Applied Probability
Control
Economics
Information and Communication
Modeling and Identification
Numerical Techniques
Optimization

Applications of Mathematics

13

Gopinath Kallianpur

Stochastic Filtering Theory

Springer-Verlag
New York Heidelberg Berlin

G. Kallianpur

Department of Statistics
University of North Carolina
Chapel Hill, NC 27514
USA

Editor

A. V. Balakrishnan

Systems Science Department
University of California
Los Angeles, CA 90024
USA

AMS Subject Classification (1980): 60G35, 60Hxx, 93E11

Library of Congress Cataloging in Publication Data

Kallianpur, Gopinath
Stochastic filtering theory.

(Applications of mathematics; v. 13)
"Based on a seminar given at the University of California at Los Angeles in the Spring of 1975."
Bibliography: p.
includes index.
1. Stochastic processes. 2. Filters (Mathematics)
3. Prediction theory. I. Title.
QA274.K34 519.2 80-13486

Printed in the United States of America.

9 8 7 6 5 4 3 2 1

ISBN 0-387-90445-X Springer-Verlag New York
ISBN 3-540-90445-X Springer-Verlag Berlin Heidelberg

To Krishna

Betake yourselves to no external refuge,
Hold fast to the truth as a lamp.
Hold fast as a refuge to the truth.

The Buddha (Mahā-Parinibbāna—Sutta, 480 B.C.)

[From *Buddhist Suttas*, translated by T. W. Rhys Davids, Dover, New York, 1969.]

Preface

This book is based on a seminar given at the University of California at Los Angeles in the Spring of 1975. The choice of topics reflects my interests at the time and the needs of the students taking the course. Initially the lectures were written up for publication in the Lecture Notes series. However, when I accepted Professor A. V. Balakrishnan's invitation to publish them in the Springer series on Applications of Mathematics it became necessary to alter the informal and often abridged style of the notes and to rewrite or expand much of the original manuscript so as to make the book as self-contained as possible. Even so, no attempt has been made to write a comprehensive treatise on filtering theory, and the book still follows the original plan of the lectures.

While this book was in preparation, the two-volume English translation of the work by R. S. Liptser and A. N. Shiryaev has appeared in this series. The first volume and the present book have the same approach to the subject, viz. that of martingale theory. Liptser and Shiryaev go into greater detail in the discussion of statistical applications and also consider interpolation and extrapolation as well as filtering.

The main purpose of this book is to present, in a compact form, the techniques now collectively known as Ito's stochastic calculus, which is the basic tool for studying nonlinear problems of filtering and prediction involving continuous stochastic processes. It is a field which has seen rapid growth in the last two decades but is not usually included in courses on probability theory or stochastic processes. The book is best suited, perhaps, for students familiar with properties of the Wiener process and with continuous parameter martingales, although the essential results pertaining to these are summarized in the first two chapters. Chapter 3 develops stochastic integrals with respect to continuous square integrable and continuous local martingales. The Ito formula for continuous local semi-martingales is treated in full detail in Chapter 4. This chapter and Chapters 5–7 constitute the core of the Ito theory: Chapter 5 gives a brief but self-contained account of the Ito theory of stochastic differential equations leading up to the result that the solution (under conditions ensuring existence and uniqueness) is a Markov process. It also discusses functional stochastic differential equations which are needed for nonlinear filtering. A unified presentation of Wiener's homogeneous chaos, the Cameron–Martin theory of Fourier–Hermite functional expansions, and multiple Wiener–Ito integrals is given in Chapter 6. The stochastic integral representation of martingales living on Wiener space is put in the perspective of this nonlinear setup.

Chapter 7 is centered around the Cameron–Martin–Girsanov theory of transformations of the Wiener process which induce measures that are absolutely continuous with respect to Wiener measure.

The central results of nonlinear filtering theory—the derivation of the stochastic equations satisfied by the optimal nonlinear filter—are obtained in Chapter 8. A separate chapter (Chapter 10) is devoted to the linear theory because of its intrinsic importance. The Kalman–Bucy filter is derived as a special case of the nonlinear theory. A derivation independent of Chapter 8 is also given.

Chapter 9 deals with Gaussian stochastic equations and is intended to serve as an introduction to the linear filtering theory. Martingale methods are avoided in this chapter in favor of operator-theoretic techniques which are natural in this context. Though not strictly necessary for the main theme of the book, it goes into the related question of non-anticipative representations of Gaussian processes via the Gohberg–Krein factorization theorem. In Chapter 11, a special case of the nonlinear filtering problem is considered, in which the signal is independent of noise in the observation process model. A Bayes formula is proved for this problem and used in obtaining the unique solution of the filter equation in a special case considered by Kunita.

The pervading influence of K. Ito's ideas on this book is obvious. The specialist will discern also in these pages the influence of the martingale and stochastic integral theory, developed over the last decade and a half by P. A. Meyer and his Strasbourg school. I have drawn freely on their work and would like here to acknowledge my scientific debt to them.

The book was written when I was at the University of Minnesota and later at the Indian Statistical Institute. My young friends and colleagues at the latter institution took it upon themselves to run a seminar based on the manuscript. This they did with great gusto, giving me invaluable assistance with their constructive criticism and suggestions for improvement. Among these friends I must mention and thank J. C. Gupta, B. V. Rao, J. Vishwanathan, and C. Bromley. The last two have also helped with the tedious job of proofreading.

I must express my special gratitude to Clayton Bromley. He has read the entire manuscript with meticulous care and weeded out numerous errors.

He has generously devoted his time and effort to discussing portions of the manuscript with me. The final form and content of the book owe much to his substantial contributions.

My thanks are due to Arun Das and Peggy Gendron for the patience and care with which they did the typing; to John Verghese and his Reprography Unit at the Indian Statistical Institute for photocopying the typescript under difficult conditions; to K. K. Kundu and Suhas Dasgupta for their unfailing courtesy in meeting my often unreasonable demands in getting the material ready for publication.

It is doubtful whether this book would ever have been written but for Professor Balakrishnan's invitation and his constant encouragement through all stages of the work. It is a pleasure to thank him not only for this but for the many hours of stimulating conversation I have had with him.

Finally, I would like to thank the staff of Springer-Verlag for their courtesy and cooperation.

Chapel Hill, N.C. Gopinath Kallianpur

Contents

Standard Notation

$\mathbf{R}^n$	n-dimensional Euclidean space
$\lvert x \rvert = \left(\sum_{i=1}^{n} x_i^2 \right)^{1/2}$	the Euclidean norm on $\mathbf{R}^n$
$\lvert A \rvert = \left(\sum_{i=1}^{m} \sum_{j=1}^{n} (a^{ij})^2 \right)^{1/2}$	the norm of an $m \times$ n matrix $A = (a^{ij})$
$A^{i\cdot}$, $A^{\cdot j}$, $\mathrm{Tr}(A)$	the i-th row-vector, j-th column-vector, and trace of a matrix A
A^*	the adjoint of a matrix or linear operator A
$a \wedge b$, $a \vee b$	the minimum, maximum of two real numbers or functions
f^+, f^-	the positive, negative parts of a real-valued function f
C^k	the space of real-valued functions which are k-times continuously differentiable
C_b^k	the subspace of functions in C^k which are bounded along with their derivatives of all orders up to and including k
C_0^k	the subspace of functions in C^k which have compact support
$L^p(\Omega,\mathscr{A},\mu)$, $1 \leq p \leq \infty$	the Lebesgue space of all measurable real- or complex-valued functions f on a given measure space with $\lVert f \rVert_p = (\int_\Omega \lvert f \rvert^p \, d\mu)^{1/p} < \infty$ for $1 \leq p < \infty$ and $\lVert f \rVert_\infty = \text{ess sup} \lvert f \rvert$

$\lvert B \rvert_t$	the total variation of a function of bounded variation B_s, $s \in [0,t]$
E	mathematical expectation
$E(\ \mid\ )$	conditional expectation
$I_A, 1_A$	the indicator function of a set A
$\perp\!\!\!\perp$	independence of random variables or sigma-fields
$\perp$	orthogonality in a Hilbert space
CONS	a complete orthonormal system in a Hilbert space

Stochastic Processes: Basic Concepts and Definitions 1

In this chapter and the next, we state a number of important results which are necessary for the work of the later chapters. Some of them might not be explicitly referred to in the later work, but they all form essential links in the chain of reasoning. To present the proofs of all of these results here would require preparatory background material which would considerably increase both the size and scope of this book. We therefore adopt the following approach with the aim of making the development of the text as self-contained as possible. We omit the proofs of those theorems which are treated in detail in well-known standard textbooks, such as P. A. Meyer's book, *Probability and Potentials* [41]. However, those proofs will be presented which are not available in existing books and are to be found scattered in the literature, or which discuss ideas specially relevant to our purpose.

1.1 Notation and Basic Definitions

$(\Omega,\mathscr{A},P)$ will denote a complete probability space. If $\{\mathscr{F}_i\}$ is a family of sub-σ-fields of $\mathscr{A}$ indexed by an arbitrary set I, $\vee\,\mathscr{F}_i$ will denote the smallest σ-field containing $\mathscr{F}_i$ for all $i \in I$. Let $\mathbf{T}$ stand for $\mathbf{R}_+ = [0,\infty)$ or a finite closed interval $[0,T]$. The choice of $\mathbf{T}$ will be made explicit in each case.

A family $(\mathscr{F}_t)$, $t \in \mathbf{T}$, is an increasing family of σ-fields if $\mathscr{F}_s \subset \mathscr{F}_t \subset \mathscr{A}$ for $s \leq t$. Let us now define the following σ-fields: $\mathscr{F}_{t-} = \vee_{s<t}\mathscr{F}_s$ if $t > 0$, $\mathscr{F}_{t+} = \bigcap_{s>t}\mathscr{F}_s$ if $t \geq 0$, $\mathscr{F}_{0-} = \mathscr{F}_0$, $\mathscr{F}_\infty = \vee_{\mathbf{R}_+}\mathscr{F}_t$, and $\mathscr{F}_{T+} = \mathscr{F}_T$ in case $\mathbf{T} = [0,T]$. An increasing family $(\mathscr{F}_t)$ is *right-continuous* if

(i) $\mathscr{F}_{t+} = \mathscr{F}_t$ for each t.

For many purposes we need the following assumption:

(ii) $\mathscr{F}_0$ contains all P-null sets in $\mathscr{A}$.

Unless otherwise stated, completeness of $(\Omega,\mathscr{A},P)$ *and condition* (ii) *will always be assumed to be in force. The assumption of right-continuity, however, will not be made in general.* To avoid possible confusion, the right-continuity assumption about the increasing family $(\mathscr{F}_t)$ will be restated in each case.

Definition 1.1.1. With respect to an increasing family of sub-σ-fields of the type discussed above, call a function $\tau: \Omega \to \bar{\mathbf{R}}_+$ ($\bar{\mathbf{R}}$ is the extended real line) a *stopping time* for $(\mathscr{F}_t)$, $t \in \mathbf{R}_+$, if $[\omega: \tau(\omega) \leq t] \in \mathscr{F}_t$ for every $t \in \mathbf{R}_+$. A similar notion is defined for a family $(\mathscr{F}_t)$, $t \in [0,T]$, with the assumption that the range of τ is now in $[0,T]$.

Note that in the presence of right-continuity of the family $(\mathscr{F}_t)$, the condition $[\omega: \tau(\omega) \leq t] \in \mathscr{F}_t, t \in \mathbf{R}_+$ (or $[0,T]$), is equivalent to requiring $[\omega: \tau(\omega) < t] \in \mathscr{F}_t$, $t \in \mathbf{R}_+$ (or $[0,T]$). Trivial examples of stopping times are obtained by setting $\tau = a \in \mathbf{R}_+$ (or $[0,T]$). Since by (ii) $\mathscr{F}_0$ contains all P-null sets, if there are two functions τ and σ on Ω with $\tau = \sigma$ (a.s.), and τ is a stopping time, then σ will also be a stopping time. Call τ a finite stopping time if $\tau(\omega) < \infty$ for all $\omega \in \Omega$.

Define the *σ-field of events prior to* τ, denoted by $\mathscr{F}_\tau$, as

$$\mathscr{F}_\tau = \{A \in \mathscr{F}_\infty: A \cap \{\tau \leq t\} \in \mathscr{F}_t \text{ for every } t\}.$$

(Here $\mathscr{F}_\infty$ is replaced by $\mathscr{F}_T$ if $t \in [0,T]$.)

Proposition 1.1.1. *For σ and τ stopping times relative to* $(\mathscr{F}_t)$,

(a) *$\mathscr{F}_\tau$ is a σ-field.*
(b) *τ is $\mathscr{F}_\tau$-measurable.*
(c) $\sigma \leq \tau$ (a.s.) $\Rightarrow \mathscr{F}_\sigma \subset \mathscr{F}_\tau$.
(d) *$\sigma \wedge \tau$ and $\sigma \vee \tau$ are also stopping times with respect to* $(\mathscr{F}_t)$.

There are a number of properties of stopping times with regard to their behavior under operations of sup, lim sup, inf, lim inf, and lim which will be used later [stated here for $(\mathscr{F}_t)$, $t \in \mathbf{R}_+$]:

1. If τ_n is a stopping time for each $n \geq 1$, then $\sup_n \tau_n$ is a stopping time, and if the assumption of right-continuity is made on $(\mathscr{F}_t)$, then $\inf_n \tau_n$ is also a stopping time.
2. If $(\mathscr{F}_t)$ is right-continuous, then both $\liminf_{n\to\infty} \tau_n$ and $\limsup_{n\to\infty} \tau_n$ are stopping times.
3. Under right-continuity of $(\mathscr{F}_t)$, if $\tau = \inf_n \tau_n$ then $\mathscr{F}_\tau = \bigcap_{n \geq 1} \mathscr{F}_{\tau_n}$.

For a general discussion of stopping times and their properties see [41]. We will find particularly useful the following result concerning the approximation of a stopping time from above by a sequence of discrete stopping times. The statement and proof below concern the case of a family $(\mathscr{F}_t)$, $t \in \mathbf{R}_+$, but a slight modification yields a similar result for $(\mathscr{F}_t)$, $t \in [0,T]$.

Proposition 1.1.2. *For τ a stopping time with respect to $(\mathscr{F}_t)$, $t \in \mathbf{R}_+$, there exists a sequence $(\tau_n)_{n\geq 1}$ of $(\mathscr{F}_t)$-stopping times with discrete ranges which approximates τ from above, i.e. such that*

(i) $\tau_n(\omega) \geq \tau(\omega)$.
(ii) $\tau_n(\omega) \downarrow \tau(\omega)$ as $n \to \infty$, for each ω.
(iii) τ_n has discrete range (a discrete subset of $\mathbf{R}_+$ and, possibly, $+\infty$).

PROOF. Let $\tau_n(\omega) = k/2^n$ if $(k-1)/2^n \leq \tau(\omega) < k/2^n$, $k = 1,2,\ldots, = \infty$ if $\tau(\omega) = \infty$. Then for $t \geq 0$.

$$\{\tau_n \leq t\} = \bigcup_{k/2^n \leq t} \{\tau_n = k/2^n\}$$

$$= \bigcup_{k/2^n \leq t} \left\{\frac{k-1}{2^n} \leq \tau < \frac{k}{2^n}\right\} \in \mathscr{F}_t.$$

So each τ_n is a stopping time and clearly $\tau_n \geq \tau$. For each ω such that $\tau(\omega) < \infty$, $0 \leq \tau_n(\omega) - \tau(\omega) < 1/2^n$. Hence $\lim_{n\to\infty} \tau_n(\omega) = \tau(\omega)$ for every ω. If $\tau_n(\omega) = k/2^n$, then $\tau_{n+1}(\omega) = (2k-1)/2^{n+1}$ or $2k/2^{n+1}$. So $\tau_n(\omega) \geq \tau_{n+1}(\omega)$ on $\{\omega: \tau(\omega) < \infty\}$, hence on all Ω. (iii) is obvious. □

Remark 1.1.1. It is not true that every stopping time can be approximated from below by an increasing sequence (τ_n) of stopping times, that is, such that $\tau_n \uparrow \tau$ and $\tau_n < \tau$ on the set $\{0 < \tau < \infty\}$. A stopping time τ is *predictable* if there exists a sequence (τ_n) of stopping times such that $\tau_n < \tau$ (a.s.) and $\tau_n \uparrow \tau$ (a.s.).

Definition 1.1.2. Let $(\Omega,\mathscr{A},P)$ be a complete probability space and $(S,\mathscr{B})$, an arbitrary measurable space.

(i) A function $\xi: \Omega \to S$ is called *$\mathscr{B}/\mathscr{A}$-measurable* if $\xi^{-1}(B) \in \mathscr{A}$ for every $B \in \mathscr{B}$.

From now on we assume S to be a complete, separable, metric space and $\mathscr{B} = \mathscr{B}(S)$, the σ-field of Borel sets of S (also called the *topological Borel σ-field of* S). Then

(ii) $\xi: \Omega \to S$ is called an *S-valued random variable* (mapping or function) if ξ is $\mathscr{B}(S)/\mathscr{A}$-measurable.
(iii) $X = (X_t)$ is called an *S-valued stochastic process* if, for each $t \in \mathbf{T}$, X_t is an S-valued random variable. S is called the *state space* of X.

An S-valued stochastic process $X = (X_t)$ with $S = \mathbf{R}^d$, $d \geq 2$, will be referred to as a d-dimensional process. By a stochastic process (or simply, process)

$X = (X_t)$, we shall always mean a real-valued process. We will often write X or (X_t) to denote a real or S-valued process.

Definition 1.1.3. The process $X = (X_t)$, $t \in \mathbf{T}$ *is continuous in probability* (or *stochastically continuous*) at $t_0 \in \mathbf{T}$ if for any $\varepsilon > 0$,

$$P\{\omega: |X_t(\omega) - X_{t_0}(\omega)| > \varepsilon\} \to 0 \quad \text{as } t \to t_0.$$

X is stochastically continuous on $\mathbf{T}$ if the above property holds for every t_0 in $\mathbf{T}$.

A process $X = (X_t)$, $t \in T$ for which $E(X_t^2) < \infty$ for every $t \in \mathbf{T}$ is said to be *continuous in quadratic mean* (c.q.m.) at $t_0 \in \mathbf{T}$ if $E[X_t - X_{t_0}]^2 \to 0$ as $t \to t_0$.

X is c.q.m. on $\mathbf{T}$ if the above property holds for all $t_0 \in \mathbf{T}$.

Definition 1.1.4. (i) The processes $X = (X_t)$ and $X' = (X'_t)$, $t \in \mathbf{T}$, defined on the probability spaces $(\Omega,\mathscr{A},P)$ and $(\Omega',\mathscr{A}',P')$ are called *equivalent* if for every $\{t_1, \ldots, t_n\} \subset \mathbf{T}$ and sets $B_i \in \mathscr{B}(\mathbf{R})$, $i = 1, \ldots, n$,

$$P\{\omega: X_{t_1}(\omega) \in B_1, \ldots, X_{t_n}(\omega) \in B_n\} = P'\{\omega': X'_{t_1}(\omega') \in B_1, \ldots, X'_{t_n}(\omega') \in B_n\}.$$

A stronger notion of equivalence is available for processes defined on the same probability space.

(ii) Let $X = (X_t)$ and $Y = (Y_t)$ be two processes defined on $(\Omega,\mathscr{A},P)$. X and Y will be called *modifications* or *versions* of each other if

$$P\{\omega: Y_t(\omega) = X_t(\omega)\} = 1 \quad \text{for each } t.$$

For each $\omega \in \Omega$, the function $t \to X_t(\omega)$ is called the *sample path*, or *trajectory* associated with ω. A process $X = (X_t)$ is called *continuous, right-continuous*, or *left-continuous* if for P-almost all ω, the trajectory of ω has this property.

Definition 1.1.5. (i) A stochastic process $X = (X_t)$, $t \in \mathbf{R}_+$, is *measurable* (*jointly measurable*, or *(t,ω)-measurable*) if the function $(t,\omega) \to X_t(\omega)$ is measurable with respect to $\mathscr{B}(\mathbf{R}_+) \times \mathscr{A}$. If $t \in [0,T]$, the measurability is with respect to $\mathscr{B}[0,T] \times \mathscr{A}$.

An S-valued process $X = (X_t)$ is called *measurable* if the function $(t,\omega) \to X_t(\omega)$ is

$$\mathscr{B}(S)/[\mathscr{B}(\mathbf{R}_+) \times \mathscr{A}]\text{-measurable.}$$

(ii) Let $(\mathscr{F}_t)$ be an increasing family. The process $X = (X_t)$ is *$(\mathscr{F}_t)$-adapted* if each X_t is measurable with respect to $\mathscr{F}_t$. If X is S-valued, it is *$(\mathscr{F}_t)$-adapted* if X_t is $\mathscr{B}(S)/\mathscr{F}_t$-measurable.

All stochastic process considered in this book will be assumed to be measurable.

Note that if X and Y are continuous (right-continuous, or left-continuous) processes, and if X is a version of Y, then a stronger statement holds:

$$P\{\omega: Y_t(\omega) = X_t(\omega) \quad \text{for all } t \in \mathbf{R}_+\} = 1.$$

For (X_i), $i \in I$, any family of random variables, $\sigma(X_i, i \in I)$ will denote the smallest σ-field with respect to which all X_i are measurable. Given a process $X = (X_t)$, $t \in \mathbf{R}_+$, clearly, $\{\sigma(X_s, s \leq t)\}_{t \geq 0}$ is the minimal increasing family with respect to which X is adapted. We shall refer to this family as the *natural family* of the process X and reserve the symbol $(\mathscr{F}_t^{X,0})$ to denote it. Let us also define

$$\mathscr{F}_t^X = \mathscr{F}_t^{X,0} \vee \{P\text{-null sets of } \mathscr{A}\}.$$

This notation will also be used for S-valued processes.

Definition 1.1.6. Suppose $X = (X_t)$ is a process and $(\mathscr{F}_t)$ is an increasing family of sub-σ-fields. We call X *progressively measurable* [with respect to $(\mathscr{F}_t)$] if for each $t \in \mathbf{R}_+$, the map

$$(s,\omega) \to X_s(\omega)$$

from $[0,t] \times \Omega$ into $\mathbf{R}$ is measurable with respect to $\mathscr{B}[0,t] \times \mathscr{F}_t$. Progressive measurability of an S-valued process is defined similarly.

Proposition 1.1.3. *Let $X = (X_t)$, $t \in \mathbf{R}_+$ be a d-dimensional process which is $(\mathscr{F}_t)$-adapted and right-continuous. Then it is progressively measurable. (A similar statement holds for left-continuous processes.)*

It is obvious that if X is progressively measurable with respect to $(\mathscr{F}_t)$, then it must be both measurable and $(\mathscr{F}_t)$-adapted. The converse of this statement is known to be false. However, if we are concerned only with versions of a given process, this converse holds in the following sense:

Proposition 1.1.4. *Let $X = (X_t)$, $t \in \mathbf{R}_+$ be a measurable, $(\mathscr{F}_t)$-adapted d-dimensional process. Then there exists a modification $Y = (Y_t)$ of X which is progressively measurable.*

Given a process $X = (X_t)$, $t \in \mathbf{R}_+$, we will call a function $\tau: \Omega \to \mathbf{R}_+$ a *stopping time for* X if τ is a stopping time with respect to the family $(\mathscr{F}_t^X)$.

Let $X = (X_t)$, $t \in \mathbf{R}_+$, be a process and $\sigma: \Omega \to \mathbf{R}_+$ be a random variable. It is desirable for many applications that the function $X_\sigma: \Omega \to S$ defined by $X_\sigma(\omega) = X_{\sigma(\omega)}(\omega)$ be also measurable. This will be true if X is a measurable process, although it is not generally true otherwise. More information on this point is provided by the following proposition.

Proposition 1.1.5. *Let $X = (X_t)$, $t \in \mathbf{R}_+$, be a progressively measurable process with respect to $(\mathscr{F}_t)$, $t \in \mathbf{R}_+$, and let τ be a finite stopping time. Then the random variable X_τ is $\mathscr{F}_\tau$-measurable.*

Using some of the notions defined above for processes, we can construct interesting examples of stopping times as follows. Let X be a process with state space $\mathbf{R}^d$ which is right-continuous and $(\mathscr{F}_t)$-adapted, where the family $(\mathscr{F}_t)$ is assumed to be right-continuous. For any Borel set B in $\mathbf{R}^d$, define τ_B, called the *hitting time* for the set B, by the rule

$$\tau_B(\omega) = \begin{cases} \inf\{t \geq 0\colon X_t(\omega) \in B\} & \\ \infty & \text{if this set is empty.} \end{cases}$$

Then τ_B is in fact a stopping time relative to the family $(\mathscr{F}_t^X)$. For $B = G$ an open set, the proof of this assertion lies in the remark that it suffices in the presence of right-continuity of $(\mathscr{F}_t)$ to show $[\tau_G < t] \in \mathscr{F}_t$ for each $t \in \mathbf{R}_+$, and that for a right-continuous process, neglecting a P-null set, $[\tau_G < t] = \bigcup_{s<t}[X_s \in G] = \bigcup_{r<t,(r \text{ rational})}[X_r \in G] \in \mathscr{F}_t$.

Because of its great importance in this book, we recall the definition of a Gaussian (or normal) distribution. A real-valued random variable ξ is said to have a *Gaussian* or *normal* distribution with mean a and variance σ^2 if for every real number x,

$$P\{\xi \leq x\} = \frac{1}{\sqrt{2\pi}\,\sigma} \int_{-\infty}^{x} e^{-(1/2\sigma^2)(u-a)^2}\,du.$$

It is easily seen that $E(\xi) = a$ and $E(\xi - a)^2 = \sigma^2$. Alternatively, ξ is called a Gaussian, normal, or $N(a,\sigma^2)$ random variable. A stochastic process $X = (X_t)$ is called a *Gaussian* process if every finite linear combination with real coefficients, $\sum c_i X_{t_i}$, is a Gaussian random variable. For a Gaussian process X, $EX_t^2 < \infty$ for each t. Since we always assume (X_t) to be a measurable process, the mean function $a(t) = E(X_t)$ and the covariance function (or kernel) $R(s,t) = E(X_s - a(s))(X_t - a(t))$ are Borel measurable functions of t and (s,t), respectively.

The symbol $\perp\!\!\!\perp$ will be used to denote stochastic independence of random variables or σ-fields.

1.2 Probability Measures Associated with Stochastic Processes

In the next chapter we introduce the Wiener process and Wiener measure in the space of continuous functions. In speaking of the Wiener process, one generally has in mind sample path properties (such as continuity, the law of the iterated logarithm, etc.), whereas Wiener measure is the appropriate tool for studying the statistical properties of the process. Both these aspects are of fundamental importance in the work of the later chapters on martingales and stochastic differential equations.

We shall preface our construction of the Wiener process with some background material concerning probability measures corresponding to stochastic processes. A convenient starting point is the well-known theorem of Kolmogorov.

Let $\mathbf{R}$ be the real line, $\mathbf{T}$ a set of arbitrary cardinality, and $\Omega^* = \mathbf{R}^{\mathbf{T}}$, the set of all functions $\omega^*: \mathbf{T} \to \mathbf{R}$. Let $\mathscr{C}(\Omega^*)$ be the σ-field generated by all finite-dimensional Borel cylinder subsets of Ω^*, that is, by sets of the form

$$\{\omega^* \in \Omega^*: (\omega^*(t_1), \ldots, \omega^*(t_n)) \in B\},$$

where $t_i \in \mathbf{T}$ $(i = 1, \ldots, n)$ are arbitrary and B ranges over all Borel sets in $\mathbf{R}^n$. Let $\mathscr{L} = \{F_{t_1, \ldots, t_n}\}$ be a family of distribution functions indexed by all finite subsets $\{t_1, \ldots, t_n\}$ of $\mathbf{T}$, $F_{t_1, \ldots, t_n}$ being a distribution function in $\mathbf{R}^n$. $\mathscr{L}$ will be called a *consistent family* if it satisfies the following consistency conditions:

(i) For $\{t_1, \ldots, t_n\} \subset \mathbf{T}$, $(a_1, \ldots, a_n) \in \mathbf{R}^n$, and $(\pi_1, \ldots, \pi_n)$ any permutation of $(1, \ldots, n)$,

$$F_{t_1, \ldots, t_n}(a_1, \ldots, a_n) = F_{t_{\pi_1}, \ldots, t_{\pi_n}}(a_{\pi_1}, \ldots, a_{\pi_n}).$$

(ii) For $1 \leq j \leq n$,

$$F_{t_1, \ldots, t_n}(a_1, \ldots, a_n) \to F_{t_1, \ldots, t_{j-1}, t_{j+1}, \ldots, t_n}(a_1, \ldots, a_{j-1}, a_{j+1}, \ldots, a_n)$$

as $a_j \to +\infty$.

The family $\mathscr{L}$ represents a stochastic process in the sense of the result stated below.

Theorem 1.2.1 (Kolmogorov). *A consistent family $\mathscr{L}$ determines a unique probability measure P on $(\Omega^*, \mathscr{C}(\Omega^*))$ such that $\mathscr{L}$ coincides with the finite-dimensional distributions of the process of coordinate random variables $(\tilde{X}_t)$ defined by*

$$\tilde{X}_t(\omega^*) = \omega^*(t) \quad \textit{for all } \omega^* \in \Omega, t \in T,$$

that is, such that

$$P\{\omega^* \in \Omega: \tilde{X}_{t_i}(\omega^*) \leq a_i, i = 1, \ldots, n\} = F_{t_1, \ldots, t_n}(a_1, \ldots, a_n).$$

Let $\mathscr{A}(\Omega^*) = \mathscr{C}'(\Omega^*)$, the completion of $\mathscr{C}(\Omega^*)$ with respect to P. The process $\tilde{X} = (\tilde{X}_t)$ defined on $(\Omega^*, \mathscr{A}(\Omega^*), P)$ is sometimes called the *canonical* process.

From Kolmogorov's theorem it follows that to every stochastic process $X = (X_t)$ given on some probability space, there corresponds a probability measure on a function space and a canonical process $\tilde{X}$ on it which is equivalent to X. The σ-field $\mathscr{C}(\Omega^*)$ is, however, too small, and, when $\mathbf{T}$ is nondenumerable, many of the interesting sets whose probabilities are required fail to lie in it. The remark below and the examples that follow shed some light on this point.

Remark 1.2.1. Let $\Lambda \in \mathscr{C}(\Omega^*)$. Then there exists a countable set $\mathbf{T}' = (t_1,t_2,\ldots,\} \subset \mathbf{T}$ such that $\Lambda = \{\omega^*(t_1),\omega^*(t_2),\ldots,) \in D\}$ for some $D \in \mathscr{B}(\mathbf{R}^\infty)$, $\mathbf{R}^\infty$ being a countable product of real lines. Call any set Λ with this property a set with a countable base. $\mathbf{T}$ is assumed to be nondenumerable.

To prove the remark, it suffices to show that the class $\mathscr{D}$ of all sets with a countable base forms a σ-field. Clearly it contains all finite-dimensional Borel cylinders. Let $\Lambda_1,\Lambda_2,\ldots,$ belong to $\mathscr{D}$, and let $\mathbf{T}'_k$ be a countable base for Λ_k. Then $\mathbf{T}' = \bigcup_k \mathbf{T}'_k$ is a countable base for all the sets Λ_k, and the latter can be written in the form $\Lambda_k = \{\omega^*:(\omega^*(t_1),\ldots) \in D_k\}$, where $\mathbf{T}' = \{t_i\}$ and $D_k \in \mathscr{B}(\mathbf{R}^\infty)$. Hence $\bigcup_k \Lambda_k \in \mathscr{D}$. Clearly $\Omega^* \in \mathscr{D}$ and $\Lambda^c \in \mathscr{D}$ if $\Lambda \in \mathscr{D}$. Thus $\mathscr{D}$ is a σ-field, and obviously, $\mathscr{D} = \mathscr{C}(\Omega^*)$. Take $\mathbf{T} = [0,1]$ in Examples 1.2.1 through 1.2.3. Let P be any probability measure on $(\Omega^*,\mathscr{C}(\Omega^*))$.

EXAMPLE 1.2.1. Let Ω_c be the set of all continuous functions in Ω^*. Then $\Omega_c \notin \mathscr{C}(\Omega^*)$.

PROOF. If $\Omega_c \in \mathscr{C}(\Omega^*)$, by the above remark,

$$\Omega_c = \{\omega^* \in \Omega^*: (\omega^*(t_1),\omega^*(t_2),\ldots) \in D\}$$

for some $D \in \mathscr{B}(\mathbf{R}^\infty)$. But clearly this is not possible, for one can obviously find discontinuous functions in the right-hand-side set. □

Let A be any subset of Ω^* and $A^c = \Omega^* \backslash A$. Define the outer measure

$$\bar{P}(A) = \inf\left\{P(\Gamma): A \subset \Gamma = \bigcup_j \Gamma_j, \Gamma_j, \text{ a finite-dimensional Borel cylinder}\right\}$$

and the inner measure $\underline{P}(A) = 1 - \bar{P}(A^c)$. Then $\mathscr{A} = \{A \subset \Omega^*: \underline{P}(A) = \bar{P}(A)\}$ is the completion of $\mathscr{C}(\Omega^*)$ with respect to P.

EXAMPLE 1.2.2. $\underline{P}(\Omega_c) = 0$.

PROOF. Suppose $(\Omega_c)^c \subset \Gamma = \bigcup_j \Gamma_j$, the Γ_j being finite-dimensional Borel cylinders. Let $\mathbf{T}'$ be a countable base set for Γ. Let $\omega^* \in \Omega^*$ and define ω_0 to be equal to ω^* on D and to be a discontinuous function of t on $\mathbf{T}$. This implies $\omega_0 \in (\Omega_c)^c$; hence $\omega_0 \in \Gamma$. Thus $\omega^* \in \Gamma$. Hence $P(\Gamma) = P(\Omega^*) = 1$ and $\bar{P}(\Omega_c)^c = 1$. □

EXAMPLE 1.2.3. Let $\Omega_m = \{\omega \in \Omega^*: \omega(t)$ is a Lebesgue measurable function of $t\}$. Then $\underline{P}(\Omega_m) = 0$.

The proof is similar to the one given in Example 1.2.2. We shall show that $\bar{P}(\Omega_m^c) = 1$. Let $\Omega_m^c \subset \Gamma \in \mathscr{C}(\Omega^*)$ and $\mathbf{T}'$ be a countable base of Γ. Suppose $\omega^* \in \Omega^*$. Let f be a non-Lebesgue-measurable function on $\mathbf{T}$. Define $\omega_1(t) = \omega^*(t)$ on $\mathbf{T}'$ and $\omega_1(t) = f(t)$ on $\mathbf{T}\backslash\mathbf{T}'$. Since $\mathbf{T}'$ is countable, $\omega_1(t)$ is still nonmeasurable. Hence $\omega_1 \in \Omega_m^c$ so $\omega_1 \in \Gamma$, and since ω^* and ω_1 agree on $\mathbf{T}'$, we have $\omega^* \in \Gamma$.

The above examples show that, whatever the family of finite-dimensional distributions which determines P on Ω^*, if Ω_c and Ω_m are $\mathscr{A}$-measurable, then they must have P-measure zero. Now take $\mathbf{T} = \mathbf{R}_+$ or $[0,T]$.

Definition 1.2.1. Let $M \subset \mathbf{R}^{\mathbf{T}}$ and let $\mathscr{C}(M)$ be the σ-field generated by the finite-dimensional Borel cylinder subsets of M. A stochastic process $X = (X_t)$ has a *realization in* M if there exists a probability measure μ on $(M,\mathscr{C}(M))$ such that the coordinate process $\hat{X}$ defined by $\hat{X}_t(\omega) = \omega(t)$ for $\omega \in M$, is equivalent to X.

The proof of the following observation is left to the reader.

Proposition 1.2.1. *The canonical process $\tilde{X} = (\tilde{X}_t)$ has a realization in M if and only if $\bar{P}(M) = 1$.*

Remark 1.2.2. When P on $(\Omega^*,\mathscr{A})$ is the measure determined by the finite-dimensional distributions of the Wiener process, it is known that $\bar{P}(\Omega_c) = 1$. Since $\underline{P}(\Omega_c) = 0$, it follows that Ω_c cannot belong to $\mathscr{A}$. Proposition 1.2.1 thus provides an alternative way of showing the existence of the Wiener process which is established by a different method in the next chapter.

Proposition 1.2.2. *If $X = (X_t)$, defined on a probability space $(\Omega,\mathscr{A},P)$ has a continuous modification, then it has a realization in the space of continuous functions.*

Proof. Take $\mathbf{T} = [0,T]$ and let $C = C[0,T]$ be the space of real, continuous functions x on $[0,T]$. C is a separable, Banach space under the norm $\|x\| = \max_{0 \le t \le T} |x(t)|$. Let $Y = (Y_t)$ be a continuous modification of X. Define $\xi : \Omega \to C$ by $\xi(\omega)(t) = Y_t(\omega)$. If $\tilde{B} = \{x \in C : (x(t_1), \ldots, x(t_n)) \in B\}$, where $B \in \mathscr{B}(\mathbf{R}^n)$, then clearly $\xi^{-1}(\tilde{B}) \in \mathscr{A}$. Hence $\xi^{-1}(E) \in \mathscr{A}$ for every $E \in \mathscr{C}(C)$, the σ-field generated by the cylinder sets of C. But it is easy to see that the latter σ-field coincides with $\mathscr{B}(C)$, proving that ξ is a C-valued random variable. The conclusion now follows if we let $\mu = P\xi^{-1}$ and define the process $\hat{X}$ on $(C,\mathscr{B}(C),\mu)$ by $\hat{X}_t(x) = x(t)$. □

A useful sufficient condition for a process $X = (X_t)$ to have a realization in $C[0,T]$ is given by the following theorem.

Theorem 1.2.2 (Kolmogorov). *Suppose that the process $X = (X_t)$, $t \in [0,T]$, satisfies the following condition: there exist positive constants p, r, and A such that*

$$E|X_t - X_{t'}|^p \le A|t - t'|^{1+r}$$

for all t,t' in $[0,T]$. *Then X has a continuous modification.*

Before proceeding to the definition and construction of the Wiener process, it may be of interest to consider briefly what might at first appear to be the simplest concrete example of a stochastic process, viz., the process $X = (X_t)$ of mutually independent random variables. That such a process cannot form the basis for a fruitful theory of stochastic processes is seen from the two examples given below, especially Example 1.2.5. We take $\mathbf{T} = [0,1]$ for convenience and assume that the random variables X_t are identically distributed and that the range of X_t contains at least two distinct points (to avoid the trivial case of degenerate random variables).

EXAMPLE 1.2.4. The process X does not have a realization in $C[0,1]$.

PROOF. Assume the contrary. Suppose μ is the probability measure in $C[0,1]$ induced by the distribution of (X_t). Letting $E = \{x \in C: x(1) > a\}$, we choose a such that $\mu(E) = 1 - \delta$, where $0 < \delta < 1$. Define the events $F_k = \{x \in C: x(1 - 1/n) > a, n = k, k+1, \ldots, 2k\}$. Since $\mu(F_k) = (1-\delta)^k$, $\sum_{k=1}^{\infty} \mu(F_k) < \infty$. By the Borel-Cantelli lemma, $\mu(\limsup_k F_k) = 0$. But this implies $\mu(E^c) = 1$, which is impossible because $\mu(E) > 0$.

For the process $X = (X_t)$ considered above, assume in addition that $EX_t = 0$ and $EX_t^2 = 1$. □

EXAMPLE 1.2.5. Let $X = (X_t)$ be a process of mutually independent random variables such that $EX_t = 0$ and $EX_t^2 = 1$ for each t. Then there is no measurable process equivalent to X.

PROOF. Suppose that on some probability space $(\Omega, \mathscr{A}, P)$ there exists a (t,ω) measurable family $\{X_t(\omega)\}$ with the specified distributions. For each t, $EX_t = 0$, $E(X_t X_s) = 0$ if $t \neq s$ and $= 1$ if $t = s$. If I is any subinterval of $[0,1]$, it is easy to see that $\int_\Omega \int_I \int_I |X_t(\omega) X_s(\omega)| P(d\omega)\, dt\, ds < \infty$. Hence, using Fubini's theorem we have

$$E\left(\int_I X_t(\omega)\, dt\right)^2 = E \iint_{II} X_t(\omega) X_s(\omega)\, dt\, ds = \iint_{II} E(X_t X_s)\, dt\, ds = 0.$$

So, $\int_I X_t(\omega)\, dt = 0$ for $\omega \notin N_I$, where $P(N_I) = 0$. Now consider all subintervals $I = [r', r'']$ with rational endpoints r' and r'' and write $N = \bigcup_I N_I$. Then $P(N) = 0$ and for all $\omega \in N^c$, we have $\int_a^b X_t(\omega)\, dt = 0$ for all subintervals $[a,b]$ of $[0,1]$. Hence, for $\omega \in N^c$, $X_t(\omega) = 0$ for all t except possibly for a set of Lebesgue measure zero. It then follows by Fubini's theorem that $\int_\Omega \int_0^1 X_t^2(\omega) P(d\omega)\, dt = 0$. This is impossible since the left-hand-side integral is $\int_0^1 E(X_t^2)\, dt = 1$. □

In statistical problems of prediction and filtering it is important to have a useful mathematical model of a "white noise" process. In the case of problems involving discrete time (that is, when $\mathbf{T}$ is at most countably infinite), white noise (or Gaussian white noise) can simply be taken to be a sequence of independent, identically distributed (Gaussian) random variables.

In view of Example 1.2.5, one has to look elsewhere for an appropriate model of a white noise process. The Wiener process turns out to be a convenient and versatile model when the noise is Gaussian. Other models of white noise which have been considered in the literature (such as generalized processes or finitely additive white noise measures) are beyond the scope of this book.

2 Martingales and the Wiener Process

2.1 The Wiener Process

In the following definition $\mathbf{T}$ is taken to be either $\mathbf{R}_+$ or $[0,T]$.

Definition 2.1.1. A process $X = (X_t)$ defined on a complete probability space $(\Omega,\mathscr{A},P)$ is called a *Wiener process* with *variance parameter* σ^2 if it is a Gaussian process with the following properties:

1. $X_0(\omega) = 0$ (a.s.).
2. For every s and t $(s \leq t)$, $X_t - X_s$ has a Gaussian distribution with zero mean and variance equal to $\sigma^2(t - s)$.
3. For all $t_i \in \mathbf{T}$ $(i = 1,2,3,4)$ such that $t_1 \leq t_2 \leq t_3 \leq t_4$, the random variables $X_{t_4} - X_{t_3}$ and $X_{t_2} - X_{t_1}$ are independent.
4. For a.a. ω, the trajectories $t \to X_t(\omega)$ are continuous.

The following equivalent definition has the advantage of carrying over directly to the case of many parameters.

Definition 2.1.1′. $X = (X_t)$ is a Wiener process with variance parameter σ^2 if X is a continuous Gaussian process with $EX_t = 0$ for all t and covariance function given by

$$E(X_t X_s) = \sigma^2 \min(t,s). \tag{2.1.1}$$

A Wiener process with $\sigma^2 = 1$ is called a *standard* Wiener process (or *standard Brownian motion*).

Remark 2.1.1. Conditions 1 to 3 suffice to determine via Theorem 1.2.1 a probability measure on $(\Omega^*,\mathscr{A}(\Omega^*))$ with the appropriate finite-dimensional distributions. But as we have seen in Section 1.2, the canonical process on

Ω^* cannot satisfy condition 4. (In this connection see Remark 2 of Section 1.2).

From among the several methods of constructing a Wiener process available in the literature we give below one which appears to be particularly simple. First let $\mathbf{T}$ be the interval $[0,1]$ for convenience. Let $\Omega = \mathbf{R}^\infty$, a countable product of real lines, let $\mathscr{B}(\mathbf{R}^\infty)$ be the σ-field generated by the Borel cylinder sets in Ω, and let P be the countable product of $N(0,1)$ measures on $\mathbf{R}$. Denote by $\mathscr{A}$ the completion with respect to P of $\mathscr{B}(\mathbf{R}^\infty)$. Let g_{00}, g_{nj} (where $n = 1,2, \ldots,$ and $j = 0,1, \ldots, 2^{n-1} - 1$) be the Haar family of functions on $[0,1]$, known to form a complete orthonormal system (CONS) in $L^2[0,1]$. They are given as follows:

$$g_{00} = 1;$$

$$g_{nj}(s) = \begin{cases} 2^{(n-1)/2} & \text{if } s \in \left[\dfrac{j}{2^{n-1}}, \dfrac{j+\frac{1}{2}}{2^{n-1}}\right), \\ -2^{(n-1)/2} & \text{if } s \in \left[\dfrac{j+\frac{1}{2}}{2^{n-1}}, \dfrac{j+1}{2^{n-1}}\right), \\ 0 & \text{otherwise.} \end{cases}$$

Let $G_{nj}(t) = \int_0^t g_{nj}(s)\,ds = (1_t, g_{nj})$, where 1_t is the indicator function of $[0,t]$ and $(\ ,\ )$ is L^2-inner product. Let $\{y_{nj}\}$ be mutually independent $N(0,1)$ random variables on $(\Omega,\mathscr{A},P)$. Consider the series

$$\sum_{n=0}^{\infty} \sum_{j \in S_n} y_{nj}(\omega) G_{nj}(t) \qquad (\omega \in \Omega,\ t \in \mathbf{T}), \tag{2.1.2}$$

where $S_n = \{j: 0 \le j \le 2^{n-1} - 1\}$ if $n \ge 1$ and $S_0 = \{0\}$. It is helpful to observe that if $n > 1$, the functions G_{nj} for different values of j have disjoint supports. Also $G_{00}(u) = u$, $G_{10}(u) = u$ in $[0,\frac{1}{2}) = 1 - u$ in $[\frac{1}{2},1)$.

$$G_{nj}(u) = \begin{cases} 0 & \text{if } 0 \le u < \dfrac{j}{2^{n-1}}, \\ 2^{(n-1)/2}\left(u - \dfrac{j}{2^{n-1}}\right) & \text{if } \dfrac{j}{2^{n-1}} \le u < \dfrac{j+\frac{1}{2}}{2^{n-1}}, \\ 2^{(n-1)/2}\left(\dfrac{j+1}{2^{n-1}} - u\right) & \text{if } \dfrac{j+\frac{1}{2}}{2^{n-1}} \le u < \dfrac{j+1}{2^{n-1}}, \\ 0 & \text{if } \dfrac{j+1}{2^{n-1}} \le u \le 1. \end{cases}$$

The maximum value of $G_{nj}(u)$ occurs at $u = (j + \frac{1}{2})/2^{n-1}$ and

$$\max_{0 \le u \le 1} G_{nj}(u) = 2^{-(n+1)/2}.$$

Set

$$f_n(t,\omega) = \sum_{j \in S_n} y_{nj}(\omega) G_{nj}(t)$$

and $Y_n(\omega) = \max_{j \in S_n} |y_{nj}(\omega)|$. Letting a_n be any positive number, for $n \geq 1$, we have

$$P(Y_n > a_n) \leq \sum_{j \in S_n} P(|y_{nj}| > a_n)$$

$$= 2^{n-1} \frac{2}{\sqrt{2\pi}} \int_{a_n}^{\infty} e^{-x^2/2}\, dx$$

$$\leq C \frac{2^n}{a_n^4} e^{-a_n^2/4}, \quad \text{where } C = \frac{1}{\sqrt{2\pi}} \int_0^{\infty} x^4 e^{-x^2/4}\, dx.$$

Now choosing $a_n = 2(n \log 2)^{\frac{1}{2}}$, we see that

$$\sum_{n=1}^{\infty} P(Y_n(\omega) > a_n) \leq C \sum_{n=1}^{\infty} \frac{1}{16(\log 2)^2 n^2} < \infty.$$

Hence by the Borel-Cantelli lemma, $P(\Omega_0) = 1$, where

$$\Omega_0 = \{\omega: Y_n(\omega) \leq 2(n \log 2)^{\frac{1}{2}} \text{ for all sufficiently large } n\}. \tag{2.1.3}$$

It is clear from its definition that $\max_{0 \leq t \leq 1} |f_n(t,\omega)|$ is a random variable. Furthermore, if $\omega \in \Omega_0$, then $\max_{0 \leq t \leq 1} |f_n(t,\omega)| \leq 2^{-(n+1)/2} 2(n \log 2)^{\frac{1}{2}}$ for all sufficiently large n. Hence $P\{\omega: \sum_{n=0}^{\infty} \max_{0 \leq t \leq 1} |f_n(t,\omega)| < \infty\} = 1$. Thus $\sum_{n=0}^{\infty} f_n(t,\omega)$ converges uniformly in t for $\omega \in \Omega_0$ [hence, (a.s.)]. Finally, define $W_t(\omega)$ to be the sum of this series when $\omega \in \Omega_0$ and to be equal to 0 for all t when $\omega \notin \Omega_0$. We shall now show that $W = (W_t)$ is a Wiener process on $(\Omega, \mathscr{A}, P)$. Since for $\omega \in \Omega_0$, $W_t(\omega)$ is a continuous function of t, condition 4 of our definition is satisfied. Also it is obvious that the finite-dimensional distributions of W are Gaussian. Since $\sum G_{nj}^2(t) = \sum (1_t, g_{nj})^2 < \infty$,

$$E\left[W_t(\omega) - \sum_{n=0}^{m} \sum_{j \in S_n} y_{nj}(\omega) G_{nj}(t)\right]^2 \to 0,$$

so that

$$E[W_t(\omega) W_s(\omega)] = \lim_{m \to \infty} E\left[\left\{\sum_{n=0}^{m} \sum_{j \in S_n} y_{nj}(\omega) G_{nj}(t)\right\}\left\{\sum_{n=0}^{m} \sum_{j \in S_n} y_{nj}(\omega) G_{nj}(s)\right\}\right]$$

$$= \sum_{n=0}^{\infty} \sum_{j \in S_n} (1_t, g_{nj})(1_s, g_{nj})$$

$$= (1_t, 1_s) = \min(t,s).$$

Also $EW_t = 0$ for all t, and so the conditions of Definition 2.1.1′ are satisfied.

Remark 2.1.2. The following, in fact, is true. Let (φ_j) be *any* CONS in $L^2[0,1]$, and let $\Phi_j(t) = \int_0^t \varphi_j(u)\, du$. If $\xi_j(\omega)$, $j \geq 1$, are independent, $N(0,1)$ random variables on $(\Omega, \mathscr{A}, P)$, then

$$P\left\{\omega: \sum_{j=1}^{\infty} \xi_j(\omega) \Phi_j(t) \text{ converges uniformly in } t \in [0,1]\right\} = 1.$$

The choice of the Haar family for (φ_j) simplifies the proof.

Remark 2.1.3. The existence of the Wiener process also follows easily from Kolmogorov's criterion (Theorem 1.2.2), since we have $E[X_{t'} - X_t]^4 = 3\sigma^4(t-s)^2$ from condition 2 of Definition 2.1.1.

Recalling Proposition 1.2.2, defining $\xi: \Omega \to C$ ($C = C[0,1]$) by $\xi(\omega)(t) = W_t(\omega)$ and setting $\mu_w = P\xi^{-1}$, it is easy to see that the coordinate process on $(C,\mathscr{B}(C),\mu_w)$ is a Wiener process. The measure space $(C,\mathscr{A}(C),\mu_w)$, where $\mathscr{A}(C)$ is the completion of $\mathscr{B}(C)$ with respect to μ_w, is often called *Wiener (function) space*, and μ_w is called *Wiener measure*.

In some problems of filtering theory and in the theory of stochastic processes generally, it is necessary to work with the Wiener process on the interval $\mathbf{R}_+$. The definition is the same as the one given above but with $\mathbf{R}_+$ instead of $[0,T]$.

We now show how the method described above can be extended to prove the existence of such a process.

The Wiener Process on $\mathbf{R}_+$

Using the Haar family $\{g_{nj}\}$ defined above, we define functions $\{h_{nj}\}$ in $L^2(\mathbf{R}_+)$. Let

$$h_{nj}(t) = \left(\frac{2}{\pi}\right)^{\frac{1}{2}}(1+t^2)^{-\frac{1}{2}}g_{nj}\left(\frac{2}{\pi}\arctan t\right) \quad \text{for } 0 \leq t < \infty,$$

where $j = 0$ if $n = 0$ and $j = 0,1,\ldots,2^{n-1}-1$ for $n \geq 1$.

Lemma 2.1.1. *(h_{nj}) is a* CONS *in $L^2(\mathbf{R}_+)$.*

PROOF. $\displaystyle\int_0^\infty h_{nj}(t)h_{mk}(t)\,dt = \frac{2}{\pi}\int_0^\infty (1+t^2)^{-1}g_{nj}\left(\frac{2}{\pi}\arctan t\right)g_{mk}\left(\frac{2}{\pi}\arctan t\right)dt$

$$= \int_0^1 g_{nj}(t)g_{mk}(t)\,dt = \delta_{mn}\delta_{jk}.$$

The following argument shows that $\{h_{nj}\}$ is also complete in $L^2[0,\infty)$. Let $L_0^1[0,\infty)$ be the set of measurable functions $x(t)$ such that

$$\int_0^\infty \frac{|x(t)|}{1+t}\,dt < \infty.$$

Then

$$\begin{aligned}\int_0^\infty x(t)^2\,dt &= \frac{\pi}{2}\int_0^1 \frac{|x(\tan(\pi s/2))|^2}{\cos^2(\pi s/2)}\,ds \\ &\geq \frac{\pi}{2}\left(\int_0^1 \frac{|x(\tan(\pi s/2))|}{\cos(\pi s/2)}\,ds\right)^2 \\ &= \frac{\pi}{2}\left(\int_0^\infty \frac{2|x(t)|}{\pi(1+t^2)^{\frac{1}{2}}}\,dt\right)^2 \\ &\geq \frac{2}{\pi}\left(\int_0^\infty \frac{|x(t)|}{1+t}\,dt\right)^2.\end{aligned}$$

Thus $L^2[0,\infty)$ is a subset of $L_0^1[0,\infty)$. Also, using the above calculations and

$$\int_0^\infty \frac{|x(t)|}{1+t}\,dt = \frac{\pi}{2}\int_0^1 \frac{|x(\tan(\pi s/2))|}{\cos(\pi s/2)} \frac{ds}{\cos(\pi s/2)+\sin(\pi s/2)}$$
$$\geq \frac{\pi}{4}\int_0^1 \frac{|x(\tan(\pi s/2))|}{\cos(\pi s/2)}\,ds,$$

we see that $x(t) \in L_0^1[0,\infty)$ if and only if

$$\int_0^1 \frac{|x(\tan(\pi s/2))|}{\cos(\pi s/2)}\,ds < \infty.$$

Let $x(t) \in L_0^1[0,\infty)$ be such that $\int_0^\infty x(t)h_{nj}(t)\,dt = 0$ for $n = 0$, $j = 0$, and $j = 0,1,\ldots,2^{n-1}-1$ for $n \geq 1$. Then

$$\int_0^\infty x(t)h_{nj}(t)\,dt = \int_0^\infty x(t)\left(\frac{2}{\pi}\right)^{\frac{1}{2}}(1+t^2)^{-\frac{1}{2}}g_{nj}\left(\frac{2}{\pi}\arctan t\right)dt$$
$$= \left(\frac{\pi}{2}\right)^{\frac{1}{2}}\int_0^1 \frac{x(\tan(\pi s/2))}{\cos(\pi s/2)}\,g_{nj}(s)\,ds = 0.$$

It is easy to see that $\{g_{nj}\}$ is complete in $L^1[0,1]$, that is, if $x \in L^1[0,1]$ and $\int_0^1 x(t)g_{nj}(t)\,dt = 0$ for each g_{nj}, then $x = 0$ a.e. This implies the completeness of $\{h_{nj}\}$ in $L_0^1[0,\infty)$. Since $L_0^1[0,\infty)$ contains $L^2[0,\infty)$, it follows that $\{h_{nj}\}$ is complete in $L^2[0,\infty)$. □

Let $H_{nj}(t) = \int_0^t h_{nj}(u)\,du$.

Lemma 2.1.2. $\sum_{j=0}^{2^{n-1}-1} |H_{nj}(t)| \leq \sqrt{\pi}2^{-n/2}(a+\frac{1}{2})$ *for* $t \in [0,a]$, $a < \infty$.

PROOF. Let $\delta = (2/\pi)\arctan t$. Then

$$H_{nj}(t) = \int_0^t h_{nj}(u)\,du = \sqrt{\frac{\pi}{2}}\int_0^\delta \left(1+\tan^2\frac{\pi s}{2}\right)^{-\frac{1}{2}} \frac{g_{nj}(s)\,ds}{\cos^2(\pi s/2)}$$
$$= \sqrt{\frac{\pi}{2}}\int_0^\delta \frac{g_{nj}(s)\,ds}{\cos(\pi s/2)}$$
$$= \sqrt{\frac{\pi}{2}}\frac{G_{nj}(s)}{\cos(\pi s/2)}\bigg|_0^\delta - \left(\frac{\pi}{2}\right)^{\frac{3}{2}}\int_0^\delta G_{nj}(s)\frac{\sin(\pi s/2)}{\cos^2(\pi s/2)}\,ds$$
$$= \sqrt{\frac{\pi}{2}}\frac{G_{nj}(\delta)}{\cos(\pi s/2)} - \left(\frac{\pi}{2}\right)^{\frac{3}{2}}\int_0^\delta G_{nj}(s)\frac{\sin(\pi s/2)}{\cos^2(\pi s/2)}\,ds.$$

It is easy to see that $\sum_{j=0}^{2^{n-1}-1} G_{nj}(s) \leq 2^{-(n+1)/2}$, and hence

$$\sum_{j=0}^{2^{n-1}-1} |H_{nj}(t)| \leq \sqrt{\frac{\pi}{2}}\, 2^{-(n+1)/2}\left(\frac{1}{\cos(\pi\delta/2)} + \frac{\pi}{2}\int_0^{\delta} \frac{\sin(\pi s/2)}{\cos^2(\pi s/2)}\, ds\right)$$

$$= \sqrt{\frac{\pi}{2}}\, 2^{-(n+1)/2}\left(\frac{2}{\cos(\pi\delta/2)} - 1\right)$$

$$= \sqrt{\frac{\pi}{2}}\, 2^{-(n+1)/2}[2(t^2+1)^{\frac{1}{2}} - 1] < \sqrt{\pi}2^{-n/2}\left(a + \frac{1}{2}\right). \quad \square$$

Consider now the series

$$\sum_{n=0}^{\infty} \sum_{j \in S_n} y_{nj}(\omega) H_{nj}(t). \tag{2.1.4}$$

Let $f_n(t,\omega) = \sum_{j \in S_n} y_{nj}(\omega) H_{nj}(t)$. For $K_a = [0,a]$, we have

$$\max_{t \in K_a} |f_n(t,\omega)| \leq Y_n(\omega) \sum_{j \in S_n} |H_{nj}(t)|$$

$$\leq Y_n(\omega) \cdot [\sqrt{\pi} 2^{-n/2}(a + \tfrac{1}{2})].$$

From Equation (2.1.3) we have $P(\Omega_0) = 1$, and for $\omega \in \Omega_0$, there exists $n_0(\omega)$ such that for all $n \geq n_0(\omega)$,

$$\max_{K_a} |f_n(t,\omega)| \leq \sqrt{\pi}(2a+1)2^{-n/2}(n \log 2)^{\frac{1}{2}}.$$

Hence again,

$$\sum_{n=0}^{\infty} \max_{K_a} |f_n(t,\omega)| < \infty \qquad \text{(a.s.)},$$

and the series (2.1.4) converges uniformly on each interval K_a for P-a.a. ω. Define $W_t(\omega)$ to be the sum of the series (2.1.4) if $\omega \in \Omega_0$ and to be zero for all $t \in \mathbf{R}_+$ if $\omega \notin \Omega_0$. Then $W = (W_t)$, $t \in \mathbf{R}_+$, is a Wiener process on $(\Omega, \mathscr{A}, P)$.

Let $C(\mathbf{R}_+)$ be the space of real-valued functions on $\mathbf{R}_+$ and $\mathscr{B}(C)$ the σ-field generated by the finite-dimensional Borel cylinder sets in $C(\mathbf{R}_+)$. Wiener measure in $C(\mathbf{R}_+)$ is defined in the same way as in the case of $C[0,T]$. Let $\mu_W = P\xi^{-1}$, the measure induced on $\mathscr{B}(C)$ by the $\mathscr{B}(C)/\mathscr{A}$-measurable map $\xi\colon \Omega \to C(\mathbf{R}_+)$ given by $\xi(\omega)(t) = W_t(\omega)$.

If $\mathscr{A}(C)$ denotes the completion of $\mathscr{B}(C)$ with respect to μ_W, the coordinate process on $C(\mathbf{R}_+)$ is then a Wiener process on $(C(\mathbf{R}_+), \mathscr{A}(C), \mu_W)$ and μ_W on $\mathscr{A}(C)$ is Wiener measure. It may also be noted that $C(\mathbf{R}_+)$ can be regarded as a complete, separable, metric space under the topology of uniform convergence on compact sets of $\mathbf{R}_+$ and that the topological Borel field coincides with $\mathscr{B}(C)$.

2.2 Martingales and Supermartingales

The definitions given below are for processes on the parameter interval $\mathbf{R}_+$. The analogous statements for the cases $\bar{\mathbf{R}}_+$ and $[0,T]$ are left to the reader.

The real-valued process (X_t) will be called a *martingale relative* to $(\mathscr{F}_t)$ if

(i) For each t, X_t is integrable and $\mathscr{F}_t$-measurable.
(ii) For $t > s$, $E(X_t|\mathscr{F}_s) = X_s$ a.s.

Alternatively, we call the system $(X_t,\mathscr{F}_t)$ a *martingale*. When other probability measures on $(\Omega,\mathscr{A})$ are under consideration, we may call (X_t) an $(\mathscr{F}_t,P)$-*martingale*. Usually the family $(\mathscr{F}_t)$ is clear from context, so that (X_t) is simply called a *martingale*. (See Section 1.1.)

Similarly, the real-valued process (X_t) is called a *supermartingale* (*submartingale*) with respect to $(\mathscr{F}_t)$ if (i) above is satisfied and (ii) holds with "$=$" replaced by "$\leq$" (respectively, by "$\geq$"). The following are some important facts which we state here for convenience and future reference. (See [41].)

1. Let (X_t) be a supermartingale. Then (X_t) is a martingale if and only if $E(X_t)$ is a constant independent of t.

2. Let (X_t) be a martingale (supermartingale) relative to $(\mathscr{F}_t)$, and let f be a real-valued concave (concave increasing) function such that the random variables $f(X_t)$ are integrable. Then the process $(f(X_t))$ is a supermartingale.

3. Let (X_t) be an arbitrary supermartingale. Then almost every sample path $t \to X_t(\omega)$ has left and right limits at every point of $\mathbf{R}_+$ in the following sense. Let S be any countable dense set in $\mathbf{R}_+$. Then $\lim_{s\to t,\, s\in S,\, s<t} X_s(\omega)$ and $\lim_{s\to t,\, s\in S,\, s>t} X_s(\omega)$ exist for each $t \in \mathbf{R}_+$, for a.e. ω.

4. Assume that the family $(\mathscr{F}_t)$ is right-continuous. Then a supermartingale (X_t) admits a right-continuous modification if and only if the function $E(X_t)$ is right-continuous. In particular, a martingale (X_t) always has a right-continuous modification.

5. The following statements are equivalent:

(i) A family $\mathscr{H}$ of integrable random variables [that is, $\mathscr{H} \subset L^1 = L^1(\Omega,\mathscr{A},P)$] is uniformly integrable.
(ii) $\mathscr{H}$ is relatively compact in L^1 in the weak topology $\sigma(L^1,L^\infty)$.
(iii) Every sequence of elements of $\mathscr{H}$ contains a subsequence that converges in the sense of the topology $\sigma(L^1,L^\infty)$.

Note that by definition, (i) holds if and only if

$$\lim_{n\to\infty} \sup_{f\in\mathscr{H}} \int_{[|f|\geq n]} |f|\, dP = 0.$$

Equivalently, for any $\varepsilon > 0$, there is a $\delta > 0$ such that $P(A) < \delta$ implies $\int_A |f|\, dP < \varepsilon$ for all $f \in \mathscr{H}$ and $E|f|$ is bounded for $f \in \mathscr{H}$.

6. Let (X_t), $t \in \mathbf{R}_+$, be a right-continuous supermartingale. Let $I = [a,b]$ be a compact interval of $\mathbf{R}_+$ and λ a positive constant. Then

(i) $P\left\{\sup_{t \in I} X_t \geq \lambda\right\} \leq \frac{1}{\lambda}[E(X_a) + E(X_b^-)]$

(ii) $P\left\{\inf_{t \in I} X_t \leq -\lambda\right\} \leq \frac{E|X_b|}{\lambda}$

7. Let (X_t) be a positive right-continuous submartingale. Let p and q be numbers such that $1 < p < \infty$ and $1/q + 1/p = 1$. Then if $I = [a,b]$, we have the inequality,

$$\left\|\sup_{t \in I} X_t\right\|_p \leq q\|X_b\|_p,$$

that is,

$$E\left(\sup_{t \in I} X_t\right)^p \leq q^p E(X_b)^p.$$

In particular, let (M_t) be a right-continuous martingale. Then we have the following inequality due to Doob. Suppose $E|M_t|^p < \infty$ for each t, where $1 < p < \infty$, then

$$E\left(\sup_{t \in I} |M_t|\right)^p \leq q^p E|M_b|^p.$$

It follows that the right-hand side of the above inequality majorizes $\lambda^p P[\sup_{t \in I} |M_t| \geq \lambda]$.

8. Let (X_t) be a right-continuous supermartingale.

(i) Assume that $\sup_t E(X_t^-) < \infty$, X_t^- being $-X_t$ if $X_t \leq 0$ and 0 otherwise. Then $\lim_{t \to \infty} X_t = X_\infty$ exists (a.s.) and X_∞ is integrable.
(ii) If $X_t \geq 0$ for each t, it follows from (i) that (X_t), $t \in \bar{\mathbf{R}}_+$, is a supermartingale.
(iii) If the random variables (X_t) are uniformly integrable, again (i) applies and (X_t), $t \in \bar{\mathbf{R}}_+$, is a supermartingale. Furthermore, $E|X_t - X_\infty| \to 0$ as $t \to \infty$.
(iv) If (M_t), $t \in \mathbf{R}_+$, is a right-continuous martingale and the M_t are uniformly integrable, $\lim_{t \to \infty} M_t = M_\infty$ exists (a.s.), (M_t), $t \in \bar{\mathbf{R}}_+$, is a martingale, and $M_t \to M_\infty$ in L_1 norm.

9. *Optional Sampling Theorem.* Let (X_t), $t \in \mathbf{R}_+$, be a right-continuous supermartingale. Suppose that $(\mathscr{F}_t)$ is right-continuous and that (X_t) has the following property:

There exists an integrable random variable Y such that for each t

$$X_t \geq E(Y|\mathscr{F}_t) \qquad \text{(a.s.).}$$

Let τ_1 and τ_2 be two stopping times such that $\tau_1 \leq \tau_2$. Then X_{τ_1}, X_{τ_2} are integrable and

$$X_{\tau_1} \geq E(X_{\tau_2}|\mathscr{F}_{\tau_1}) \qquad \text{(a.s.)}$$

10. Let (X_t), $t \in \mathbf{R}_+$, be a right-continuous, uniformly integrable martingale, and let $(\mathscr{F}_t)$ be right-continuous. Since $X_t = E(X_\infty|\mathscr{F}_t)$, the condition stated in 9 holds for (X_t) as well as for $(-X_t)$ by taking Y to be X_∞ and $-X_\infty$, respectively. The optional sampling theorem (9) applied to (X_t) and $(-X_t)$ then yields the following result: for τ_1 and τ_2 two stopping times such that $\tau_1 \leq \tau_2$, we have

$$E(X_{\tau_2}|\mathscr{F}_{\tau_1}) = X_{\tau_1} \qquad \text{(a.s.)}.$$

In particular, if τ is a stopping time, then $(X_{t\wedge\tau}, \mathscr{F}_{t\wedge\tau})$, $t \in \mathbf{R}_+$, is a martingale. It will be seen in Section 2.6 (Lemma 2.6.1) that $(X_{t\wedge\tau}, \mathscr{F}_t)$ is also a martingale.

11. Let (X_t), $t \in [0,T]$, be a right-continuous, $(\mathscr{F}_t)$-adapted process such that $E|X_\tau| < \infty$ and $E(X_\tau) = 0$ for every stopping time τ of $(\mathscr{F}_t)$. Then $(X_t, \mathscr{F}_t)$ is a uniformly integrable martingale.

Proof. Fix $t \in [0,T]$ and let $A \in \mathscr{F}_t$. Define $\tau(\omega) = t$ if $\omega \in A$ and $\tau(\omega) = T$ if $\omega \in A^c$. Note that the right-continuity of (X_t) ensures that X_τ is a random variable. By hypothesis,

$$0 = E(X_\tau) = \int_A X_t\,dP + \int_{A^c} X_T\,dP.$$

Also

$$\int_A X_T\,dP + \int_{A^c} X_T\,dP = E(X_T) = 0.$$

Hence

$$\int_A X_T\,dP = \int_A X_t\,dP.$$

Since this equality holds for all $A \in \mathscr{F}_t$, we have

$$X_t = E(X_T|\mathscr{F}_t). \qquad \square$$

2.3 Properties of Wiener Processes—Wiener Martingales

Definition 2.3.1. A stochastic process (X_t), $t \in \mathbf{R}_+$, is called a *Wiener martingale* with respect to an increasing σ-field family $(\mathscr{F}_t)$ if

(i) (X_t) is a Wiener process.
(ii) $(X_t, \mathscr{F}_t)$ is a martingale.

A Wiener process $W = (W_t)$ is a Wiener martingale with respect to the family $(\mathscr{F}_t^W)$. It is easy to find $(\mathscr{F}_t)$ such that $\mathscr{F}_t \neq \mathscr{F}_t^W$ and for which $(W_t, \mathscr{F}_t)$ is a martingale.

Definition 2.3.2. $W = (W_t)$, where $W_t = (W_t^1, \ldots, W_t^d)$, is a *d-dimensional (standard) Wiener process* if each (W_t^j) $(j = 1, \ldots, d)$ is a (standard) Wiener process and the σ-fields $\mathscr{F}_\infty^{W_j}$ are independent.

We state here some important and well-known properties of Wiener processes. Let (W_t), $t \in \mathbf{R}_+$, be a (standard) Wiener process.

1.

(i) $P\{\omega$: the trajectory $t \to W_t(\omega)$ is differentiable$\} = 0$.

(ii) $P\{\omega$: the trajectory $t \to W_t(\omega)$ is of bounded variation in any finite interval$\} = 0$.

2.

(i) Law of the iterated logarithm:

$$P\left\{\omega: \limsup_{t\to\infty} \frac{W_t(\omega)}{\sqrt{2t\log\log t}} = 1\right\} = 1.$$

(ii) Local law of the iterated logarithm:

$$P\left\{\omega: \limsup_{t\downarrow 0} \frac{W_t(\omega)}{\sqrt{2t\log\log(1/t)}} = 1\right\} = 1.$$

(iii)
$$P\left\{\omega: \limsup_{\substack{0\le s<t\\ t-s\downarrow 0}} \frac{|W_t(\omega) - W_s(\omega)|}{\sqrt{2(t-s)\log 1/(t-s)}} = 1\right\} = 1.$$

Let us now assume that (W_t) is a d-dimensional Wiener process.

3. Let $\mathscr{B}$ be the family of all real-valued, bounded Borel measurable functions on $\mathbf{R}^d$. For each $t > 0$ and $f \in \mathscr{B}$, define the function $P_t f$ on $\mathbf{R}^d$ by

$$(P_t f)(x) = (2\pi t)^{-d/2} \int_{\mathbf{R}^d} f(y) \exp\left[-\frac{|y-x|^2}{2t}\right] dy.$$

(i) $P_t f \in \mathscr{B}$.

(ii) For $0 < s < t$ and $f \in \mathscr{B}$,

$$(P_{t-s}f)(x) = E(f(W_t)\,|\,W_s = x) \qquad \text{(a.e.)}$$

with respect to Lebesgue measure on $\mathbf{R}^d$.

(iii) $E(f(W_t)\,|\,\mathscr{F}_s^W) = E(f(W_t)\,|\,W_s) = (P_{t-s}f)(W_s)$ (a.s.).

The equation in (iii) shows that (W_t) is a Markov process. (See Section 5.4 for the definition of a Markov process.)

4. $\mathscr{F}_{t+}^W = \mathscr{F}_t^W$ for each $t \ge 0$.

Proof of 4. Fix $t \ge 0$ and let H be the family of all bounded, real-valued, $\mathscr{F}_t^{W,0}$-measurable random variables such that

$$E(h\,|\,\mathscr{F}_{t+}^W) = E(h\,|\,\mathscr{F}_t^W) \qquad \text{(a.s.)} \tag{2.3.1}$$

For $f \in \mathscr{B}$ and $s \leq t$, Equation (2.3.1) clearly holds for $h = f(W_s)$. Suppose $s > t$ and let $s > u > t$. By property 3(iii) we have

$$E(f(W_s)|\mathscr{F}_u^W) = (P_{s-u}f)(W_u)$$
$$= (2\pi(s-u))^{-d/2} \int_{\mathbf{R}^d} f(y) \exp\left(-\frac{|y - W_u|^2}{2(s-u)}\right) dy.$$

For all ω whose trajectories are continuous, as $u \downarrow t$, this last expression converges to $(P_{s-t}f)(W_t)$ evaluated at ω. Hence as $u \downarrow t$, $E(f(W_s)|\mathscr{F}_u^W)$ converges (a.s.) to $(P_{s-t}f)(W_t) = E(f(W_s)|\mathscr{F}_t^W)$. On the other hand, according to a standard martingale theorem, $E(f(W_s)|\mathscr{F}_u^W) \to E(f(W_s)|\mathscr{F}_{t+}^W)$ (a.s.) as $u \downarrow t$. Hence Equation (2.3.1) holds for $f(W_s)$ if $s > t$. Next, let $h = f_1(W_{s_1})f_2(W_{s_2})$, where f_1 and f_2 are in $\mathscr{B}$ and $0 \leq s_1 < s_2$. We now show that $h \in H$. The case $s_2 \leq t$ is trivial, and the case $s_1 \leq t < s_2$ follows from the first step.

Suppose then that $t < s_1 < s_2$. For any u such that $t \leq u < s_1$, we have

$$\begin{aligned} E[(f_1 \circ W_{s_1})(f_2 \circ W_{s_2})|\mathscr{F}_u^W] &= E[(f_1 \circ W_{s_1})E(f_2 \circ W_{s_2}|\mathscr{F}_{s_1}^W)|\mathscr{F}_u^W] \\ &= E[(f_1 \circ W_{s_1})[(P_{s_2-s_1}f_2) \circ W_{s_1}]|\mathscr{F}_u^W] \\ &= E[g \circ W_{s_1}|\mathscr{F}_u^W], \end{aligned}$$

where $g = f_1(P_{s_2-s_1}f_2)$.

Considering the limit of both sides as $u \downarrow t$, we obtain

$$E[h|\mathscr{F}_{t+}^W] = E[g(W_{s_1})|\mathscr{F}_{t+}^W].$$

By the same equation for $u = t$,

$$E[h|\mathscr{F}_t^W] = E[g(W_{s_1})|\mathscr{F}_t^W].$$

Since $g \in \mathscr{B}$, we conclude from the first step

$$E[g(W_{s_1})|\mathscr{F}_{t+}^W] = E[g(W_{s_1})|\mathscr{F}_t^W].$$

Hence $h \in H$.

This argument immediately extends to show that all functions h of the form $(f_1 \circ W_{s_1}) \cdots (f_n \circ W_{s_n})$ for arbitrary n and $s_j \geq 0$ must also belong to H. Then by an argument using the monotone class theorem, H contains all bounded random variables measurable with respect to the σ-field $\mathscr{F}_\infty^{W,0}$. In particular, if h is the indicator function of a set A in $\mathscr{F}_{t+}^W$, $E(h|\mathscr{F}_t^W) = h$ (P-a.s.), since h is (a.s.) equal to a random variable measurable with respect to $\mathscr{F}_\infty^{W,0}$. Since $\mathscr{F}_t^W$ contains all P-null sets, $A \in \mathscr{F}_t^W$. □

5. Let τ be a stopping time for W, that is, a stopping relative to the family $(\mathscr{F}_t^W)$. Assume that τ is finite. Then

(a) $Y_t = W_{t+\tau} - W_\tau$ for $t \geq 0$ is a Wiener process.
(b) $\sigma[Y_t: t \geq 0] \perp\!\!\!\perp \mathscr{F}_\tau$.

The proof below is given for the case $d = 1$. The same argument works for any Wiener martingale $(W_t, \mathscr{F}_t)$ with τ a stopping time relative to $(\mathscr{F}_t)$; so in the proof we have written $\mathscr{F}_t$ instead of $\mathscr{F}_t^W$.

PROOF. First, note that since W is continuous, W_τ and $W_{t+\tau}$ are random variables.

(i) Assume that τ has countable range (s_j). For $B \in \mathscr{F}_\tau$, $0 \leq t_1 < t_2 < \cdots < t_k$, and $A_1, \ldots, A_k$ any Borel sets,

$$\begin{aligned} P\{Y_{t_1} \in A_1, \ldots, Y_{t_k} \in A_k, B\} &= \sum_j P\{Y_{t_1} \in A_1, \ldots, Y_{t_k} \in A_k, B, \tau = s_j\} \\ &= \sum_j P\{W_{t_1+s_j} - W_{s_j} \in A_1, \ldots, W_{t_k+s_j} \\ &\qquad - W_{s_j} \in A_k, B, \tau = s_j\}. \end{aligned}$$

Since $B \cap [\tau = s_j] = (B \cap [\tau \leq s_j]) \cap (\tau = s_j) \in \mathscr{F}_{s_j}$ and $\sigma[W_{t+s_j} - W_{s_j}: t \geq 0] \perp\!\!\!\perp \mathscr{F}_{s_j}$, the right-hand side above equals

$$\begin{aligned} &\sum_j P\{W_{t_1+s_j} - W_{s_j} \in A_1, \ldots, W_{t_k+s_j} - W_{s_j} \in A_k\} P\{\tau = s_j, B\} \\ &\qquad = \sum_j P\{W_{t_1} \in A_1, \ldots, W_{t_k} \in A_k\} P\{\tau = s_j, B\} \\ &\qquad = P\{W_{t_1} \in A_1, \ldots, W_{t_k} \in A_k\} P(B). \end{aligned}$$

We thus obtain

$$P\{Y_{t_1} \in A_1, \ldots, Y_{t_k} \in A_k, B\} = P\{W_{t_1} \in A_1, \ldots, W_{t_k} \in A_k\} P(B). \quad (2.3.2)$$

Taking $B = \Omega$ in Equation (2.3.2), it follows that Y_t is a Wiener process, continuous by definition. Now allowing B to be any element of $\mathscr{F}_\tau$, (b) follows.

(ii) Let $\tau_n \downarrow \tau$, where each τ_n has discrete range, and define $Y_t^n = W_{t+\tau_n} - W_{\tau_n}$. Note that since $\tau_n \geq \tau$, $\mathscr{F}_\tau \subset \mathscr{F}_{\tau_n}$. Taking $B \in \mathscr{F}_\tau$ in Equation (2.3.2) (with Y_t replaced by Y_t^n), we obtain

$$P[Y_{t_1}^n \leq a_1, \ldots, Y_{t_k}^n \leq a_k, B] = P[W_{t_1} \leq a_1, \ldots, W_{t_k} \leq a_k] P(B).$$

Observe that for every $t \geq 0$, $Y_t^n \to Y_t$ (P-a.s.), and therefore $(Y_{t_i}^n, Y_{t_2}^n, \ldots, Y_{t_k}^n) \to (Y_{t_1}, \ldots, Y_{t_k})$ (P-a.s.) (using continuity of W). We would like to prove that for all $a_1, \ldots, a_k$, and $B \in \mathscr{F}_\tau$,

$$P[Y_{t_1} \leq a_1, \ldots, Y_{t_k} \leq a_k, B] = P[W_{t_1} \leq a_1, \ldots, W_{t_k} \leq a_k] P(B). \quad (2.3.3)$$

Notice that Equation (2.3.3) holds if $P(B) = 0$. Therefore fix $B \in \mathscr{F}_\tau$ such that $P(B) > 0$. Let m be the probability measure $m(A) = P(A \cap B)/P(B)$. Since $m \ll P$, we have $(Y_{t_1}^n, \ldots, Y_{t_k}^n) \to (Y_{t_1}, \ldots, Y_{t_k})$ (a.s.) m. Hence for any $\alpha_1, \ldots, \alpha_k$ real, $\int_\Omega \exp i[\alpha_1 Y_{t_1}^n + \cdots + \alpha_k Y_{t_k}^n]\, dm$ converges by the bounded convergence theorem to $\int_\Omega \exp i[\alpha_1 Y_{t_1} + \cdots + \alpha_k Y_{t_k}]\, dm$. It follows that for all points $(a_1, \ldots, a_k)$ of continuity of $\mathscr{L}_m(Y_{t_1}, \ldots, Y_{t_k})$ (the distribution law of this vector with respect to m), $m[Y_{t_1}^n \leq a_1, \ldots, Y_{t_k}^n \leq a_k] \to m[Y_{t_1} \leq a_1, \ldots, Y_{t_k} \leq a_k]$. Then by definition of m, Equation (2.3.3) holds for such $(a_1, \ldots, a_k)$. But then Equation (2.3.3) holds for all vectors $(a_1, \ldots, a_k)$, and $B \in \mathscr{F}_\tau$. Taking $B = \Omega$ in Equation (2.3.3) shows that $Y = (Y_t)$ is a Wiener process. To get (b), write first from (a) $\mathscr{L}[Y_{t_1}, \ldots, Y_{t_k}] = \mathscr{L}[W_{t_1}, \ldots, W_{t_k}]$. Then

Equation (2.3.3) can be rewritten in the form

$$P\{Y_{t_1} \leq a_1, \ldots, Y_{t_k} \leq a_k, B\} = P\{Y_{t_1} \leq a_1, \ldots, Y_{t_k} \leq a_k\}P(B).$$

This completes the proof of (5). □

The above property (5) defines the strong Markov property of the Wiener process. This proof also gives the strong Markov property for (W_t) replaced by a right-continuous process (X_t) which has stationary, independent increments, that is, such that for each $t > s \geq 0$, $(X_t - X_s) \perp\!\!\!\perp \sigma[X_u: u \leq s]$ and the distribution of $X_t - X_s$ depends only on $t - s$.

2.4 Decomposition of Supermartingales

The main aim of this section is to establish the Doob-Meyer decomposition for right-continuous potentials. This problem is first considered for discrete parameter supermartingales, for which all the necessary concepts must be introduced. The existence of Doob's decomposition of a potential in the continuous parameter case is then obtained as a natural generalization from the discrete parameter case.

Let $(\Omega, \mathscr{A}, P)$ be a complete probability space. The condition of right-continuity and assumption (ii) stated at the beginning of Chapter 1 will be in force throughout this section and Sections 2.5 and 2.6. We will first confine our attention to discrete parameter supermartingales.

Definition 2.4.1. Call $(X_n, \mathscr{F}_n)$, $n \geq 0$, a (discrete parameter) *supermartingale* if each X_n is a real-valued, integrable, $\mathscr{F}_n$-measurable random variable, where $\mathscr{F}_n \subset \mathscr{F}_{n+1}$ for $n \geq 0$, and $E(X_m | \mathscr{F}_n) \leq X_n$ (a.s.) for $m \geq n \geq 0$. Such a system is called a *martingale* if, instead, $E(X_m | \mathscr{F}_n) = X_n$. (These notions can in fact be viewed as special cases of their continuous parameter analogs.)

Theorem 2.4.1 (Doob Decomposition in Discrete Case). *A supermartingale* $(X_n, \mathscr{F}_n)$, $n \geq 0$, *has exactly one decomposition*:

$$X_n = M_n - A_n \qquad \text{(a.s.)}, \tag{2.4.1}$$

where (M_n, F_n) *is a martingale and* $(A_n)_{n \geq 0}$ *is a sequence of random variables such that*

(i) $A_0 = 0$ (a.s.).
(ii) $A_n \leq A_{n+1}$ (a.s.).
(iii) A_{n+1} is $\mathscr{F}_n$*-measurable* $(n \geq 0)$.

PROOF. Define the random variables M_n inductively as

$$M_0 = X_0,$$
$$M_n = M_{n-1} + [X_n - E(X_n | \mathscr{F}_{n-1})] \quad \text{for } n \geq 1.$$

Clearly M_n is $\mathscr{F}_n$-measurable, and

$$E(M_n|\mathscr{F}_{n-1}) = M_{n-1}.$$

Thus $(M_n, \mathscr{F}_n)_{n \geq 0}$ is a martingale. Now define A_n by

$$A_n = M_n - X_n.$$

Note that for each $n \geq 1$,

$$A_n = A_{n-1} + [X_{n-1} - E(X_n|\mathscr{F}_{n-1})]. \tag{2.4.2}$$

Since $(X_n, \mathscr{F}_n)$ is a supermartingale, $A_n \geq A_{n-1}$ (a.s.) for every $n \geq 1$. Also it follows that $A_n \geq 0$ (a.s.) and that A_n is $\mathscr{F}_{n-1}$-measurable. This concludes the proof of existence of (M_n) and (A_n) such that the above conditions hold.

To show uniqueness, let (B_n), $n \geq 0$, be a sequence with properties (i) to (iii) above and $(Z_n, \mathscr{F}_n)_{n \geq 0}$ a martingale such that $X_n = Z_n - B_n$. It suffices to note that B_n satisfies Equation (2.4.2).

Let $(Y_n, \mathscr{F}_n)$, $n \geq 0$, be any bounded martingale and consider

$$\begin{aligned} E\left[\sum_{k=1}^{n} (Y_k - Y_{k-1})(A_k - A_{k-1})\right] &= \sum_{k=1}^{n} E[E((Y_k - Y_{k-1})(A_k - A_{k-1})|\mathscr{F}_{k-1})] \\ &= \sum_{k=1}^{n} E[(A_k - A_{k-1})E[(Y_k - Y_{k-1})|\mathscr{F}_{k-1}]]. \end{aligned}$$

Since $(Y_k, \mathscr{F}_k)$ is a martingale, this last expression equals 0. Hence,

$$E\sum_{k=1}^{n} Y_k(A_k - A_{k-1}) = E\sum_{k=1}^{n} Y_{k-1}(A_k - A_{k-1}).$$

But on the left side, we have the simplification

$$\begin{aligned} E\sum_{k=1}^{n} Y_k(A_k - A_{k-1}) &= \sum_1^n EY_kA_k - \sum_1^n E[E(Y_kA_{k-1}|\mathscr{F}_{k-1})] \\ &= \sum_1^n EY_kA_k - \sum_1^n E[A_{k-1}E(Y_k|\mathscr{F}_{k-1})] \\ &= \sum_1^n EY_kA_k - \sum_1^n E(A_{k-1}Y_{k-1}) \\ &= EY_nA_n - EA_0Y_0 = EY_nA_n. \end{aligned}$$

Therefore,

$$EY_nA_n = E\sum_1^n Y_{k-1}(A_k - A_{k-1}).$$

Conversely, let this equation hold for each $n \geq 1$ and any bounded martingale $(Y_n, \mathscr{F}_n)$, $n \geq 0$, but with A_n in Theorem 2.4.1 only supposed to be $\mathscr{F}_n$-adapted. It then follows that $E(ZA_n) = E(E(Z|\mathscr{F}_{n-1})A_n)$ for any $n \geq 1$ and for any bounded $\mathscr{F}_n$-measurable random variable Z. For $n = 1$ this is simply a restatement of the above equation with $Y_1 = Z$ and $Y_0 = E(Z|\mathscr{F}_0)$.

For $n > 1$, let $Y_k = E(Z|\mathscr{F}_k)$ for $k \geq 0$. Assume by induction, $EY_kA_k = E(E(Y_k|\mathscr{F}_{k-1})A_k) = E(Y_{k-1}A_k)$ for $1 \leq k < n$. Then

$$\begin{aligned} EY_nA_n &= E\sum_1^n Y_{k-1}(A_k - A_{k-1}) \\ &= E\sum_1^n Y_{k-1}A_k - E\sum_1^n Y_{k-1}A_{k-1} \\ &= EY_{n-1}A_n + \sum_1^{n-1} E(Y_{k-1} - Y_k)A_k - EY_0A_0 \\ &= EY_{n-1}A_n. \end{aligned}$$

Therefore, $EZA_n = E(E(Z|\mathscr{F}_{n-1})A_n)$, completing the induction. Since for any $n \geq 1$ $E(E(Z|\mathscr{F}_{n-1})A_n) = E(ZE(A_n|\mathscr{F}_{n-1}))$, it follows that $A_n = E(A_n|\mathscr{F}_{n-1})$ a.s. and therefore that each A_n is in fact $\mathscr{F}_{n-1}$-measurable for any $n \geq 1$.

Definition 2.4.2. A supermartingale $(X_n, \mathscr{F}_n)$, $n \geq 0$, is called a *potential* if

(a) $X_n \geq 0$ (a.s.) for all n.
(b) $\lim_{n\to\infty} EX_n = 0$.

Remark 2.4.1. If the supermartingale $(X_n, \mathscr{F}_n)$, $n \geq 0$, is nonnegative, then the A_n's of Theorem 2.4.1 satisfy the following inequality:

$$EA_n \leq EM_0.$$

Therefore, $A_\infty = \lim_{n\to\infty} A_n$ exists (a.s.) and has finite expectation.

Remark 2.4.2. If $(X_n, \mathscr{F}_n)$, $n \geq 0$, is a potential, then the martingale $M_n = A_n + X_n$ is uniformly integrable because both $\{A_n\}$, $n \geq 0$, and $\{X_n\}$, $n \geq 0$, are uniformly integrable. By the martingale convergence theorem,

$$M_\infty = \lim_{n\to\infty} M_n \quad \text{and} \quad X_\infty = \lim_{n\to\infty} X_n$$

exist (a.s.) and in L^1, and

$$M_n = E(M_\infty|\mathscr{F}_n).$$

Since $(X_n, \mathscr{F}_n)$ is a potential, Fatou's lemma implies that $X_\infty = 0$. Thus

$$A_\infty = M_\infty.$$

We state this now as

Theorem 2.4.2. *A potential $(X_n, \mathscr{F}_n)$, $n \geq 0$, can be written uniquely in the form*

$$X_n = E(A_\infty|\mathscr{F}_n) - A_n \quad \textit{for all } n \geq 0,$$

where $A_0 = 0$, $A_n \leq A_{n+1}$ (a.s.), A_{n+1} is $\mathscr{F}_n$-measurable, and A_∞ is integrable, with $A_\infty = \lim_{n\to\infty} A_n$ (a.s.) and in the L^1 sense.

Remark 2.4.3. If (A_n), $n \geq 0$, is a sequence satisfying (i) and (ii) of Theorem 2.4.1, and if $A_\infty = \lim A_n$ has finite expectation, then A_{n+1} is $\mathscr{F}_n$-measurable for $n \geq 0$ if and only if

$$E\left[\sum_{1}^{\infty} Y_{k-1}(A_k - A_{k-1})\right] = EY_\infty A_\infty$$

for every bounded martingale $(Y_n, \mathscr{F}_n)$, $n \geq 0$.

We now turn to the task of considering the analogous theory in the case of supermartingales which are defined on the parameter intervals $\mathbf{R}_+$ or $[0,T]$. The essential features of the decomposition carry over in this case, but only for a certain class of supermartingales to be defined later. In the rest of this section and in Sections 2.5 and 2.6 we suppose that the family $(\mathscr{F}_t)$ is right-continuous.

Definition 2.4.3. A real stochastic process (A_t), $t \in \mathbf{R}_+$, is called an *increasing process* with respect to an increasing family $(\mathscr{F}_t)$ if

(a) (A_t) is $(\mathscr{F}_t)$-adapted.
(b) $A_0 = 0$ and $A_t(\omega)$ is an increasing right-continuous function of t, for a.a. ω.
(c) $E(A_t) < \infty$ for each $t \in \mathbf{R}_+$.

The increasing process (A_t) is said to be *integrable* if $\sup_t E(A_t) < \infty$. If the index set is the interval $[0,T]$, then the definition of an increasing process implies that it is integrable.

Remark 2.4.4. Let (A_t) be an increasing process and (X_t) a measurable, non-negative process. For each fixed $\omega \in \Omega$, the function $t \to X_t(\omega)$ is measurable. We can consider the Lebesgue-Stieltjes integral on $\mathbf{R}_+$ for each ω:

$$\int_0^\infty X_t(\omega)\, dA_t(\omega).$$

If, in particular, (X_t) is progressively measurable with respect to an increasing family $(\mathscr{F}_t)$ of sub-σ-fields of $\mathscr{A}$, then

$$Y_t(\omega) = \int_{[0,t]} X_s(\omega)\, dA_s(\omega)$$

is $\mathscr{F}_t$-measurable for each $t \in \mathbf{R}_+$. In the future we will use the notation $\int_0^t X_s(\omega)\, dA_s(\omega)$ always with integration on the interval $[0,t]$ in mind. The process (Y_t) then is right-continuous and therefore progressively measurable. Consequently, for each stopping time τ,

$$Y_\tau = \int_0^\tau X_s\, dA_s$$

is $\mathscr{F}_\tau$-measurable.

Definition 2.4.4. An increasing process is called *natural* if

$$E\int_0^t Y_s\,dA_s = E\int_0^t Y_{s-}\,dA_s$$

for every $t \in \mathbf{R}_+$ and each nonnegative, bounded right-continuous martingale (Y_t).

Theorem 2.4.3. *Let (A_t) be an integrable increasing process. Then A is natural if and only if*

$$E\left[\int_0^\infty Y_s\,dA_s\right] = E\left[\int_0^\infty Y_{s-}\,dA_s\right]$$

for every nonnegative, bounded right-continuous martingale (Y_t).

For the proof of this fact, we refer the reader to Theorem 19 on page 112 of [41].

Definition 2.4.5. Let $(X_t)_{t \in \mathbf{R}_+}$ be a right-continuous supermartingale relative to the family $(\mathscr{F}_t)_{t \in \mathbf{R}_+}$, and let $\mathscr{T}$ be the collection of all the finite stopping times relative to this family (respectively, $\mathscr{T}_c$ the collection of all stopping times bounded by a positive number c). (X_t) is said to belong to the class (D) (respectively, belong to the class (D) on $[0,c]$) if the collection $\{X_\tau, \tau \in \mathscr{T}\}$ (respectively $\{X_\tau, \tau \in \mathscr{T}_c\}$) is uniformly integrable.

(X_t) is said to belong to the class (DL) or locally to the class (D), if (X_t) belongs to the class (D) on every interval $[0,c]$ $(0 \leq c < \infty)$.

Remark 2.4.5.

a. Let $(X_n)_0^\infty$ be a uniformly integrable supermartingale. Then (X_n) is of class (D) if we define $X_t = X_n$ for $n \leq t < n + 1$.
b. There exist uniformly integrable supermartingales that do not belong to the class (D).
c. Every right-continuous martingale belongs to the class (DL).
d. A right-continuous, uniformly integrable martingale belongs to the class (D).
e. A negative right-continuous supermartingale belongs to the class (DL).

Definition 2.4.6. A positive, right-continuous supermartingale is called a *potential* if $\lim_{t \to \infty} E(X_t) = 0$.

Theorem 2.4.4 (Doob-Meyer Decomposition for Potentials). *A potential $(X_t, \mathscr{F}_t)_{t \in \mathbf{R}_+}$ can be decomposed in the form*

$$X_t = E[A_\infty | \mathscr{F}_t] - A_t, \qquad t \in \mathbf{R}_+,$$

where (A_t) is an integrable increasing process if and only if (X_t) is of class (D).

PROOF. For each integer $n \geq 1$ and $i \geq 0$, consider $(X_{i/2^n}, \mathscr{F}_{i/2^n})$, $i \geq 0$. Since (X_t) is a potential, for each n this will be a discrete-parameter potential. By Theorem 2.4.2, we may write

$$X_{i/2^n} = E[A(\infty,n)|\mathscr{F}_{i/2^n}] - A\left(\frac{i}{2^n}, n\right),$$

where $A(0,n) = 0$, $A(i/2^n,n) \leq A((i+1)/2^n,n)$, $A(i/2^n,n)$ is $\mathscr{F}_{(i-1)/2^n}$-measurable, and $A(\infty,n)$ is integrable with

$$A(\infty,n) = \lim_{i\to\infty} A\left(\frac{i}{2^n}, n\right).$$

Assume for the moment that the family $\{A(\infty,n)\}_{n\geq 1}$ is uniformly integrable. This will be shown in Lemma 2.4.1. By property 5 of Section 2.2, there is an integrable random variable $A(\infty)$ such that for some subsequence,

$$A(\infty,n_k) \to A(\infty)$$

weakly [in the $\sigma(L^1,L^\infty)$ topology] as $k \to \infty$.

Consider any two dyadic rationals $r \leq s$. Then there is an $N \geq 1$ such that for $n \geq N$, both $A(r,n)$ and $A(s,n)$ are defined and $A(r,n) \leq A(s,n)$. Since

$$X_r = E[A(\infty,n)|\mathscr{F}_r] - A(r,n) \quad \text{and} \quad X_s = E[A(\infty,n)|\mathscr{F}_s] - A(s,n),$$

we have

$$E[A(\infty,n_k)|\mathscr{F}_r] - X_r \leq E[A(\infty,n_k)|\mathscr{F}_s] - X_s$$

for $n_k \geq N$. Letting k tend to ∞, and as the operation of conditional expection is continuous in the weak topology, we obtain

$$E[A(\infty)|\mathscr{F}_r] - X_r \leq E[A(\infty)|\mathscr{F}_s] - X_s.$$

Denote by M_t a right-continuous version of the uniformly integrable martingale

$$(E[A(\infty)|\mathscr{F}_t],\mathscr{F}_t), \qquad t \in \mathbf{R}_+,$$

and let

$$A_t = M_t - X_t.$$

Then (A_t) is right-continuous and increasing on the dyadic rationals and thus on all of $\mathbf{R}_+$. Since $X_t \to 0$ as $t \to \infty$, we have, moreover,

$$\lim_{t\to\infty} A_t = A_\infty \qquad \text{(a.s.) and in } L^1.$$

It therefore remains only to verify the claim that $\{A(\infty,n)\}$, $n \geq 1$, is uniformly integrable.

Lemma 2.4.1. *$\{A(\infty,n)\}$, $n \geq 1$ is uniformly integrable if and only if (X_t) belongs to class* (D).

PROOF. Suppose that $\{A(\infty,n)\}$ is uniformly integrable. Then the argument above yields

$$X_t = E[A(\infty)|\mathscr{F}_t] - A_t = M_t - A_t.$$

Hence $0 \le X_\tau \le M_\tau$ for τ any finite stopping time. Since (M_t) is a right-continuous, uniformly integrable martingale, it is of class (D). It now follows easily that the family $\{X_\tau: \tau \text{ a finite stopping time}\}$ is uniformly integrable and (X_t) is of class (D).

Conversely, suppose that (X_t) belongs to class (D). For every $\lambda > 0$ and $n \ge 1$, define

$$T_{n,\lambda} = \begin{cases} \inf\left\{\dfrac{i}{2^n}: A\left(\dfrac{i+1}{2^n}, n\right) > \lambda\right\} \\ \infty & \text{if this set is empty.} \end{cases}$$

Observe that $T_{n,\lambda}$ is a stopping time with respect to $\{\mathscr{F}_{i/2^n}\}$, and also

a. $\{T_{n,\lambda} \le k/2^n\} \in \mathscr{F}_{k/2^n}$.
b. For all $t \in [k/2^n,(k+1)/2^n)$, $\{T_{n,\lambda} \le t\} \in \mathscr{F}_t$.
c. $A(T_{n,\lambda},n) \le \lambda$.
d. $A(\infty,n) > \lambda \Leftrightarrow T_{n,\lambda} < \infty$.

We have

$$X_{T_{n,\lambda}} = E[A(\infty,n)|\mathscr{F}_{T_{n,\lambda}}] - A(T_{n,\lambda},n).$$

Consider

$$\int_{\{A(\infty,n)>\lambda\}} A(\infty,n)\,dP = \int_{\{T_{n,\lambda}<\infty\}} A(T_{n,\lambda},n)\,dP + \int_{\{T_{n,\lambda}<\infty\}} X_{T_{n,\lambda}}\,dP.$$

Since $A(T_{n,\lambda},n) \le \lambda$ by the definition of $T_{n,\lambda}$, by item d we have

$$\int_{\{A(\infty,n)>\lambda\}} A(\infty,n)\,dP \le \lambda P\{A(\infty,n) > \lambda\} + \int_{\{T_{n,\lambda}<\infty\}} X_{T_{n,\lambda}}\,dP. \tag{2.4.3}$$

Now

$$\begin{aligned} \lambda P\{A(\infty,n) > 2\lambda\} &\le \int_{\{A(\infty,n)>2\lambda\}} [A(\infty,n) - \lambda]\,dP \\ &\le \int_{\{A(\infty,n)>\lambda\}} [A(\infty,n) - \lambda]\,dP \\ &\le \int_{\{T_{n,\lambda}<\infty\}} X_{T_{n,\lambda}}\,dP. \end{aligned}$$

Replacing λ by 2λ in (2.4.3) gives

$$\begin{aligned} \int_{\{A(\infty,n)>2\lambda\}} A(\infty,n)\,dP &\le 2\lambda P\{A(\infty,n) > 2\lambda\} + \int_{\{T_{n,2\lambda}<\infty\}} X_{T_{n,2\lambda}}\,dP \\ &\le 2\int_{\{T_{n,\lambda}<\infty\}} X_{T_{n,\lambda}}\,dP + \int_{\{T_{n,2\lambda}<\infty\}} X_{T_{n,2\lambda}}\,dP. \end{aligned} \tag{2.4.4}$$

Note that

$$\lambda P\{T_{n,\lambda} < \infty\} = \lambda P\{A(n,\infty) > \lambda\} \leq E[A(n,\infty)]$$
$$= E[E(A(n,\infty)|\mathscr{F}_0) - A_0] = EX_0.$$

Thus $P\{T_{n,\lambda} < \infty\}$ is small for all sufficiently large λ.

The stopping times $T_{n,\lambda}$ are not finite (a.s.), so that the hypothesis that (X_t) is of class (D) cannot be applied directly to the family $\{X_{T_{n,\lambda}}\}$. Nevertheless the family $\{X_{T_{n,\lambda}} I_{[T_{n,\lambda}<\infty]}: n \geq 1, \lambda > 0\}$ is uniformly integrable because the family $\{X_{T_{n,\lambda}\wedge a}: n \geq 1, \lambda > 0, a > 0\}$ being uniformly integrable, $\{X_{T_{n,\lambda}\wedge a} \cdot I_{[T_{n,\lambda}<\infty]}: n \geq 1, \lambda > 0, a > 0\}$ is also uniformly integrable, and each $X_{T_{n,\lambda}} \cdot I_{[T_{n,\lambda}<\infty]}$ is (a.s.) the limit of the sequence $\{X_{T_{n,\lambda}\wedge m} \cdot I_{[T_{n,\lambda}<\infty]}\}_{m\geq 1}$. Therefore, given $\varepsilon > 0$, there exists $\delta > 0$ such that if $P(A) < \delta$, $\int_A |X_{T_{n,\lambda}}| \cdot I_{[T_{n,\lambda}<\infty]}\, dP < \varepsilon$. For this δ, find λ_0 such that if $\lambda > \lambda_0$, then

$$P\{T_{n,\lambda} < \infty\} < \delta.$$

From (2.4.4) we then conclude that $\int_{\{A(\infty,n)>2\lambda\}} A(\infty,n)\, dP \leq 3\varepsilon$, for all $n \geq 1$, $\lambda > \lambda_0$.

This concludes the proof of Lemma 2.4.1, and so also that of Theorem 2.4.4. □

Theorem 2.4.5. *The integrable increasing process (A_t) associated with the potential (X_t) in the proof of Theorem 2.4.4 is natural. It is the only natural integrable increasing process for which*

$$X_t = E(A_\infty|\mathscr{F}_t) - A_t, \qquad t \in \mathbf{R}_+.$$

PROOF. We shall assume here a fact proved in [41], page 110: Let (Y_t) be a positive right-continuous martingale and let (A_t) be an increasing process. If (Y_t) is uniformly integrable, then

$$E\left(\int_0^\infty Y_s\, dA_s\right) = E(A_\infty Y_\infty).$$

Using the criterion given in Theorem 2.4.3, it is sufficient to show that

$$E\left(\int_0^\infty Y_{s-}\, dA_s\right) = E(A_\infty Y_\infty),$$

for (Y_t) a nonnegative, bounded, right-continuous martingale. By the Lebesgue dominated convergence theorem,

$$E\int_0^\infty Y_{s-}\, dA_s = \lim_{n\to\infty} \sum_{i=0}^{\infty} E\left[Y_{i/2^n}\left(A\left(\frac{i+1}{2^n}\right) - A\left(\frac{i}{2^n}\right)\right)\right].$$

Because $Y_{i/2^n}$ is $\mathscr{F}_{i/2^n}$-measurable,

$$\begin{aligned}\sum_{i=0}^{\infty} E&\left[Y_{i/2^n}\left(A\left(\frac{i+1}{2^n}\right)-A\left(\frac{i}{2^n}\right)\right)\right]\\
&=\sum_{i=0}^{\infty} E\left[Y_{i/2^n}E\left\{A\left(\frac{i+1}{2^n}\right)-A\left(\frac{i}{2^n}\right)\middle|\mathscr{F}_{i/2^n}\right\}\right]\\
&=\sum_{i=0}^{\infty} E\left[Y_{i/2^n}E\left\{X\left(\frac{i}{2^n}\right)-X\left(\frac{i+1}{2^n}\right)\middle|\mathscr{F}_{i/2^n}\right\}\right]\\
&=\sum_{i=0}^{\infty} E\left[Y_{i/2^n}E\left\{A\left(\frac{i+1}{2^n},n\right)-A\left(\frac{i}{2^n},n\right)\middle|\mathscr{F}_{i/2^n}\right\}\right]\\
&=\sum_{i=0}^{\infty} E\left[Y_{i/2^n}\left\{A\left(\frac{i+1}{2^n},n\right)-A\left(\frac{i}{2^n},n\right)\right\}\right].\end{aligned}$$

Since $A((i+1)/2^n, n)$ is $\mathscr{F}_{i/2^n}$-measurable,

$$E\left[Y_{i/2^n}A\left(\frac{i+1}{2^n},n\right)\right]=E\left[Y_{(i+1)/2^n}A\left(\frac{i+1}{2^n},n\right)\right].$$

Therefore,

$$\sum_{i=0}^{\infty} E\left[Y\left(\frac{i}{2^n}\right)\left\{A\left(\frac{i+1}{2^n}\right)-A\left(\frac{i}{2^n}\right)\right\}\right]=E[A(\infty,n)Y_\infty].$$

Making $n\to\infty$ along the subsequence n_k, we have

$$E\int_0^\infty Y_{s-}\,dA_s=E[A_\infty Y_\infty].$$

Hence (A_t) is natural.

Let (B_t) be a natural, integrable increasing process associated with (X_t), that is,

$$X_t=E[B_\infty|F_t]-B_t.$$

Since it is natural, we have

$$E\int_0^\infty Y_{s-}\,dB_s=E[Y_\infty B_\infty]$$

for every nonnegative, bounded, right-continuous martingale (Y_t).

Note that for $s<t$,

$$E(Y_s(B_t-B_s))=E(Y_s(A_t-A_s)).$$

Therefore,

$$\begin{aligned}E(Y_\infty B_\infty)&=\lim_{n\to\infty}\sum_{i=0}^{\infty} E\left[Y_{i/2^n}\left(B\left(\frac{i+1}{2^n}\right)-B\left(\frac{i}{2^n}\right)\right)\right]\\
&=\lim_{n\to\infty}\sum_{i=0}^{\infty} E\left[Y_{i/2^n}\left(A\left(\frac{i+1}{2^n}\right)-A\left(\frac{i}{2^n}\right)\right)\right]\\
&=E(Y_\infty A_\infty).\end{aligned}$$

Since we may take Y_∞ to be the indicator function of any set in $\mathscr{F}_\infty$, we conclude

$$A_\infty = B_\infty \qquad \text{(a.s.)}.$$

Hence $A_t = B_t$ (a.s.) for each $t \geq 0$. □

The following theorem is based on Theorem 2.4.5. For the proof, see [41], page 122.

Theorem 2.4.6. *A right-continuous supermartingale* (X_t) *has a Doob-Meyer decomposition*:

$$X_t = M_t - A_t, \qquad t \in \mathbf{R}_+,$$

where (M_t) *denotes a right-continuous martingale and* (A_t) *is an increasing process if and only if* (X_t) *is of class* (DL). *In this case there is exactly one pair* (M_t, A_t) *which yields such a decomposition, where* (A_t) *is also natural.*

For later use, we cite here the following result which may be found in [41].

Theorem 2.4.7. *Let* (X_t) *be a potential of class* (D) *with* (A_t) *its associated integrable, natural, increasing process. Then*

$$E(A_\infty^2) = E\left[\int_0^\infty (X_t + X_{t-})\,dA_t\right].$$

2.5 The Quadratic Variation of a Square Integrable Martingale

Definition 2.5.1. A martingale $M = (M_t, \mathscr{F}_t)$, $t \in \mathbf{R}_+$, is said to be *square-integrable* (or called an *L^2-martingale*) if

$$\sup_t E(M_t^2) < \infty. \tag{2.5.1}$$

If we consider martingales on $[0,T]$ for some $T > 0$, then (2.5.1) may be replaced by the requirement

$$E(M_T^2) < \infty. \tag{2.5.1a}$$

[That (2.5.1a) implies $\sup_{0 \leq t \leq T} E(M_t^2) < \infty$ follows from the fact that $(-M_t^2, \mathscr{F}_t)_{[0,T]}$ is a supermartingale.]

We now associate with a given L^2-martingale M a certain natural integrable increasing process called the *quadratic variation process*, or simply *variation process* of M. In what follows, $\mathbf{R}_+$ is taken as the parameter interval, and those changes necessary for intervals $[0,T]$ will be left to the reader.

First, replace (M_t) by a right-continuous version. (Since two different right-continuous versions agree for all t, except on a P-null set, future steps will be seen to be independent of our choice of version.) Condition (2.5.1) implies that the family $(M_t)_{t \geq 0}$ is uniformly integrable. From 8(iv) of Section 2.2, $M_\infty = \lim_{t \to \infty} M_t$ exists (a.s.) and in the L^1 sense, and $M_t = E(M_\infty | \mathscr{F}_t)$ (a.s.). Also, by Fatou's lemma, $E(M_\infty^2) < \infty$. Then, since $(M_t, \mathscr{F}_t)$, $t \in \bar{\mathbf{R}}_+$, is a martingale, $(-M_t^2, \mathscr{F}_t)$, $t \in \bar{\mathbf{R}}_+$, is a uniformly integrable supermartingale. The inequality of property 7 of Section 2.2 with $p = 2$ can now be extended. Taking $I = [0,b]$ and letting $b \to \infty$, we obtain from property 7, Doob's martingale inequality

$$E\left(\sup_{0 \leq t} |M_t|\right)^2 \leq 4E(M_\infty^2).$$

Next, define

$$\xi_t = E(M_\infty^2 | \mathscr{F}_t) - M_t^2$$

where we first take a right-continuous version for the uniformly integrable martingale

$$(E(M_\infty^2 | \mathscr{F}_t), \mathscr{F}_t).$$

Then $(\xi_t, \mathscr{F}_t)$ is a nonnegative right-continuous supermartingale, and

$$\xi_\infty = \lim_{t \to \infty} \xi_t$$

exists (a.s.). Since $E(M_\infty^2 | \mathscr{F}_t) \to M_\infty^2$ (a.s.) as $t \to \infty$ and $M_t^2 \to M_\infty^2$ (a.s.) as $t \to \infty$,

$$\xi_\infty = 0 \qquad \text{(a.s.)}.$$

Hence $E(\xi_t) \to 0$, and $(\xi_t, \mathscr{F}_t)$ is a potential. In fact, (ξ_t) belongs to class (D) because it is dominated by a right-continuous, uniformly integrable martingale [which is therefore of class (D)], that is,

$$0 \leq \xi_t \leq E(M_\infty^2 | \mathscr{F}_t).$$

We may now apply the Doob decomposition theorem for potentials to obtain a unique natural increasing process (A_t) such that

$$\xi_t = E(A_\infty | \mathscr{F}_t) - A_t.$$

The quadratic variation process of M, denoted henceforth by $\langle M \rangle$ or $\langle M,M \rangle$ is defined by

$$\langle M \rangle_t - \langle M \rangle_0 = A_t,$$

where we set $\langle M \rangle_0 = M_0^2$.

In fact, this process is unique (up to a version) in that it is the only natural integrable increasing process for which $M_t^2 - \langle M \rangle_t$ is a martingale. In particular, if $0 \leq s \leq t$,

$$\begin{aligned} E((M_t - M_s)^2 | \mathscr{F}_s) &= E((M_t^2 - M_s^2) | \mathscr{F}_s) \\ &= E(\langle M \rangle_t - \langle M \rangle_s | \mathscr{F}_s). \end{aligned}$$

Remark 2.5.1. In the special case where the L^2-martingale M is a continuous process, the variation $\langle M \rangle$ will also be continuous. This follows from Theorem 37 of [41] where continuity of $\langle M \rangle$ is related to "regularity" of (ξ_t). Since $\xi_t = E(M_\infty^2|\mathscr{F}_t) - M_t^2$ is the sum of a right-continuous, uniformly integrable martingale, and a nonpositive continuous supermartingale, it can be shown to be regular.

EXAMPLE 2.5.1. Consider the Wiener process restricted to any interval $[0,T]$. Then W is an L^2-martingale relative to $(\mathscr{F}_t^W)$. Let $A_t(\omega) = t$ for all ω and $t \in [0,T]$. Then A is a natural increasing process, and for $0 \leq s < t \leq T$,

$$E[(W_t - W_s)^2|\mathscr{F}_s^W] = t - s = E[A_t - A_s|\mathscr{F}_s^W].$$

Hence $\langle W \rangle_t = t$.

EXAMPLE 2.5.2. (Poisson Process). Let $(N_t)_{t \geq 0}$ be a right-continuous version of the Poisson process with parameter $\lambda > 0$. Since N is a process of independent increments and $E(N_t - N_s) = \lambda(t - s)$, it follows that $((N_t - \lambda t), \mathscr{F}_t^N)$ is a martingale on $\mathbf{R}_+$. Over the interval $[0,T]$, $Y_t = N_t - \lambda t$ is then an L^2-martingale, and we find that $E[(Y_t - Y_s)^2|\mathscr{F}_s^N] = \lambda(t - s)$ for $0 \leq s < t \leq T$. In this case then, $A_t = \lambda t$ will be the variation of (Y_t).

Remark 2.5.2. If in Example 2.5.2 we take $\lambda = 1$, we then have two L^2-martingales W_t and N_t on the same interval $[0,T]$ and the same quadratic variation process $A_t = t$. This should draw attention to the fact that the quadratic variation process does not uniquely determine the L^2-martingale and, in particular, continuity of the variation process does not imply continuity of the martingale.

Theorem 2.5.1. *Let (A_t), $t \in \mathbf{R}_+$, be a continuous, integrable increasing process, and let Π be a partition of $[0,t]$. Define*

$$A_t^\Pi = \sum_{j=0}^{n-1} E[(A_{t_{j+1}} - A_{t_j})|\mathscr{F}_{t_j}].$$

Then we have

$$E|A_t^\Pi - A_t| \to 0 \quad \textit{as } |\Pi| \to 0, \textit{ for all } t.$$

Here $|\Pi| = \max_j |t_{j+1} - t_j|$.

We consider two cases:

Case I. $E(A_t^2) < \infty$.

$$\begin{aligned} E(A_t - A_t^\Pi)^2 &= E\left[\sum_j (A_{t_{j+1}} - A_{t_j}) - \sum_j E[(A_{t_{j+1}} - A_{t_j})|\mathscr{F}_{t_j}]\right]^2 \\ &= E\left\{\sum_j [(A_{t_{j+1}} - A_{t_j}) - E[(A_{t_{j+1}} - A_{t_j})|\mathscr{F}_{t_j}]]\right\}^2. \end{aligned}$$

Let us denote $A_{t_{j+1}} - A_{t_j}$ by Δ_j.

$$\begin{aligned} E(A_t - A_t^\Pi)^2 &= E\left\{\sum_j \{\Delta_j - E(\Delta_j|\mathscr{F}_{t_j})\}\right\}^2 \\ &= E\sum_j [\Delta_j - E(\Delta_j|\mathscr{F}_{t_j})]^2 \\ &\quad + 2\sum_{t_i<t_j} E\{\Delta_i - E(\Delta_i|\mathscr{F}_{t_i})\}\{\Delta_j - E(\Delta_j|\mathscr{F}_{t_j})\}. \end{aligned}$$

The cross product vanishes because $\mathscr{F}_{t_i} \subset \mathscr{F}_{t_j}$ for $t_i < t_j$. Hence the right-hand side

$$\begin{aligned} &= E\sum_j [\Delta_j - E(\Delta_j|\mathscr{F}_{t_j})]^2 \\ &= E\sum_j \Delta_j^2 - E\sum_j [E(\Delta_j|\mathscr{F}_{t_j})]^2 \\ &\leq E\sum_j (A_{t_{j+1}} - A_{t_j})^2 \\ &\leq E\left[\sup_\Pi (A_{t_{j+1}} - A_{t_j})A_t\right]. \end{aligned}$$

By the uniform continuity of A_s on $[0,t]$ and because $A_t \in L^2$, it follows by the dominated convergence theorem that the right-hand side goes to 0. Hence

$$A_t^\Pi \to A_t \quad \text{in} \quad L^2.$$

Case II. $A_t \in L^1$. Define, for $s \in [0,t]$,

$$B_s = A_s \wedge n \quad \text{and} \quad C_s = A_s - B_s.$$

Then for $B_t^\Pi = \sum_j E(B_{t_{j+1}} - B_{t_j}|\mathscr{F}_{t_j})$ and $C_t^\Pi = \sum_j E(C_{t_{j+1}} - C_{t_j}|\mathscr{F}_{t_j})$, we have

$$A_t^\Pi = B_t^\Pi + C_t^\Pi$$

and

$$\begin{aligned} |A_t - A_t^\Pi| &\leq |B_t - B_t^\Pi| + |C_t - C_t^\Pi| \\ &\leq |B_t - B_t^\Pi| + |C_t| + |C_t^\Pi|. \end{aligned}$$

Therefore

$$E|A_t - A_t^\Pi| \leq E|B_t - B_t^\Pi| + E|C_t| + E|C_t^\Pi|.$$

Note that

$$EC_t^\Pi = EC_t.$$

Now

$$\begin{aligned} EC_t = E(A_t - B_t) &= \int_{[A_t>n]} (A_t - n)\,dP \\ &\leq \int_{[A_t>n]} A_t\,dP < \frac{\varepsilon}{2}, \end{aligned}$$

if n is taken large enough. The term $E|B_t - B_t^\Pi|$ can be made sufficiently small by making the partition fine enough. Thus we have proved $A_t^\Pi \to A_t$ in L^1. □

Let us apply this result to the case of the increasing process $A_t = \langle M \rangle_t - \langle M \rangle_0$ which is the variation process of an L^2-martingale. We then have the following useful result.

Theorem 2.5.2. *Let (M_t) be an L^2-martingale and suppose its variation process $\langle M \rangle$ is continuous. Then if $\Pi_n = \{t_j^n\}$ is a finite partition of $[0,t]$, $\sum_{\Pi_n} E[(M_{t_{j+1}^n} - M_{t_j^n})^2 | \mathscr{F}_{t_j^n}] \to \langle M \rangle_t - \langle M \rangle_0$ as $|\Pi| \to 0$ in the L^1 sense.*

As part of our notation we define

$$\langle M,N \rangle = \tfrac{1}{4}[\langle M + N \rangle - \langle M - N \rangle]$$

for any two L^2-martingales M and N.

Remark 2.5.3. $E[(M_t - M_s)(N_t - N_s)|\mathscr{F}_s] = E[\langle M,N \rangle_t - \langle M,N \rangle_s | \mathscr{F}_s]$ for $0 \leq s < t$.

Remark 2.5.4. If $\langle M \rangle$ and $\langle N \rangle$ are continuous,

$$\sum_{\Pi_n} E[(M_{t_{j+1}^n} - M_{t_j^n})(N_{t_{j+1}^n} - N_{t_j^n})|\mathscr{F}_{t_j^n}] \to \langle M,N \rangle_t - M_0 N_0$$

in the L^1 sense as $|\Pi_n| \to 0$.

Remark 2.5.5. If $(M_t,\mathscr{F}_t)$ and $(N_t,\mathscr{F}_t)$ are continuous L^2-martingales, then

$$\sum_{\Pi_n} [(M_{t_{j+1}^n} - M_{t_j^n})(N_{t_{j+1}^n} - N_{t_j^n})] \to \langle M,N \rangle_t - M_0 N_0$$

in L^1 as $|\Pi_n| \to 0$.

A more general statement and proof of this last fact can be found in Meyer's article "Integrales Stochastique I," *Seminaire de Probabilities I*, [45].

Remark 2.5.6. If (M_t) and (N_t) are continuous L^2-martingales, then $M_t N_t - \langle M,N \rangle_t$ is a (continuous) martingale.

The proof follows immediately if we observe that $M_t N_t - \langle M,N \rangle_t = \frac{1}{4}[(M+N)_t^2 - \langle M + N \rangle_t] - \frac{1}{4}[(M-N)_t^2 - \langle M - N \rangle_t]$.

2.6 Local Martingales

Definition 2.6.1. (a) A real-valued process (M_t), $t \in \mathbf{R}_+$, which is $(\mathscr{F}_t)$-adapted is a *local martingale* with respect to $(\mathscr{F}_t)$ if there is a sequence of stopping times $(\tau_n)_{n \geq 1}$ such that $\tau_n \uparrow \infty$ (a.s.) [or, in the case of the interval $[0,T]$,

$\tau_n \uparrow T$ (a.s.)] and for each n, $(M_t^n, \mathscr{F}_t)$ is a martingale, where

$$M_t^n = M_{t \wedge \tau_n}.$$

(b) A real-valued process (M_t), $t \in \mathbf{R}_+$, is called a *local L^2-martingale* if it satisfies the conditions of (a) with the stronger requirement that $(M_t^n, \mathscr{F}_t)$ be an L^2-martingale.

Remark 2.6.1. The condition above that $(M_t^n, \mathscr{F}_t)$ be a martingale (or L^2-martingale) is equivalent to requiring that $(M_t^n, \mathscr{F}_{t \wedge \tau_n})$ be a martingale. This is shown in the following result.

Lemma 2.6.1. *Suppose the process (M_t) is $(\mathscr{F}_t)$-adapted and that τ is a stopping time with respect to $(\mathscr{F}_t)$. Then*

(i) *$(M_{t \wedge \tau})$ is $(\mathscr{F}_t)$-adapted if and only if $(M_{t \wedge \tau})$ is $(\mathscr{F}_{t \wedge \tau})$-adapted.*
(ii) *$(M_{t \wedge \tau})$ is a martingale with respect to $(\mathscr{F}_t)$ if and only if it is a martingale with respect to $(\mathscr{F}_{t \wedge \tau})$.*

PROOF. (i) Since $t \wedge \tau \leq t$, $\mathscr{F}_{t \wedge \tau} \subset \mathscr{F}_t$. Suppose $(M_{t \wedge \tau})$ is $(\mathscr{F}_t)$-adapted. Fix $t_0 \geq 0$. Let $A = \{M_{t_0 \wedge \tau} \in B\}$, for B a Borel set. Then $A \in \mathscr{F}_{t_0}$. We must show $A \cap \{t_0 \wedge \tau \leq s\} \in \mathscr{F}_s$ for all $s \geq 0$. If $s \geq t_0$, then this set is all of A, hence belongs to $\mathscr{F}_{t_0} \subset \mathscr{F}_s$. Suppose therefore that $0 \leq s < t_0$. Then $t_0 \wedge \tau \leq s$ if and only if $t_0 \wedge \tau = \tau = s \wedge \tau$. Thus

$$\begin{aligned} A \cap \{t_0 \wedge \tau \leq s\} &= \{\omega\colon M_{t_0 \wedge \tau} \in B \text{ and } t_0 \wedge \tau \leq s\} \\ &= \{\omega\colon M_{s \wedge \tau} \in B \text{ and } \tau \leq s\} \\ &= \{M_{s \wedge \tau} \in B\} \cap \{\tau \leq s\}. \end{aligned}$$

Both the last two sets belong to $\mathscr{F}_s$.

(ii) Suppose $(M_{t \wedge \tau})$ is a martingale relative to $(\mathscr{F}_t)$. By (i), each $M_{t \wedge \tau}$ is $\mathscr{F}_{t \wedge \tau}$-measurable. Also, for $0 \leq s < t$,

$$\begin{aligned} E(M_{t \wedge \tau} | \mathscr{F}_{s \wedge \tau}) &= E[E(M_{t \wedge \tau} | \mathscr{F}_s) | \mathscr{F}_{s \wedge \tau}] \\ &= E(M_{s \wedge \tau} | \mathscr{F}_{s \wedge \tau}) \\ &= M_{s \wedge \tau}. \end{aligned}$$

Conversely, suppose $(M_{t \wedge \tau})$ is a martingale with respect to $(\mathscr{F}_{t \wedge \tau})$. Again, by (i), it remains only to check $E(M_{t \wedge \tau} | \mathscr{F}_s) = M_{s \wedge \tau}$, which we may write in the form

$$\int_\Lambda M_{t \wedge \tau}\, dP = \int_\Lambda M_{s \wedge \tau}\, dP \tag{2.6.1}$$

for all $\Lambda \in \mathscr{F}_s$, where $0 \leq s < t$ are fixed. Note that we are given that Equation (2.6.1) holds for all $\Lambda \in \mathscr{F}_{s \wedge \tau}$. Let

$$\Lambda_1 = \Lambda \cap \{\tau > s\} \qquad \Lambda_2 = \Lambda \cap \{\tau \leq s\}.$$

For $s' < s$, $\Lambda_1 \cap \{s \wedge \tau \leq s'\} = \varnothing \in \mathscr{F}_{s'}$, and for $s' \geq s$,

$$\Lambda_1 \cap \{s \wedge \tau \leq s'\} = \Lambda \cap \{\tau > s \text{ and } s \wedge \tau \leq s'\} = \Lambda_1 \in \mathscr{F}_s \subset \mathscr{F}_{s'}.$$

Therefore, $\Lambda_1 \in \mathscr{F}_{s\wedge\tau}$. Hence Equation (2.6.1) holds for Λ_1. On the other hand,

$$\begin{aligned}\int_{\Lambda_2} M_{t\wedge\tau}\,dP &= \int_{\Lambda\cap\{\tau\le s\}} M_{t\wedge\tau}\,dP\\ &= \int_{\Lambda\cap\{\tau\le s\}} M_{s\wedge\tau}\,dP\\ &= \int_{\Lambda_2} M_{s\wedge\tau}\,dP.\end{aligned}$$

Therefore,

$$\begin{aligned}\int_{\Lambda} M_{t\wedge\tau}\,dP &= \int_{\Lambda_1} M_{t\wedge\tau}\,dP + \int_{\Lambda_2} M_{t\wedge\tau}\,dP\\ &= \int_{\Lambda_1} M_{s\wedge\tau}\,dP + \int_{\Lambda_2} M_{s\wedge\tau}\,dP\\ &= \int_{\Lambda} M_{s\wedge\tau}\,dP.\end{aligned}$$

□

Remark 2.6.2. A bounded local martingale relative to $(\mathscr{F}_t)$ is a martingale.

PROOF. Suppose $|X_t| \le B$ for all (t,ω), and suppose $\tau_n \uparrow \infty$ (a.s.) is a family of stopping times such that $(X_{t\wedge\tau_n})_{t\ge 0}$ is a martingale for each $n \ge 1$. By Lemma 2.6.1, it makes no difference whether we say relative to $(\mathscr{F}_t)$ or $(\mathscr{F}_{t\wedge\tau_n})$.

$$E(X_{t\wedge\tau_n}|\mathscr{F}_s) = X_{s\wedge\tau_n}, \qquad s \le t.$$

Fix $s \le t$. Then $X_{t\wedge\tau_n} \to X_t$ (a.s.) as $n \to \infty$ and $X_{s\wedge\tau_n} \to X_s$ (a.s.) as $n \to \infty$. By bounded convergence, these are also limits in L^1. Hence $E(X_t|\mathscr{F}_s) = X_s$. □

Remark 2.6.3. Let (M_t) be a continuous local martingale. Define

$$\sigma_n(\omega) = \begin{cases}\inf\{t\colon |M_t(\omega)| \ge n\} & \\ \infty \text{ (or } T) & \text{if this is an empty set.}\end{cases}$$

It is easy to verify by the optional sampling theorem (9) of Section 2.2 that $(M_{t\wedge\sigma_n})$ will again be a local martingale, but clearly each $M_{t\wedge\sigma_n}$ is bounded. Therefore, by Equation (2.6.1), $(M_{t\wedge\sigma_n},\mathscr{F}_t)$ is a bounded continuous martingale (hence also an L^2-martingale). (In the future, when continuous local martingales are discussed, the stopping times (σ_n) will always be some variant of those defined above.)

Lemma 2.6.2. *Let $(M_t,\mathscr{F}_t)$, $t \in \mathbf{R}_+$, be a continuous L^2-martingale with associated increasing process (A_t). Then for any stopping time τ, the increasing process associated with the martingale $(M_{t\wedge\tau})$ is given by $(A_{t\wedge\tau})$.*

PROOF. Recall that $Y_t = M_t^2 - A_t$ is a martingale and $E|Y_t| \le E(M_\infty^2) + E(A_\infty) < \infty$. Furthermore, since $(M_t^2, 0 \le t \le \infty)$ is a submartingale we have, for $C > 0$,

$$\int_{[|Y_t|>C]} |Y_t|\,dP \le \int_{[|Y_t|>C]} M_\infty^2\,dP + \int_{[|Y_t|>C]} A_\infty\,dP.$$

Since $P(|Y_t| > C)$ tends to zero when $C \to \infty$, (Y_t) is uniformly integrable. By the optional sampling theorem, if $s \le t$, $E(Y_{t\wedge\tau}|\mathscr{F}_{s\wedge\tau}) = Y_{s\wedge\tau}$. Hence $(Y_{t\wedge\tau}, \mathscr{F}_{t\wedge\tau})$ is a martingale, and from the preceding lemma it follows that $(Y_{t\wedge\tau}, \mathscr{F}_t)$ is a martingale. Then for $s \le t$, we have $E(M_{t\wedge\tau}^2 - A_{t\wedge\tau}|\mathscr{F}_s) = M_{s\wedge\tau}^2 - A_{s\wedge\tau}$. This means that the process (B_t), where $B_t = A_{t\wedge\tau}$, is the increasing process of the L^2-martingale $(M_{t\wedge\tau})_{t\ge 0}$. Note that naturality of (B_t) follows from its continuity. □

Theorem 2.6.1. *To every continuous local martingale (M_t) there corresponds a process (A_t) with these properties:*

(i) $A_0 = 0$ (a.s.).
(ii) $A_t \ge A_s$ (a.s.) for $t \ge s$.
(iii) *(A_t) is continuous.*
(iv) *$(M_t^2 - A_t)$ is a local martingale.*

Furthermore, this process is unique in the sense that if (B_t) is any process with properties (i) *through* (iv), *then $A_t = B_t$ for every t* (a.s.).

PROOF. Since $M_0 \in L^1$, we may assume $M_0 = 0$ by replacing (M_t) by $(M_t - M_0)$. Let σ_n be the sequence of stopping times introduced in Remark 2 above. Then for each n, the bounded continuous martingale

$$M_t^n = M_{t\wedge\sigma_n}$$

has a unique increasing integrable process which we denote by $A_t^n = \langle M^n \rangle_t$. The continuous martingale $M_{t\wedge\sigma_n}^{n+1}$ has the variation process $A_{t\wedge\sigma_n}^{n+1}$. But for all $t \in \mathbf{R}_+$,

$$M_{t\wedge\sigma_n}^{n+1} = M_{t\wedge\sigma_n\wedge\sigma_{n+1}} = M_{t\wedge\sigma_n} = M_t^n \qquad \text{(a.s.)}.$$

Hence, by uniqueness of the variation process,

$$A_{t\wedge\sigma_n}^{n+1} = A_t^n \qquad \text{(a.s.)}$$

Since these are continuous processes, we actually have $A_{t\wedge\sigma_n}^{n+1} = A_t^n$ for all t, (a.s.). Therefore, on a set of probability 1,

$$A_t^{n+1}(\omega) = A_t^n(\omega) \quad \text{for } t \le \sigma_n(\omega).$$

From this it follows that if we define

$$A_t(\omega) = A_t^n(\omega) \quad \text{on } \{\omega: \sigma_n(\omega) \ge t\},$$

then A_t is consistently defined except on a P-null set. Furthermore, A_t inherits the properties (i) through (iii) from those for A_t^n. Finally, from $A_{t\wedge\sigma_n}^{n+1} = A_t^n$ it is easily deduced that $A_{t\wedge\sigma_n} = A_t^n$. Since $M_{t\wedge\sigma_n}^2 = (M_t^n)^2$, it then follows that $(M_t^2 - A_t)$ is a local martingale with regard to the stopping times (σ_n). The uniqueness of A_t follows from that of the processes A_t^n. □

As in the case of L^2-martingales, we will use the notation $\langle M \rangle$ for the process $(A + M_0^2)$, calling it the *variation process* of M.

Remark 2.6.4. For convenience later we define a *continuous locally increasing process* to be an $(\mathscr{F}_t)$-adapted process which has properties (i), (ii), and (iii) in Theorem 2.6.1. Then A can be referred to as the unique continuous locally increasing process for which $(M_t^2 - A_t)$ is a local martingale.

Remark 2.6.5. In the above proof, any sequence (τ_n) of stopping times such that

$$\tau_n \leq \sigma_n \quad \text{and} \quad \tau_n \uparrow \infty\ (T) \qquad \text{(a.s.)}$$

could be used to obtain a sequence M^n of bounded continuous martingales such that

$$\langle M \rangle_{t \wedge \tau_n} = \langle M^n \rangle_t \qquad \text{(a.s.)}$$

for all t and $n \geq 1$.

Lemma 2.6.3. *Let $(M_t, \mathscr{F}_t)$ be a continuous local martingale with variation process $(\langle M \rangle_t)$. If $E\langle M \rangle_t < \infty$ for each t in $[0,T]$, then $(M_t, \mathscr{F}_t)$ is an L^2-martingale for $t \in [0,T]$.*

PROOF. By definition, since $M_t^2 - A_t$ is a local martingale, there exists a sequence (τ_n) of stopping times with $\tau_n \uparrow T$ (a.s.) and such that for $s \leq t$, $E(M_{t \wedge \tau_n}^2 - A_{t \wedge \tau_n} | \mathscr{F}_s) = M_{s \wedge \tau_n}^2 - A_{s \wedge \tau_n}$. Taking $s = 0$, we have

$$E(M_{t \wedge \tau_n}^2 - A_{t \wedge \tau_n}) = E(M_0^2 - A_0).$$

Since for all n, $E(\langle M \rangle_{t \wedge \tau_n}) \leq E(\langle M \rangle_t) < \infty$ by hypothesis, it follows that $E(M_{t \wedge \tau_n}^2)$ is a bounded sequence and uniform integrability of the sequence $\{M_{t \wedge \tau_n}\}$ follows. $(M_t, \mathscr{F}_t)$ is thus a martingale on $[0,T]$ and, in fact, square-integrable. □

As in the case of square-integrable martingales, if $M = (M_t)$ and $N = (N_t)$ are continuous local martingales, we define the process $\langle M,N \rangle$ by

$$\langle M,N \rangle = \tfrac{1}{4}\{\langle M + N, M + N \rangle - \langle M - N, M - N \rangle\}.$$

If (τ_n) is a stopping time sequence such that $(M_t^n) = (M_{t \wedge \tau_n})$ and $(N_t^n) = (N_{t \wedge \tau_n})$ are L^2-martingales (clearly such a sequence (τ_n) can be found), then it follows that

$$\langle M,N \rangle_t(\omega) = \langle M^n,N^n \rangle_t(\omega) \quad \text{if } t \leq \tau_n(\omega).$$

The following result is very useful in applications.

Theorem 2.6.2. *Suppose that $M = (M_t)$ $(M_0 = 0)$ is a continuous local martingale which is of bounded variation in $[0,t]$ for every t. Then $M_t = 0$ (a.s.).*

PROOF. We first assume that M is a continuous L^2-martingale. Let $\pi_n = \{t_j^n\}$ be a finite partition of $[0,t]$ such that $|\pi_n| \to 0$ as $n \to \infty$. Write

$$S_{\pi_n} = \sum_{\pi_n} [M_{t_{j+1}^n} - M_{t_j^n}]^2.$$

Then it is easy to see that

$$S_{\pi_n} \le \max_{\pi_n} \left| M_{t_{j+1}^n} - M_{t_j^n} \right| \cdot |M|_t,$$

where $|M|_t$ is the total variation of M_s in $[0,t]$. The continuity of M_t then yields $S_{\pi_n} \to 0$ (a.s.). However, from Remark 5 of Section 2.5, $S_{\pi_n} \to \langle M \rangle_t$ in L^1. Hence $\langle M \rangle_t = 0$ (a.s.) for every t. It follows that $E(M_t - M_0)^2 = 0$ and so $M_t = 0$ (a.s.).

Suppose now that M is a continuous local martingale with the stated property. Let σ_N be an increasing sequence of stopping times tending to ∞ and such that $(M_t^N) = (M_{t \wedge \sigma_N})$ is an L^2-martingale. For π, an arbitrary finite partition of $[0,t]$, it is easy to verify that for almost all ω,

$$\sum_{\pi} |M_{t_{j+1}}^N(\omega) - M_{t_j}^N(\omega)| \le |M|_t(\omega) + |M_{\sigma_N(\omega)}(\omega)| + \sup_{0 \le s \le t} |M_s(\omega)|.$$

Hence, for each N, $M_s^N(\omega)$ is of bounded variation in $[0,t]$. By what has been shown above, we have $M_t^N = 0$ (a.s.) Making $N \to \infty$, we obtain the desired conclusion. □

Obviously, the above theorem holds if the index set is taken to be a finite interval $[0,T]$ instead of $\mathbf{R}_+$.

As an application of Theorem 2.6.2, we prove the following uniqueness property of the process $\langle M,N \rangle$ associated with the continuous local martingales M and N.

Theorem 2.6.3. *Let $M = (M_t)$ and $N = (N_t)$ be continuous local martingales relative to $(\mathcal{F}_t)$ with $M_0 = N_0 = 0$. Suppose that B_t $(B_0 = 0)$ is a continuous, $(\mathcal{F}_t)$-adapted process such that* (i) *for almost all ω, $B_s(\omega)$ is of bounded variation on $[0,t]$ and* (ii) *$M_tN_t - B_t$ is a local martingale for each t. Then*

$$B_t = \langle M,N \rangle_t \qquad \text{(a.s.)}.$$

Proof. Write $Z_t = M_tN_t - B_t$. By the definition of $\langle M,N \rangle$ we have

$$\begin{aligned} Z_t &= \tfrac{1}{4}[(M+N)_t^2 - \langle M+N \rangle_t - \{(M-N)_t^2 - \langle M-N \rangle_t\}] \\ &\quad + \langle M,N \rangle_t - B_t \\ &= Y_t + \langle M,N \rangle_t - B_t, \end{aligned}$$

say, where (Y_t) is a continuous local martingale because

$$\{(M+N)_t^2 - \langle M+N \rangle_t\} \quad \text{and} \quad \{(M-N)_t^2 - \langle M-N \rangle_t\}$$

are continuous local martingales. Thus we have the relation

$$Z_t - Y_t = \langle M,N \rangle_t - B_t,$$

where the left-hand side is a continuous local martingale and the right-hand side is a continuous process of bounded variation in every finite interval.

Moreover, $Z_0 - Y_0 = 0$. Hence, from Theorem 2.6.2 it follows that the $Z_t - Y_t = 0$ (a.s.) for each t, that is, $\langle M,N\rangle_t - B_t = 0$ (a.s.) for each t. □

Let $M = (M_t)$ be a continuous, L^2-martingale and let τ be a stopping time. We introduce the notation $M_t^\tau = M_{t \wedge \tau}$ and $M^\tau = (M_t^\tau)$. As has been shown above, the process $(M_t^\tau, \mathscr{F}_t)$ is a continuous, L^2-martingale.

Lemma 2.6.4. *Let (M_t) $(M_0 = 0)$ and (N_t) $(N_0 = 0)$ be continuous, L^2-martingales relative to $(\mathscr{F}_t)$. If τ is a stopping time, then*

$$\langle M^\tau, N^\tau\rangle_t = \langle M, N^\tau\rangle_t = \langle M,N\rangle_{t\wedge\tau} \qquad \text{(a.s.)}.$$

Proof. Since (M_t), (N_t), (M_t^τ) and (N_t^τ) are continuous L^2-martingales, the martingales (ξ_t) and (η_t) given by

$$\xi_t = M_t N_t^\tau - \langle M, N^\tau\rangle_t$$

and

$$\eta_t = M_t N_t - \langle M,N\rangle_t$$

are uniformly integrable, and (λ_t) with $\lambda_t = M_t^\tau N_t^\tau - \langle M^\tau, N^\tau\rangle_t$ is a martingale. Hence (η_t^τ) is also a martingale, so that

$$\eta_t^\tau - \lambda_t = \langle M^\tau, N^\tau\rangle_t - \langle M,N\rangle_{t\wedge\tau}.$$

Theorem 2.6.2 then shows that

$$\langle M^\tau, N^\tau\rangle_t = \langle M,N\rangle_{t\wedge\tau}.$$

Now set $X_t = M_t N_t^\tau - M_t^\tau N_t^\tau$.

For $A \in \mathscr{F}_t$,

$$\int_{A\cap[\tau\le t]} X_t\,dP = \int_{A\cap[\tau\le t]} E[M_\infty N_{t\wedge\tau} - M_{t\wedge\tau}N_{t\wedge\tau}|\mathscr{F}_t]\,dP$$

$$= \int_{A\cap[\tau\le t]} (M_\infty N_\tau - M_\tau N_\tau)\,dP.$$

It is easy to see that $\int_{A\cap[\tau>t]} X_t\,dP = 0$. Since $A \cap [\tau > t] \in \mathscr{F}_\tau$, we have

$$\int_{A\cap[\tau>t]} (M_\infty - M_\tau)N_\tau\,dP = \int_{A\cap[\tau>t]} E[(M_\infty - M_\tau)N_\tau|\mathscr{F}_\tau]\,dP = 0.$$

Hence

$$\int_A X_t\,dP = \int_A (M_\infty - M_\tau)N_\tau\,dP \quad \text{for all } A \in \mathscr{F}_t,$$

that is,

$$X_t = E[(M_\infty - M_\tau)N_\tau|\mathscr{F}_t].$$

Thus (X_t) is a continuous martingale with $X_0 = 0$. Applying Theorem 2.6.2 to the relation

$$\xi_t - \eta_t^\tau = X_t - [\langle M,N^\tau\rangle_t - \langle M,N\rangle_{t\wedge\tau}]$$

we conclude that $\langle M,N^\tau\rangle_t = \langle M,N\rangle_{t\wedge\tau}$. This completes the proof of Lemma 2.6.4. □

Remark 2.6.6. Let M_i and N_i be continuous local martingales relative to $(\mathscr{F}_t)$ and a_i,b_i be any real numbers $(i = 1,2)$. Then

(a) $a_1M_1 + a_2M_2$ is a continuous local martingale relative to $(\mathscr{F}_t)$.
(b) $\langle a_1M_1 + a_2M_2, b_1N_1 + b_2N_2\rangle = \sum_{i,j=1}^2 a_ib_j\langle M_i,N_j\rangle$.

The proof is easy. To show (a), let σ_n be a sequence of stopping times increasing to ∞ such that $(M_i^n)_t = (M_i)_{t\wedge\sigma_n}$ are L^2-martingales. It is then easy to see that

$$(a_1M_1 + a_2M_2)_{t\wedge\sigma_n} = a_1(M_1^n)_t + a_2(M_2^n)_t$$

is an L^2-martingale for every n. Hence (a) follows. To prove (b) let us write $X = a_1M_1 + a_2M_2$ and $Y = b_1N_1 + b_2N_2$ for convenience. We first observe that if M_i and N_i are continuous L^2-martingales then the verification of (b) is easy using Remark 2.5.5. For the case of local martingales, set $X_t^n = X_{t\wedge\sigma_n}$ and $Y_t^n = Y_{t\wedge\sigma_n}$ where σ_n is the stopping time used in (a). Then X^n and Y^n are L^2-martingales, and on the set $[\omega\colon t \le \sigma_n(\omega)]$ we have

$$\begin{aligned}\langle X,Y\rangle_t(\omega) &= \langle X^n,Y^n\rangle_t(\omega)\\ &= \sum_{i,j=1}^2 a_ib_j\langle M_i^n,N_j^n\rangle_t(\omega)\\ &= \sum_{i,j=1}^2 a_ib_j\langle M_i,N_j\rangle_t(\omega).\end{aligned}$$

The result now follows since $\sigma_n(\omega)\uparrow\infty$ (a.s.).

Remark 2.6.7. Let $M = (M_t)$ and $N = (N_t)$ be L^2-martingales with $M_0 = N_0 = 0$. For $s < t$, we then have

(a) $|\langle M,N\rangle_t - \langle M,N\rangle_s| \le [\langle M\rangle_t - \langle M\rangle_s]^{\frac12}[\langle N\rangle_t - \langle N\rangle_s]^{\frac12}$, (a.s.).
(b) $|\langle M,N\rangle_t| \le [\langle M\rangle_t\langle N\rangle_t]^{\frac12}$, (a.s.).
(c) If M and N are continuous local martingales $(M_0 = N_0 = 0)$, then (b) holds.

PROOF. For all real λ we have

$$\langle M + \lambda N, M + \lambda N\rangle_t - \langle M + \lambda N, M + \lambda N\rangle_s \ge 0 \qquad \text{(a.s.)}$$

and (a) follows immediately. We obtain (b) upon taking $s = 0$ and noting that by definition $\langle M\rangle_0 = M_0^2 = 0$, $\langle N\rangle_0 = 0$ and $\langle M,N\rangle_0 = 0$.

To prove (c) let σ_n be a sequence of stopping times as in Remark 2.6.6. From what has just been shown we have,

$$|\langle M^n,N^n\rangle_t| \le [\langle M^n\rangle_t\langle N^n\rangle_t]^{\frac12}$$

where M^n and N^n are L^2-martingales as defined Remark 2.6.6. Hence it follows that

$$|\langle M,N\rangle_t(\omega)| \leq [\langle M\rangle_t(\omega)\langle N\rangle_t(\omega)]^{\frac{1}{2}}$$

on $[t \leq \sigma_n(\omega)]$. This yields the assertion since $\sigma_n \uparrow \infty$ (a.s.). □

2.7 Some Useful Theorems

Let $Z = (Z_t)$ $(0 \leq t \leq T)$ be a continuous process on $(\Omega,\mathscr{A},P)$, and let $\mathscr{F}_t^Z$ be the smallest σ-field with respect to which the family $\{Z_s,s \leq t\}$ is measurable and containing all P-null sets. The following result will prove very useful in the filtering theory to be developed in later chapters.

Theorem 2.7.1. *Let $h_t(\omega)$ be jointly measurable and such that $E|h_t| < \infty$ for each t. Let*

$$\int_0^T E|h_t|\,dt < \infty.$$

Then there exists a jointly measurable and $(\mathscr{F}_t^Z)$-adapted process $H(t,\omega)$ such that for a.e. *t in $[0,T]$, $E(h_t|\mathscr{F}_t^Z)(\omega) = H(t,\omega)$* (a.s.).

First we prove the following lemma. For convenience of writing, we set $T = 1$.

Lemma 2.7.1. *Let f be a real random variable with $E|f| < \infty$. Then there is a jointly measurable modification of $E(f|\mathscr{F}_t^Z)$.*

Proof. For $n \geq 1$ define

$$\phi_n(t,\omega) = E(f|\mathscr{F}^Z_{(k-1)/2^n})(\omega)$$

for $(k-1)/2^n \leq t < k/2^n$ and $\omega \in \Omega$, $k = 1,\ldots,2^n$. Clearly (ϕ_n) is a sequence of (t,ω)-measurable functions. Now for t fixed, let t^n be the unique number of the form $(k-1)/2^n$ such that $(k-1)/2^n \leq t < k/2^n$. Then (t^n) is an increasing sequence, $t^n \uparrow t$ as $n \to \infty$. Since (Z_t) is a continuous process, it is easy to see that $\mathscr{F}^Z_{t^n} \subseteq \mathscr{F}^Z_{t^{n+1}}$ and $\vee_n \mathscr{F}^Z_{t^n} = \mathscr{F}^Z_t$. Hence $E(f|\mathscr{F}^Z_t)(\omega) = \lim_{n\to\infty} E(f|\mathscr{F}^Z_{t^n})(\omega) = \lim_{n\to\infty}\phi_n(t,\omega)$ (a.s.). The set C of (t,ω) points for which $\lim_{n\to\infty}\phi_n(t,\omega)$ exists is obviously jointly (that is, $\mathscr{B}([0,T]) \times \mathscr{F}^Z_T$) measurable. Define $\phi(t,\omega) = \lim_{n\to\infty}\phi_n(t,\omega)$ if $(t,\omega) \in C$ and $= 0$ otherwise. Then ϕ is (t,ω)-measurable and, for each t, equals $E(f|\mathscr{F}^Z_t)$ (a.s.), and is hence a (t,ω)-measurable version of the conditional expectation. □

Proof of Theorem 2.7.1. The class $\mathscr{M}$ of sets of the form $M =$ a finite union of disjoint sets $[a_j,b_j) \times A_j$, where $[a_j,b_j) \subset [0,T]$ and $A_j \in \mathscr{F}^Z_T$ is a field that generates $\mathscr{B}([0,T]) \times \mathscr{F}^Z_T$. From the integrability of $E|h_t|$ assumed in the theorem it follows that $\int_0^T\int_\Omega |h_t(\omega) - h_t^n(\omega)|\,dt\,dP \to 0$, where each h^n is a finite sum of the form $\sum a_k I_{M_k}(t,\omega)$ for $M_k \in \mathscr{M}$. From the lemma proved above, there exists a jointly measurable modification $H^n(t,\omega)$ of $E(h_t^n|\mathscr{F}_t^Z)(\omega)$. Since there exists a subsequence $(h^{n'})$ such that $E|h_t^{n'} - h_t| \to 0$ for a.e. (Lebesgue) t, we have for a.e. t, $E(h_t|\mathscr{F}_t^Z)(\omega) = \lim_{n' \to \infty} E(h_t^{n'}|\mathscr{F}_t^Z)(\omega)$ (a.s.). Define $H(t,\omega) = \lim_{n' \to \infty} H^{n'}(t,\omega)$ when the limit exists and $= 0$ otherwise. $H(t,\omega)$ has the required properties, and for a.e. t, we have $E(h_t|\mathscr{F}_t^Z)(\omega) = H(t,\omega)$ (a.s.). □

The following result which is useful in our study of stochastic differential equations is taken from Liptser and Shiryaev [40].

Let $C_d = C([0,T]; \mathbf{R}^d)$ be the Banach space (with sup norm) of continuous functions x from $[0,T]$ to $\mathbf{R}^d$. Denote by $\mathscr{B}_t(C_d)$, or simply $\mathscr{B}_t$, the σ-field of subsets of C_d generated by the family $\{x(s), 0 \le s \le t\}$.

Theorem 2.7.2. *Let $\xi = (\xi_t)$, $0 \le t \le T$, be a continuous d-dimensional process defined on a complete probability space $(\Omega,\mathscr{A},P)$. Also defined on the latter is a d-dimensional measurable process $\alpha = (\alpha_t)$, $0 \le t \le T$, which is adapted to the family $(\mathscr{F}_t^\xi)$.*

Then there exists a measurable functional $\gamma = \gamma(t,x)$ defined on $([0,T] \times C_d$, $\mathscr{B}[0,T] \times \mathscr{B}_T)$ which is $\mathscr{B}_{t_+}$-measurable for each t (we set $\mathscr{B}_{T_+} = \mathscr{B}_T$) and is such that

$$L \times P\{(t,\omega): \alpha_t(\omega) \ne \gamma(t,\xi(\omega))\} = 0,$$

where L is Lebesgue measure on $[0,T]$ and $L \times P$ is the product of L and P.

PROOF. By Proposition 1.1.4, we may assume $\alpha_t(\omega)$ is progressively measurable. Also $\xi_t(\omega)$ may be assumed continuous in t for all ω. For fixed $u \in [0,T]$, let us denote by $\mathscr{B}_u[0,T]$ the σ-field of $[0,T]$ generated by all subintervals of $[0,u]$. An elementary argument now shows that there is a function $\hat{\alpha}_t(\omega)$ defined on $[0,T] \times \Omega$ and measurable with respect to $\mathscr{B}_u[0,T] \times \mathscr{F}_u^{\xi,0}$ such that

$$L \times P\{(t,\omega): \hat{\alpha}_t(\omega) \ne \alpha_{t \wedge u}(\omega)\} = 0.$$

Recall $\mathscr{F}_t^{\xi,0} = \sigma(\xi_s, 0 \le s \le t)$, that is, without null sets. Hence for every $u \in [0,T]$, there exists on $([0,T] \times C_d, \mathscr{B}_u[0,T] \times \mathscr{B}_u)$ a d-dimensional measurable functional $\gamma_u(t,x)$ such that

$$L \times P\{(t,\omega): \alpha_{t \wedge u}(\omega) \ne \gamma_u(t,\xi(\omega))\} = 0.$$

Let $u_{k,n} = kT/2^n$, $k = 0, \ldots, 2^n$ and $n = 1,2, \ldots$. Set

$$\gamma^{(n)}(t,x) = \gamma_0(0,x) I_{\{0\}}(t) + \sum_{k=1}^{2^n} \gamma_{u_{k,n}}(t,x) I_{(u_{k-1,n},u_{k,n}]}(t)$$

and

$$\gamma(t,x) = \limsup_{n \to \infty} \gamma^{(n)}(t,x).$$

The lim sup is to be taken for each component of the d-dimensional functional $\gamma^{(n)}$. The symbol $|\cdot|$ in the steps below stands for the length of a d

vector. The functional $\gamma^{(n)}(t,x)$ is (t,x)-measurable for all n and so is the functional $\gamma(t,x)$. From the construction of the functionals $\gamma^{(n)}(t,x)$ $(n = 1,2,\ldots)$, it follows that for every t, $\gamma(t,x)$ is $\mathscr{B}_{t_+}$-measurable. Furthermore, for $n \geq 1$,

$$\begin{aligned} L \times P(\{(t,\omega): \gamma^{(n)}(t,\xi(\omega)) \neq \alpha_t(\omega)\}) &\leq L \times P(\{(0,\omega): \gamma_0(0,\xi(\omega)) \neq \alpha_0(\omega)\}) \\ &\quad + \sum_{k=1}^{2^n} L \times P(\{(t,\omega): \gamma_{u_{k,n}}(t,\xi(\omega)) \neq \alpha_t(\omega), u_{k-1,n} < t \leq u_{k,n}\}) \\ &= 0. \end{aligned}$$

Hence

$$L \times P(\{(t,\omega): \gamma^{(n)}(t,\xi(\omega)) \neq \gamma^{(n+1)}(t,\xi(\omega))\}) = 0$$

and, by the definition of $\gamma(t,x)$,

$$L \times P(\{(t,\omega): \gamma(t,\xi(\omega)) \neq \alpha_t(\omega)\}) = 0. \qquad \square$$

3 Stochastic Integrals

3.1 Predictable Processes

Let **L** denote the family of all real-valued functions $Y_t(\omega)$ defined on $\mathbf{R}_+ \times \Omega$ which are measurable with respect to $\mathscr{B}(\mathbf{R}_+) \times \mathscr{A}$ and have the following properties:

1. $Y = (Y_t)$ is adapted to $(\mathscr{G}_t)$.
2. For each ω, the function $t \to Y_t(\omega)$ is left-continuous.

Here P is a complete probability measure on $(\Omega,\mathscr{A})$ and $(\mathscr{G}_t)(t \in \mathbf{R}_+)$ is a right-continuous, increasing family of sub-σ-fields of $\mathscr{A}$. It is assumed that $\mathscr{G}_0$ contains all P-null sets in $\mathscr{A}$. Let **P** be the smallest σ-field of subsets of $\mathbf{R}_+ \times \Omega$ with respect to which all the functions belonging to **L** are measurable.

Definition 3.1.1. A stochastic process $X = (X_t(\omega))$ is *predictable* (or *previsible*) relative to $(\mathscr{G}_t)$, or $(\mathscr{G}_t)$-predictable, if the function

$$(t,\omega) \to X_t(\omega)$$

is **P**-measurable.

The phrase "relative to $(\mathscr{G}_t)$" will usually be dropped when referring to predictable processes unless the context requires it. Similarly we shall say "adapted" instead of "$(\mathscr{G}_t)$-adapted." Let us first look at some simple (and obvious) examples of predictable processes.

EXAMPLE 3.1.1. All $B(\mathbf{R}_+) \times \mathscr{A}$-measurable, adapted left-continuous processes are predictable.

EXAMPLE 3.1.2. A *simple* process (ϕ_t) which is defined to be of the form

$$\phi_t(\omega) = \phi_0(\omega)I_{\{0\}}(t) + \sum_{j=0}^{n-1} \phi_j(\omega)I_{(t_j,t_{j+1}]}(t) \qquad (0 = t_0 < t_1 < \cdots < t_n) \quad (3.1.1)$$

is predictable if each ϕ_j is $(\mathscr{G}_{t_j})$-measurable. The process given by Eq. (3.1.1) is both adapted and left-continuous.

EXAMPLE 3.1.3. Let (η_t) be an adapted, right-continuous simple process given by

$$\eta_t(\omega) = \sum_{j=0}^{n} \eta_{t_j}(\omega)I_{[t_j,t_{j+1})}(t). \quad (3.1.2)$$

Let (ξ_t) be the process defined by $\xi_t = \eta_{t-}$, the left limit of η_t. Then (ξ_t) is predictable.

Lemma 3.1.1. *The σ-field* $\mathbf{P}$ *is generated by all sets of the form* $\Gamma = (u,v] \times B$, *where* $B \in \mathscr{G}_u$, *or* $\Gamma = \{0\} \times B$, *where* $B \in \mathscr{G}_0$.

PROOF. Denote by $\mathscr{U}$ the class of all (t,ω) functions of the form $I_{(u,v]}(t)I_B(\omega)$, where $B \in \mathscr{G}_0$ and $u,v \in \mathbf{R}_+$ $(u \leq v)$, or of the form $I_{\{0\}}(t)I_B(\omega)$, where $B \in \mathscr{G}_0$. Clearly, each member of $\mathscr{U}$ is $\mathbf{P}$-measurable, so that $\sigma(\mathscr{U}) \subseteq \mathbf{P}$, where $\sigma(\mathscr{U})$ is the minimal σ-field with respect to which all the functions in $\mathscr{U}$ are measurable. Next, if $\phi \in \mathbf{L}$, we can see that it is, for each (t,ω), the limit of a sequence of simple processes of the form given in Example 3.1.2. Such a sequence is given by (ϕ^n), where

$$\phi_t^n(\omega) = \phi_0(\omega)I_{\{0\}}(t) + \sum_{j=0}^{j_n-1} \phi_{t_j^n}(\omega)I_{(t_j^n,t_{j+1}^n]}(t)$$

and $0 = t_0^n < t_1^n < \cdots < t_{j_n}^n = n$ is a subdivision of $[0,n]$ such that the length of each subinterval is $\leq 1/n$. Since $\phi_{t_j^n}(\omega)$ is $\mathscr{G}_{t_j^n}$-measurable, it is the pointwise limit of a sequence of simple functions of the form $\sum_i a_i I_{B_i}(\omega)$, where $B_i \in \mathscr{G}_{t_j^n}$. Hence $\phi_t^n(\omega)$ is measurable with respect to $\sigma(\mathscr{U})$, and so the same is true of $\phi_t(\omega)$. This shows $\mathbf{P} \subseteq \sigma(\mathscr{U})$ and completes the proof. □

Denote by $\mathscr{S}_T$ the class of sets mentioned in the above lemma where we restrict u and v to $[0,T]$.

Lemma 3.1.2. *Let* $\mathscr{M}_T$ *be the family of finite disjoint unions of sets in* $\mathscr{S}_T$. *Then* $\mathscr{M}_T$ *is a field of subsets of* $[0,T] \times \Omega$.

PROOF. Clearly $\mathscr{M}_T$ is closed under finite disjoint unions, and if $M_1,M_2 \in \mathscr{M}_T$, then $M_1 \cap M_2 \in \mathscr{M}_T$. To see this, it is enough to verify it for $M_1 = (u_1,v_1] \times B_1$ and $M_2 = (u_2,v_2] \times B_2$ when $u_2 \geq u_1$. Then $M_1 \cap M_2 = (u_2,\min(v_1,v_2)] \times (B_1 \cap B_2) \in \mathscr{S}_T$ since $B_1 \cap B_2 \in \mathscr{G}_{u_2}$. Next, if $M = \{0\} \times B$ $(B \in \mathscr{G}_0)$, we have

$$([0,T] \times \Omega) - M = ((0,T] \times \Omega) \cup (\{0\} \times B^c).$$

For $M = (u,v] \times B$ ($B \in \mathscr{G}_u$), we have

$$([0,T] \times \Omega) - M = (\{0\} \times \Omega) \cup ((0,u] \times \Omega) \cup ((u,v] \times B^c) \cup ((v,T] \times \Omega).$$

Thus if $M \in \mathscr{M}_T$, its complement is an intersection of disjoint unions of sets in $\mathscr{M}_T$. It follows that $\mathscr{M}_T$ is a field. □

Before we prove the main proposition of this section we need the following well-known result which we state here without proof ([8], page 601).

Lemma 3.1.3. *Let $\mathscr{S}$ be a class of subsets of $[0,T] \times \Omega$ such that the class of finite unions of disjoint $\mathscr{S}$ sets is a field. Let $\mathscr{H}$ be a class of real-valued, $\sigma(\mathscr{S})$-measurable functions on $[0,T] \times \Omega$ such that*

(i) *$\mathscr{H}$ includes every indicator function I_M, where $M \in \mathscr{S}$.*
(ii) *$\mathscr{H}$ includes every linear combination of a finite number of its functions.*
(iii) *If $\phi^n \in \mathscr{H}$, $n \geq 1$, and if $\lim_{n\to\infty} \phi_t^n(\omega) = \phi_t(\omega)$ exists and is finite for all (t,ω), then $\phi \in \mathscr{H}$.*

Then $\mathscr{H}$ contains all functions measurable with respect to $\sigma(\mathscr{S})$.

As we shall see in the next section, predictable processes provide a natural class of processes for which stochastic integrals can be defined. In problems of stochastic filtering and control—in particular, in the formulation of the innovation process theorem—there is an additional reason for considering predictable processes. The family of σ-fields which we shall be dealing with is furnished by an observation process, and it is not convenient to assume a priori that this family is right-continuous. On the other hand, the apparatus of martingale theory and its basic theorems (such as the Doob decomposition theorem) rest on such an assumption. The use of predictable processes helps us to overcome this obstacle as we shall now explain.

Let $(\mathscr{F}_t)$, $t \in \mathbf{R}_+$, be an increasing family of sub σ-fields of $\mathscr{A}$, which is *not* assumed to be right-continuous. Let $\mathscr{G}_t = \mathscr{F}_{t+}$. Then we have the following result.

Theorem 3.1.1. *Let $\phi = (\phi_t)$ be a predictable process relative to $(\mathscr{G}_t)$. Then ϕ_t is $\mathscr{F}_t$-measurable for each $t > 0$.*

It suffices to consider the time interval $[0,T]$, where T is finite but arbitrary. The definition of predictability when $\mathbf{R}_+$ is replaced by $[0,T]$ is given as before, and it is obvious that the assumption on ϕ implies that the restriction of ϕ to $[0,T] \times \Omega$ is predictable relative to $(\mathscr{G}_t)$, $0 \leq t \leq T$. We shall only be concerned with the restriction in this proof. In Lemma 3.1.3, take $\mathscr{S} = \mathscr{S}_T$, $\mathscr{S}_T$ being the class given in Lemma 3.1.2. Define $\mathscr{H}$ to be the class of all processes f defined on $[0,T] \times \Omega$, predictable relative to $(\mathscr{G}_t)$, $0 \leq t \leq T$, and such that

$$f_t \text{ is } \mathscr{F}_t\text{-measurable for each } t \in (0,T].$$

Conditions (ii) and (iii) of Lemma 3.1.3 are obviously satisfied. Let us now verify condition (i). If $M \in \mathscr{S}_T$ is of the form $\{0\} \times B$ $(B \in \mathscr{G}_0)$, then for $t > 0$, $I_M(t,\omega) = 0$ for all ω. Hence $I_M(t,\cdot)$ is $\mathscr{F}_t$-measurable. Next let $M = (u,v] \times B$, where $0 \leq u < v \leq T$, $B \in \mathscr{G}_u$. If $u = 0$, $B \in \mathscr{F}_t$-measurable for all $t > 0$ since $\mathscr{G}_0 = \mathscr{F}_{0+} \subset \mathscr{F}_t$. For $u > 0$, $I_M(t,\omega) = 0$ (for all ω) if $0 < t \leq u$ or if $t > v$. If $u < t \leq v$, B is $\mathscr{F}_s$-measurable for every $s > u$ and is hence $\mathscr{F}_t$-measurable. Thus (i) holds, and hence it follows that $\mathscr{H}$ includes all processes measurable with respect to $\sigma(\mathscr{S}_T) = \mathbf{P}_T$, where $\mathbf{P}_T$ is the σ-field in Definition 3.1.1 with $\mathbf{R}_+$ replaced by $[0,T]$. The proof of the theorem is then complete. □

The proof of the next result we need is to be found in Dellacherie's book ([7], page 501). It will take us too far afield to give it here.

Theorem 3.1.2. *Let $A = (A_t)$ be an integrable, increasing process adapted to a right-continuous family $(\mathscr{G}_t)$. Then A is predictable if and only if*

$$E\left[\int_0^t Y_s \, dA_s\right] = E\left[\int_0^t Y_{s-} \, dA_s\right]$$

for every bounded, right-continuous, positive martingale $Y = (Y_t)$.

The above result shows that the class of predictable, increasing processes coincides with the class of natural, increasing processes defined earlier. Hence we may reformulate the Doob decomposition theorem for potentials in the following form. The right-continuous family of σ-fields will be denoted by $(\mathscr{G}_t)$.

Theorem 3.1.3. *Let $(X_t,\mathscr{G}_t)$ be a potential of class* (D). *Then there exists a unique, integrable and predictable increasing process $A = (A_t)$ such that*

$$X_t = E[A_\infty|\mathscr{G}_t] - A_t.$$

From our study of predictable processes in this section we can extract the following result which will be used in the chapter on the stochastic filtering problem.

Theorem 3.1.4. *Let $(\mathscr{F}_t)$ be an increasing family not assumed to be right-continuous. Let $\mathscr{G}_t = \mathscr{F}_{t+}$. Let (U_t) be an integrable, increasing process not necessarily adapted to $(\mathscr{F}_t)$. Then there exists an integrable increasing process (U_t^*) such that*

(i) *U_t^* is $\mathscr{F}_t$-measurable for each t.*
(ii) *$E[(U_t - U_s)|\mathscr{F}_s] = E[(U_t^* - U_s^*)|\mathscr{F}_s]$, for $s,t \in \mathbf{R}_+$, $s \leq t$.*

Proof. The process $X_t = E(U_\infty - U_t|\mathscr{G}_t)$ is a $\mathscr{G}_t$-adapted potential of class (D). From Theorem 3.1.3 it follows that there exists an integrable, predictable increasing process (U_t^*) such that

$$X_t = E[U_\infty^*|\mathscr{G}_t] - U_t^*. \tag{3.1.3}$$

Since (U_t^*) is predictable [relative to $(\mathscr{G}_t)$], from Theorem 3.1.1 U_t^* is $\mathscr{F}_t$-measurable for each $t > 0$. Thus (i) is proved since $U_0^* = 0$. The validity of (ii) is now obvious. □

We shall prove below a very useful result which will be constantly used in our work.

Theorem 3.1.5. *Let $f = (f_t(\omega))$, $(t,\omega) \in [0,T] \times \Omega$ be a predictable process relative to $(\mathscr{F}_t)$, where $(\mathscr{F}_t)$ is assumed to be right-continuous. Suppose τ is a stopping time for $(\mathscr{F}_t)$. Then f_τ is $\mathscr{F}_\tau$-measurable.*

PROOF. Apply Lemma 3.1.3, taking $\mathscr{S} = \mathscr{S}_T$, so that $\sigma(\mathscr{S}) = \mathbf{P}_T$. Define $\mathscr{H}$ to be the class of all processes ϕ defined on $[0,T] \times \Omega$, predictable relative to $(\mathscr{F}_t)$ and such that ϕ_τ is $\mathscr{F}_\tau$-measurable. Then conditions (ii) and (iii) of Lemma 3.1.3 are obviously satisfied. It remains only to verify (i). Let $M \in \mathscr{S}_T$, $M = (u,v] \times A$ $(A \in \mathscr{F}_u)$. We have to show that $I_M \in \mathscr{H}$. Now, $I_M(t,\omega) = I_{(u,v]}(t) \cdot I_A(\omega)$. We must show that

$$[\omega: I_{(u,v]}(\tau(\omega)) \cdot I_A(\omega) = 1] \in \mathscr{F}_\tau.$$

For $a \geq 0$,

$$\begin{aligned}[\omega: I_{(u,v]}(\tau(\omega)) \cdot I_A(\omega) = 1] &\cap [\omega: \tau(\omega) \leq a] \\ &= A \cap [\omega: u < \tau(\omega) \leq v] \cap [\omega: \tau(\omega) \leq a].\end{aligned}$$

It is readily verified that the set on the right-hand side is in $\mathscr{F}_a$ for $a \geq 0$. By the definition of $\mathscr{F}_\tau$ it follows that

$$[\omega: I_{(u,v]}(\tau(\omega)) \cdot I_A(\omega) = 1] \in \mathscr{F}_\tau.$$

If $M = \{0\} \times A$, where $A \in \mathscr{F}_0$, then

$$[\omega: I_M(\tau(\omega),\omega) = 1] \cap [\omega; \tau(\omega) \leq a] = A \cap [\omega: \tau(\omega) = 0] \in \mathscr{F}_0 \subset \mathscr{F}_a.$$

So again we have the $\mathscr{F}_\tau$-measurability of $I_M(\tau(\omega),\omega)$. Hence $I_M \in \mathscr{H}$ for $M \in \mathscr{S}_T$. Thus $\mathscr{H}$ contains all $\mathbf{P}_T$-measurable (that is, predictable) functions by Lemma 3.1.3 and the theorem is proved. □

3.2 Stochastic Integrals for L^2-Martingales

Let $M = (M_t)$, $0 \leq t \leq T$, be a right-continuous, square-integrable martingale with respect to a *right-continuous* family of σ-fields $(\mathscr{F}_t)$. Assume $M_0 = 0$. We first define the stochastic integral with respect to M of a simple, predictable process $f = (f_t)$ given by

$$f_t(\omega) = \sum_{j=0}^{n-1} f_j(\omega) I_{(t_j,t_{j+1}]}(t),$$

where $0 \leq t_0 < \cdots < t_n = T$, f_j is $\mathscr{F}_{t_j}$-measurable, and $E\int f_s^2\, d\langle M\rangle_s < \infty$. *Define*

$$\int_0^T f_s\, dM_s = \sum_j f_j(M_{t_{j+1}} - M_{t_j}). \tag{3.2.1}$$

Since $f_s \cdot I_{[0,t]}(s)$ is also a simple predictable process for $t \geq 0$, the integral $\int_0^t f_s\, dM_s$ is defined as $\int_0^T f_s I_{[0,t]}(s)\, dM_s$. It is easy to verify the following properties. For convenience we write $\int$ instead of $\int_0^T$.

$$E\int f_s\, dM_s = 0 \tag{3.2.2}$$

$$E\left[\int f_s\, dM_s\right]^2 = E\left[\int f_s^2\, d\langle M\rangle_s\right] \tag{3.2.3}$$

where $\langle M\rangle$ is the quadratic variation process associated with M. We shall now show that the definition [Equation (3.2.1)] extends to all integrand processes which are predictable and satisfy the condition $E[\int f_s^2\, d\langle M\rangle_s] < \infty$. It is convenient to introduce the measure ν_M on $([0,T] \times \Omega, \mathbf{P}_T)$ defined by the relation $\int \Phi\, d\nu_M = E[\int \Phi_t\, d\langle M\rangle_t]$, where Φ ranges over all nonnegative, $\mathbf{P}_T$-measurable processes. Then the stochastic integral will be defined for all $f \in L^2([0,T] \times \Omega, \mathbf{P}_T, \nu_M) = L^2(\nu_M)$. Let $E \in \mathbf{P}_T$ be a subset of $[0,T] \times \Omega$ for T as above. It has been shown earlier that the field $\mathscr{M}_T$ of finite disjoint unions of sets in $\mathscr{S}_T$ generates $\mathbf{P}_T$. Hence there exists a sequence of sets $E_n \in \mathscr{M}_T$ such that for each n, $\int [I_E - I_{E_n}]^2\, d\nu_M \leq 1/n$. Hence it follows that the class $\mathbf{L}_0$ of all (left-continuous) predictable, simple processes is dense in $L^2(\nu_M)$.

By Equation (3.2.3), *the linear map $f \to \int f_s\, dM_s$ from $\mathbf{L}_0$ to $L^2(\Omega,\mathscr{A},P)$ is an isometry which has a unique, continuous extension as an isometry from $L^2(\nu_M)$ into $L^2(\Omega,\mathscr{A},P)$.* It is clear from this definition that the stochastic integral $\int f_s\, dM_s$, $f \in L^2(\nu_M)$, is defined only up to an equivalence.

From now on we shall denote $L^2(\nu_M)$ by $L^2(M)$.

Properties of Stochastic Integrals

In addition to Equations (3.2.2) and (3.2.3), the stochastic integral $\int f_s\, dM_s$ has the following properties:

1. (Linearity) If $f,g \in L^2(M)$ and $\alpha \in \mathbf{R}$, then $f + g \in L^2(M)$, $\alpha f \in L^2(M)$, and

$$\int (f + g)_u\, dM_u = \int f_u\, dM_u + \int g_u\, dM_u,$$

$$\int \alpha f_u\, dM_u = \alpha \int f_u\, dM_u.$$

2. If $f, f^n \in L^2(M)$ and $f^n \to f$ (that is, $E\int |f_u^n - f_u|^2\, d\langle M\rangle_u \to 0$ as $n \to \infty$), then

$$\int f_u^n\, dM_u \to \int f_u\, dM_u$$

in the L^2 sense.

3. For each $t \geq 0$ $(0 \leq t \leq T)$ and $f \in L^2(M)$, we have
$$fI_{[0,t]} \in L^2(M),$$
and we then define
$$\int_0^t f_s\,dM_s = \int f_s I_{[0,t]}(s)\,dM_s.$$
Similarly, we define, for $0 \leq s < t$,
$$\int_s^t f_u\,dM_u = \int f_u I_{(s,t]}(u)\,dM_u.$$
Set $Y_t = \int_0^t f_u\,dM_u$, $f \in L^2(M)$. Clearly Y_t is $\mathscr{F}_t$-measurable.
4. (a) $(Y_t, \mathscr{F}_t)$ is an L^2-martingale.
   (b) There exists a right-continuous version of the process (Y_t).

PROOF. If f is a predictable, simple function in $L^2(M)$ (a) can be directly verified. For $f \in L^2(M)$, let (f^n) be a sequence of predictable, simple functions belonging to $L^2(M)$ such that $E\int [f_u - f_u^n]^2\,d\langle M\rangle_u \to 0$. For s and t fixed $(s \leq t)$, we have
$$E\left|E\left(\int_s^t f_u\,dM_u \middle| \mathscr{F}_s\right) - E\left(\int_s^t f_u^n\,dM_u \middle| \mathscr{F}_s\right)\right| \leq \left\{E\left[\int_s^t (f_u - f_u^n)\,dM_u\right]^2\right\}^{\frac{1}{2}}$$
$$\leq \left\{E\left(\int_0^T (f_u - f_u^n)^2\,d\langle M\rangle_u\right)\right\}^{\frac{1}{2}} \to 0.$$
Since $E(\int_s^t f_u^n\,dM_u | \mathscr{F}_s) = 0$ for all n, it follows that $E(\int_s^t f_u\,dM_u | \mathscr{F}_s) = 0$ and (a) is proved. Conclusion (b) follows from property 4 of Section 2.2. □

In working with the process Y_t, it will always be assumed that a right-continuous version is chosen.

5. Suppose that $M = (M_t)$ is a *continuous*, L^2-martingale and $f \in L^2(M)$. Then the stochastic integral $\int_0^t f_s\,dM_s$ $(0 \leq t \leq T)$ has a continuous version.

PROOF. Let (f^n) be a sequence of predictable, simple (left-continuous) processes such that for each n,
$$E\int_0^T (f_s - f_s^n)^2\,d\langle M\rangle_s < 2^{-n}.$$
From the martingale inequality,
$$E\left[\sup_{0 \leq t \leq T}\left|\int_0^t f_s\,dM_s - \int_0^t f_s^n\,dM_s\right|^2\right] \leq 4E\int_0^T (f_s - f_s^n)^2\,d\langle M\rangle_s.$$
The Borel-Cantelli lemma gives
$$P\left[\sup_{0 \leq t \leq T}\left|\int_0^t f_s\,dM_s - \int_0^t f_s^n\,dM_s\right| > \frac{1}{n}, \text{ infinitely often}\right] = 0.$$

Now, for each n, $\int_0^t f_s^n\,dM_s$ is a continuous process and

$$\sup_{0\le t\le T}\left|\int_0^t f_s^n\,dM_s \to \int_0^t f_s\,dM_s\right| \to 0 \qquad \text{(a.s.) as } n\to\infty.$$

The version of $\int_0^t f_s\,dM_s$ obtained above is, in fact, a continuous process. □

Hence, in dealing with the stochastic integral process (Y_t), $Y_t = \int_0^t f_s\,dM_s$, where M is a continuous L^2-martingale, we may and do choose a version which is a continuous L^2-martingale.

In what follows M and N are L^2-martingales with respect to $(\mathscr{F}_t)$. For convenience we assume that $M_0 = N_0 = 0$.

6. (a) Let f be a bounded, predictable process. Then

$$\left|\int f_t\,d\langle M,N\rangle_t\right| \le \left[\int f_t^2\,d\langle M\rangle_t \langle N\rangle_T\right]^{\frac{1}{2}} \qquad \text{(a.s.)}.$$

(b) If f and g are predictable processes, then

$$\int |f_t g_t|\,d|\langle M,N\rangle|_t \le \left[\int f_t^2\,d\langle M\rangle_t \int g_t^2\,d\langle N\rangle_t\right]^{\frac{1}{2}} \qquad \text{(a.s.)}.$$

(c) $E[\int|f_t g_t|\,d|\langle M,N\rangle|_t] \le [E\int f_t^2\,d\langle M\rangle_t E\int g_t^2\,d\langle N\rangle_t]^{\frac{1}{2}}$.

Proof. We shall only give an outline of the proof of (a) following [43]. For the proof of (b) the reader is referred to [36] or [43]. Inequality (c) is a direct consequence of (b).

Let $\langle M,N\rangle_s^t = \langle M,N\rangle_t - \langle M,N\rangle_s$, and let $\langle M\rangle_s^t$ and $\langle N\rangle_s^t$ be similarly defined. Then by Remark 2.6.7(a), we have

$$|\langle M,N\rangle_s^t| \le [\langle M\rangle_s^t \langle N\rangle_s^t]^{\frac{1}{2}} \qquad \text{(a.s.)}.$$

Now, referring to the proof of Remark 2.6.7(a), let Ω_0 be the set of ω's for which the above inequality holds for all rational s and t $(s < t)$. Suppose

$$f_t(\omega) = \sum_{j=0}^{n-1} f_j(\omega)[M_{t_{j+1}}(\omega) - M_{t_j}(\omega)]$$

where the t_j's $(0 \le t_j < T)$ are rational. For $\omega \in \Omega_0$, we have

$$\left[\int f_t\,d\langle M,N\rangle_t\right](\omega) = \sum_j f_j(\omega)\langle M,N\rangle_{t_j}^{t_{j+1}}(\omega).$$

Hence we obtain

$$\left|\left[\int f_t\,d\langle M,N\rangle_t\right](\omega)\right| \le \left[\sum_j f_j^2(\omega)\langle M\rangle_{t_j}^{t_{j+1}}(\omega)\right]^{\frac{1}{2}} [\langle N\rangle_T(\omega)]^{\frac{1}{2}}$$

$$= \left\{\left[\int f_t^2\,d\langle M\rangle_t\right](\omega)\langle N_T\rangle(\omega)\right\}^{\frac{1}{2}}$$

which is (a). To show that (a) holds for all bounded, predictable processes, we apply a monotone class argument which needs a variant of Lemma 3.1.3 (see Theorem 20, p. 11 in [41]). The details are left to the reader. □

7. Let $f \in L^2(M)$. Then for fixed $t \in [0,T]$,

$$E\left[\int_0^t f_s\, dM_s \cdot N_t\right] = E\left[\int_0^t f_s\, d\langle M,N\rangle_s\right].$$

Proof. It is easy to verify property 7 directly if f is a predictable simple process belonging to $L^2(M)$. Suppose $f \in L^2(M)$ is bounded, and let (f^n) be a sequence of predictable simple processes such that $E\int_0^T (f_s - f_s^n)^2\, d\langle M\rangle_s \to 0$. We have

$$E\left|\int_0^t f_s\, dM_s \cdot N_t - \int_0^t f_s^n\, dM_s \cdot N_t\right| \le \left\{E\int_0^T (f_s - f_s^n)^2\, d\langle M_s\rangle E\langle N\rangle_t\right\}^{\frac{1}{2}} \to 0$$

so that

$$E\left[\int_0^t f_s^n\, dM_s \cdot N_t\right] \to E\left[\int_0^t f_s\, dM_s \cdot N_t\right].$$

Now, from 6(a),

$$E\left|\int_0^t f_s^n\, d\langle M,N\rangle_s - \int_0^t f_s\, d\langle M,N\rangle_s\right| \le \left\{E\int_0^t (f_s - f_s^n)^2\, d\langle M\rangle_s \cdot E\langle N\rangle_t\right\}^{\frac{1}{2}}.$$

The proof for arbitrary $f \in L^2(M)$ is similar, using 6(c). □

8. Let M be a continuous L^2-martingale, $f \in L^2(M)$ and $Y_t = \int_0^t f_s\, dM_s$. If N is any continuous L^2-martingale, then

$$\langle Y,N\rangle_t = \int_0^t f_s\, d\langle M,N\rangle_s \qquad \text{(a.s.)}.$$

Proof. Define $Z_t = Y_t N_t - \int_0^t f_s\, d\langle M,N\rangle_s$, where it is understood that continuous versions of Y and N are taken. Let τ be a stopping time. Since N^τ is also a continuous L^2-martingale (see Section 2.6), it follows from property 7 that

$$E\left[Y_T N_T^\tau - \int_0^T f_s\, d\langle M,N^\tau\rangle_s\right] = 0.$$

Now,

$$E[Y_T N_T^\tau] = E[Y_T N_\tau] = E[N_\tau E(Y_T|\mathscr{F}_\tau)] = E(N_\tau Y_\tau).$$

Also, from Lemma 2.6.4, $\langle M,N^\tau\rangle_s = \langle M,N\rangle_{s\wedge\tau}$, and so $\int_0^T f_s\, d\langle M,N^\tau\rangle_s = \int_0^\tau f_s\, d\langle M,N\rangle_s$. Hence we have $E[Y_\tau N_\tau - \int_0^\tau f_s\, d\langle M,N\rangle_s] = 0$. Thus (Z_t) is a continuous, $(\mathscr{F}_t)$-adapted process such that $E(Z_\tau) = 0$ for every stopping time. From property 11 of Section 2.2, it follows that (Z_t) is a uniformly integrable martingale. From Theorem 2.6.3 it then follows that

$$\int_0^t f_s\, d\langle M,N\rangle_s = \langle Y,N\rangle_t. \qquad \square$$

9. Suppose M is a continuous, L^2-martingale ($M_0 = 0$) and $f \in L^2(M)$. Then the continuous L^2-martingale Y, where $Y_t = \int_0^t f_s \, dM_s$ has the quadratic variation

$$\langle Y \rangle_t = \int_0^t f_s^2 \, d\langle M \rangle_s.$$

PROOF. Taking $N = Y$ in property 8, we obtain

$$\langle Y \rangle_t = \int_0^t f_s \, d\langle M, Y \rangle_s.$$

Again using property 8, we have

$$\langle M, Y \rangle_s = \int_0^s f_u \, d\langle M \rangle_u,$$

and hence

$$\langle Y \rangle_t = \int_0^t f_s^2 \, d\langle M \rangle_s. \qquad \square$$

10. Let $f \in L^2(M)$ and suppose τ is a stopping time for the continuous L^2-martingale M. Then the martingale Y^τ is given by

$$Y_t^\tau = \int_0^t f_s I_{[0,\tau]}(s) \, dM_s \qquad \text{(a.s.)}.$$

PROOF. Clearly $f I_{[0,\tau]} \in L^2(M)$, so that the right-hand stochastic integral exists. Writing $\xi_t = \int_0^t f_s I_{[0,\tau]}(s) \, dM_s$ (and taking a continuous version of the L^2-martingale ξ), we have

$$\begin{aligned} E[Y_t^\tau - \xi_t]^2 &= E\langle Y^\tau - \xi \rangle_t \\ &= E\langle Y^\tau \rangle_t + E\langle \xi \rangle_t - 2E\langle Y^\tau, \xi \rangle_t. \end{aligned}$$

Now

$$E\langle Y^\tau \rangle_t = E\langle Y \rangle_{t \wedge \tau} = E \int_0^{t \wedge \tau} f_s^2 \, d\langle M \rangle_s,$$

$$E\langle \xi \rangle_t = E \int_0^t f_s^2 I_{[0,\tau]}(s) \, d\langle M \rangle_s,$$

and

$$\begin{aligned} E\langle Y^\tau, \xi \rangle_t &= E \int_0^t f_s I_{[0,\tau]}(s) \, d\langle Y^\tau, M \rangle_s \\ &= E \int_0^t f_s I_{[0,\tau]}(s) \, d\langle Y, M \rangle_{s \wedge \tau} \\ &= E \int_0^{t \wedge \tau} f_s^2 I_{[0,\tau]}(s) \, d\langle M \rangle_s \\ &= E \int_0^t f_s^2 I_{[0,\tau]}(s) \, d\langle M \rangle_s. \end{aligned}$$

Hence $E[Y_t^\tau - \xi_t]^2 = 0$ and the proof is complete. $\square$

The stochastic integral $\int_0^t f_s I_{[0,\tau]}(s) \, dM_s$ *is also written* $\int_0^{t \wedge \tau} f_s \, dM_s$.

11. (a) Let f be predictable and bounded, and let σ and τ be stopping times such that $\sigma \le \tau$. Then $f_\sigma I_{(\sigma,\tau]} \in L^2(M)$ and $\int_0^T f_\sigma I_{(\sigma,\tau]}(t) \, dM_t = f_\sigma[M_\tau - M_\sigma]$. Here and in (b), M is continuous.

PROOF. From the increasing property of $(\mathscr{F}_t)$ we have

$$[\omega: I_{(\sigma(\omega),\tau(\omega)]}(t) = 1] = [\omega: \sigma(\omega) < t] \cap [\omega: \tau(\omega) < t]^c \in \mathscr{F}_t.$$

Hence $f_\sigma I_{(\sigma,\tau]}(t)$ is $\mathscr{F}_t$-measurable and $f_\sigma I_{(\sigma,\tau]}(t)$ is clearly left-continuous. The boundedness of f shows that $f_\sigma I_{(\sigma,\tau]} \in L^2(M)$. Write

$$Z_T = \int_0^T f_\sigma I_{(\sigma,\tau]}(s)\, dM_s$$

and

$$N = f_\sigma[M_\tau - M_\sigma].$$

We have

$$E(Z_T - N)^2 = E\int_0^T f_\sigma^2 I_{(\sigma,\tau]}(s)\, d\langle M\rangle_s - 2E(Z_T N) + Ef_\sigma^2(M_\tau - M_\sigma)^2.$$

The first term on the right-hand side equals

$$Ef_\sigma^2[\langle M\rangle_\tau - \langle M\rangle_\sigma].$$

$$\begin{aligned} E[Z_T N \mid \mathscr{F}_\sigma] &= f_\sigma E\left[\int_0^T f_\sigma I_{(\sigma,\tau]}(s)\, dM_s \cdot \int_0^T I_{(\sigma,\tau]}(s)\, dM_s \Big| \mathscr{F}_\sigma\right] \\ &= f_\sigma E\left[\left\langle \int_0^T f_\sigma I_{(\sigma,\tau]}(s)\, dM_s, \int_0^T I_{(\sigma,\tau]}(s)\, dM_s \right\rangle \Big| \mathscr{F}_\sigma\right] \\ &= f_\sigma E\left[\int_0^T f_\sigma I_{(\sigma,\tau]}(s)\, d\langle M\rangle_s \Big| \mathscr{F}_\sigma\right]. \end{aligned}$$

Hence

$$\begin{aligned} E[Z_T N] &= E\left[\int_0^T f_\sigma^2 I_{(\sigma,\tau]}(s)\, d\langle M\rangle_s\right] \\ &= E\{f_\sigma^2[\langle M\rangle_\tau - \langle M\rangle_\sigma]\}. \end{aligned}$$

Finally, the last term on the right-hand side equals

$$E[f_\sigma^2 E\{\langle M\rangle_\tau - \langle M\rangle_\sigma \mid \mathscr{F}_\sigma\}] = E\{f_\sigma^2[\langle M\rangle_\tau - \langle M\rangle_\sigma]\}.$$

It follows that $E[Z_T - N]^2 = 0$ and the assertion is proved. □

11. (b) Suppose that (X_t) is a continuous, $(\mathscr{F}_t)$-adapted process, and let g be a bounded, continuous function of a real variable. Let (τ_i) be a finite sequence of stopping times with $\tau_i \leq \tau_{i+1}$ (a.s.) for each i. Then

$$f = \sum_i g(X_{\tau_i}) I_{(\tau_i,\tau_{i+1}]} \in L^2(M)$$

and

$$\int_0^T f_t\, dM_t = \sum_i g(X_{\tau_i})[M_{\tau_{i+1}} - M_{\tau_i}].$$

PROOF. Since (X_t) is continuous and $(\mathscr{F}_t)$-adapted, X_{τ_i} is $\mathscr{F}_{\tau_i}$-measurable, and hence $g(X_{\tau_i})$ is $\mathscr{F}_{\tau_i}$-measurable. From property 11(a), for each i, $g(X_{\tau_i})I_{(\tau_i,\tau_{i+1}]}$ is predictable and bounded and

$$f = \sum_i g(X_{\tau_i})I_{(\tau_i,\tau_{i+1}]} \in L^2(M).$$

Using property 11(a) and the linearity of the stochastic integral, we obtain

$$\int_0^T f_t\,dM_t = \sum_i \int_0^T g(X_{\tau_i})I_{(\tau_i,\tau_{i+1}]}(t)\,dM_t$$

$$= \sum_i g(X_{\tau_i})[M_{\tau_{i+1}} - M_{\tau_i}].$$ □

12. Let M be a continuous L^2-martingale, and $f,g \in L^2(M)$. If $A = \{\omega: f_t(\omega) = g_t(\omega), 0 \le t \le T\}$, then

$$\int_0^T f_t\,dM_t = \int_0^T g_t\,dM_t \qquad \text{(a.s.) on } A.$$

PROOF. Let

$$\tau = \begin{cases} \inf\{t: f_t \neq g_t, 0 \le t \le T\} & \\ T & \text{if this set is empty.} \end{cases}$$

Then τ is a stopping time for $(\mathscr{F}_t)$ on $[0,T]$. Let (Y_t) be a right-continuous version of the martingale $\int_0^t (f_s - g_s)\,dM_s$, $0 \le t \le T$. By property 10,

$$Y_\tau = \int_0^T (f_s - g_s)I_{[0,\tau]}(s)\,dM_s \qquad \text{(a.s.)}.$$

For each ω, $[f_s(\omega) - g_s(\omega)]^2 I_{[0,\tau(\omega)]}(s) = I_{\{\tau(\omega)\}}(s)[f_{\tau(\omega)}(\omega) - g_{\tau(\omega)}(\omega)]^2$. Since $\langle M\rangle$ is continuous, for almost all ω the measure $d\langle M\rangle(\omega)$ has no jumps. Consequently, $\int_0^T (f_t - g_t)^2 I_{[0,\tau]}(t)\,d\langle M\rangle_t = 0$ (a.s.), so that $EY_\tau^2 = 0$, that is, $Y_\tau = 0$ (a.s.). But for $\omega \in A$, $\tau(\omega) = T$, so that $Y_\tau = Y_T$ on A. From $Y_T = \int_0^T (f_s - g_s)\,dM_s$, it now follows that $\int_0^T f_s\,dM_s = \int_0^T g_s\,dM_s$ (a.s.) on A. □

3.3 The Ito Integral

The stochastic integral that concerns us most in this book is the integral with respect to the Wiener process $W = (W_t)$, where $(W_t,\mathscr{F}_t)$ is a martingale. Let us first consider the case of a finite interval $[0,T]$. Then W is a square-integrable martingale with $\langle W\rangle_t = t$. The definition of $\int_0^T f_t\,dW_t$ follows immediately from Section 3.2 for integrand processes f belonging to the class $L^2(W)$. It is instructive, however, to study Ito's method of defining the stochastic integral that now bears his name.

Ito's Definition of the Stochastic Integral

Let $\mathbf{T}$ stand for $[0,T]$ or $\mathbf{R}_+$, and let $(W_t,\mathscr{F}_t)$, $t \in \mathbf{T}$ be a Wiener martingale. For convenience we shall assume $\mathbf{T} = \mathbf{R}_+$, though the same method gives the definition of the integral over $[0,T]$. Let $\mathscr{M}^W_{2,\infty}$ be the class of all processes $f = (f_t)$ satisfying the conditions

(i) (f_t) is $\mathscr{B}(\mathbf{R}_+) \times \mathscr{A}$ measurable and $(\mathscr{F}_t)$-adapted.
(ii) $E\int f_t^2\, dt < \infty$.

Define the stochastic integral first for the class $\mathscr{E}$ of simple functions belonging to $\mathscr{M}^W_{2,\infty}$, that is, for $f \in \mathscr{M}^W_{2,\infty}$ of the form

$$f_t(\omega) = f_{t_j}(\omega) \quad \text{if } t_j \le t < t_{j+1}, j = 0, 1, \ldots, n-1,$$

where $0 = t_0 < \cdots < t_n < \infty$ is a partition of $[0,\infty)$ and f_{t_j} is $\mathscr{F}_{t_j}$-measurable. Writing $I(f)$ for $\int f_t\, dW_t$, define

$$I(f) = \sum f_{t_j}[W_{t_{j+1}} - W_{t_j}].$$

If g is another simple function in the same class it is easily seen that

$$E[I(f) - I(g)]^2 = E\int [f_t - g_t]^2\, dt.$$

From this relation it follows that the map $f \to \int f_t\, dW_t$ from $\mathscr{E}$ to $L^2(\Omega,\mathscr{A},P)$ has a unique extension to $\mathscr{M}^W_{2,\infty}$ if the following fact is established.

Proposition 3.3.1. *Let $f \in \mathscr{M}^W_{2,\infty}$. Then there exists a sequence (f^n) of processes $f^n \in \mathscr{E}$ such that*

$$\lim_{n\to\infty} E\int [f_t - f_t^n]^2\, dt = 0. \tag{3.3.1}$$

PROOF. It is sufficient to prove Equation (3.3.1) for an f which is bounded in (t,ω) and such that $f_t(\omega) = 0$ for t outside some finite interval $[a,b]$ in $\mathbf{R}_+$. Extend the definition of $f_t(\omega)$ for $t < 0$ by setting $f_t(\omega) = 0$ for all ω if $t < 0$. Condition (ii) implies $\int_{-\infty}^{\infty} f_t^2(\omega)\, dt < \infty$ (a.s.). Hence

$$\int_{-\infty}^{\infty} |f_{t+h}(\omega) - f_t(\omega)|^2\, dt \to 0 \qquad \text{(a.s.) as } h \to 0.$$

Furthermore,

$$\int_{-\infty}^{\infty} |f_{t+h}(\omega) - f_t(\omega)|^2\, dt \le 4\int_{-\infty}^{\infty} f_t^2(\omega)\, dt \in L^1(\Omega)$$

for almost all ω. So

$$\int_\Omega \int_{-\infty}^{\infty} |f_{t+h}(\omega) - f_t(\omega)|^2\, dt\, dP \to 0 \quad \text{as } h \to 0.$$

For $n \ge 1$, define $\alpha_n(t)$ by

$$\alpha_n(t) = \frac{j}{2^n}, \qquad \frac{j}{2^n} \le t < \frac{j+1}{2^n} \qquad (j = 0, \pm 1, \pm 2, \ldots).$$

Then

$$\int_\Omega \int_{-\infty}^{\infty} |f_{s+\alpha_n(t)}(\omega) - f_{s+t}(\omega)|^2 \, ds \, dP \to 0 \quad \text{as } n \to \infty.$$

Since $\int_\Omega \int_{-\infty}^{\infty} |f_{s+\alpha_n(t)}(\omega) - f_{s+t}(\omega)|^2 \, ds \, dP \leq 4 \int_\Omega \int_{-\infty}^{\infty} f_s^2(\omega) \, ds \, dP$ for every t, it follows that

$$\int_{a-1}^{b} \int_\Omega \int_{-\infty}^{\infty} |f_{s+\alpha_n(t)}(\omega) - f_{s+t}(\omega)|^2 \, ds \, dP \, dt \to 0 \quad \text{as } n \to \infty,$$

or

$$\int_{-\infty}^{\infty} ds \int_{a-1}^{b} \int_\Omega |f_{s+\alpha_n(t)} - f_{s+t}(\omega)|^2 \, dP \, dt \to 0.$$

Hence there exists a subsequence (n_i) such that

$$\int_{a-1}^{b} \int_\Omega |f_{s+\alpha_{n_i}(t)}(\omega) - f_{s+t}(\omega)|^2 \, dP \, dt \to 0$$

for almost every s. Choose and fix one such value of $s \in [0,1]$. Then

$$\int_{a-1}^{b} \int_\Omega |f_{s+\alpha_{n_i}(t)}(\omega) - f_{s+t}(\omega)|^2 \, dP \, dt \to 0 \quad \text{as } n_i \to \infty,$$

and making a change of variable, we get

$$\int_{a-1+s}^{b+s} \int_\Omega |f_{s+\alpha_{n_i}(t-s)}(\omega) - f_t(\omega)|^2 \, dP \, dt \to 0.$$

Since $s \in [0,1]$,

$$\int_a^b \int_\Omega |f_{s+\alpha_{n_i}(t-s)}(\omega) - f_t(\omega)|^2 \, dP \, dt \to 0.$$

Defining $f_t^i(\omega) = f_{s+\alpha_{n_i}(t-s)}(\omega)$, we see that (f_t^i) is a sequence of processes in $\mathscr{E}$ which satisfies Equation (3.3.1). □

Remark 3.3.1. The integrands belonging to $L^2(W)$ are predictable processes. The connection between them and the processes satisfying (i) and (ii) can be seen as follows. For convenience, let t range over a finite interval $[0,T]$. Observe first, that the approximating sequence of simple processes in Equation (3.3.1) of Proposition 3.3.1 were taken to be right-continuous but, obviously, can be taken to be left-continuous. Then from Equation (3.3.1) there exists a subsequence $(f^{n'})$ converging to f almost everywhere with respect to the product measure $dt\,dP$. Hence for any measurable f which satisfies (i) and (ii) there exists a predictable process g such that $f = g$ a.e. $(dt\,dP)$. However, the stochastic integral $\int f_t \, dM_t$, where M is not a Wiener process, cannot be extended to the class of measurable f satisfying conditions (i) and the condition $E\int_0^T f_t^2 \, d\langle M\rangle_t < \infty$. Here is an example due to P. Courrége [5].

Let $\Omega = \{\omega_1,\omega_2\}$ $(\omega_1 \neq \omega_2)$ and $\mathscr{A} = \{\varnothing,\{\omega_1\},\{\omega_2\},\Omega\}$. Define P by $P(\{\omega_1\}) = \alpha$, $0 < \alpha < 1$, $\alpha \neq \frac{1}{2}$, and the σ-fields family $(\mathscr{F}_t)$ thus: $\mathscr{F}_t = \{\varnothing,\Omega\}$ for $0 \leq t < t_0 < T$, $\mathscr{F}_t = \mathscr{A}$ for $t_0 \leq t \leq T$. Clearly, in $(\Omega,\mathscr{A},P)$, (a.s.) convergence, L^2-convergence, and pointwise convergence are all equivalent. Choose $h(\omega) = I_{\{\omega_1\}}(\omega) - \alpha$ and define the martingale $M_t = E(h|\mathscr{F}_t)$. Then

M_t is right-continuous and square-integrable. If $f \in L^2(M)$, it is easy to see that

$$\int f_t \, dM_t = f_{t_0} h. \tag{3.3.2}$$

We shall show that the process Φ given by

$$\Phi_t(\omega) = I_{\{\omega_1\}}(\omega) I_{[t_0,T]}(t) \tag{3.3.3}$$

satisfies the above conditions, but $\int \Phi_t \, dM_t$ cannot be consistently defined. Suppose the integral exists and satisfies Equations (3.2.2) and (3.2.3). We obtain a contradiction as follows. It is easy to verify that we must have $\int \Phi_t \, dM_t = \Phi_{t_0} h$. Hence

$$E \int \Phi_t \, dM_t = E(\Phi_{t_0} \cdot h) = \alpha(1 - \alpha) \neq 0 \tag{3.3.4}$$

and

$$E\left[\int \Phi_t \, dM_t\right]^2 - E\left[\int \Phi_t^2 \, d\langle M \rangle_t\right] = E(\Phi_{t_0} h)^2 - E(h^2) E(\Phi_{t_0}^2)$$

$$= \alpha(1 - \alpha)(1 - 2\alpha) \neq 0, \tag{3.3.5}$$

where $\langle M \rangle_t(\omega_1) = \langle M \rangle_t(\omega_2) = E(h^2) I_{[t_0,T]}(t)$.

Let t range over the finite interval $[0,T]$. *Then the class of integrands satisfying* (i) *and* (ii) *above will be denoted by* $\mathscr{M}_2^W[0,T]$. Consider the stochastic process $Y = (Y_t)$ where $Y_t = \int_0^t f_s \, dW_s$. For each t, the stochastic integral Y_t is defined only up to an equivalence; that is, we may choose any random variable Y_t' as the stochastic integral, which has the property $P[\omega\colon Y_t'(\omega) = Y_t(\omega)] = 1$. It will now be shown that the stochastic integral Y_t can be so chosen that the process (Y_t) is sample continuous, that is, the paths $t \to Y_t(\omega)$ are continuous for almost all ω.

First suppose f is a simple process. Then

$$\int_0^t f_s \, dW_s = \sum_{j=1}^n f_{t_j}[W_{t_{j+1} \wedge t} - W_{t_j \wedge t}].$$

Hence $\int_0^t f_s \, dW_s$ is a continuous process with probability 1 as is evident from the expression on the right-hand side. Let the sequence s_{nj} $(j = 0,1, \ldots ,n)$ of partitions of $[0,T]$ be such that for each n, f_s is constant on the intervals $[s_{nj}, s_{n,j+1})$. Since $\int_0^t f_s \, dW_s$ is a continuous process $\sup_{0 \le t \le T} \left|\int_0^t f_s \, dW_s\right|$ is a random variable. Let

$$Z_n = \sup_j \left|\int_0^{s_{nj}} f_s \, dW_s\right|. \tag{3.3.6}$$

We need the following lemma which includes the discrete version of the martingale inequalities (7) given in Section 2.2.

Lemma 3.3.1. *Let* $\xi_1, \ldots ,\xi_n$ *be random variables such that* $E|\xi_k| < \infty$ *and* $E[\xi_n - \xi_k | \xi_1, \ldots ,\xi_k] = 0$ *for each* k. *Let* $\zeta = \sup\{0,\xi_1, \ldots ,\xi_n\}$, $\xi_n^+ =$

$\frac{1}{2}(\xi_n + |\xi_n|)$, *and* $\alpha > 1$. *Then we have the following conclusions*:

1. $E(\zeta^\alpha) \le \left(\frac{\alpha}{\alpha - 1}\right)^\alpha E(\xi_n^+)^\alpha$.

2. $P(\zeta > a) \le \frac{1}{a^2} E(\xi_n^+)^2, (a > 0)$.

3. $P\left[\max_k |\xi_k| > a\right] \le \frac{E\xi_n^2}{a^2}, (a > 0)$.

4. $E\left(\max_k |\xi_k|\right)^\alpha \le \left(\frac{\alpha}{\alpha - 1}\right)^\alpha E|\xi_n|^\alpha$.

PROOF. Fix $a > 0$. Define $I_k(a)$ to be the indicator of the set $[\xi_1 \le a, \ldots, \xi_{k-1} \le a, \xi_k > a]$. Then $a\sum_1^n I_k(a) \le \sum_1^n \xi_k I_k(a)$, and $a^{\alpha-1}\sum_1^n I_k(a) \le \sum_1^n \xi_k a^{\alpha-2} I_k(a)$. Since $E(\xi_n - \xi_k)I_k(a) = EI_k(a)E[(\xi_n - \xi_k)|\xi_1, \ldots, \xi_k] = 0$, we obtain the inequality

$$Ea^{\alpha-1}\sum_1^n I_k(a) \le E\sum_1^n a^{\alpha-2} I_k(a)\xi_n \le E\sum_1^n a^{\alpha-2} I_k(a)\xi_n^+.$$

Integrating the above relations with respect to a from 0 to ∞, we have

$$E\zeta^\alpha \le \left(\frac{\alpha}{\alpha - 1}\right) E(\zeta^{\alpha-1}\xi_n^+).$$

Applying Hölder's inequality to the right-hand side, we have

$$E(\zeta^\alpha) \le \left(\frac{\alpha}{\alpha - 1}\right) [E\zeta^\alpha]^{(\alpha-1)/\alpha} [E(\xi_n^+)^\alpha]^{1/\alpha}$$

from which conclusion 1 follows. Conclusion 2 is an immediate consequence of the relations $P(\zeta > a) = \sum_{k=1}^n EI_k(a)$ and $a^2\sum_1^n I_k(a) \le \sum_1^n (\xi_k^+)^2 I_k(a)$.

Applying conclusion 1 to the random variables $(-\xi_1), (-\xi_2), \ldots, (-\xi_n)$, writing $\zeta_- = \max\{0, (-\xi_1), \ldots, (-\xi_n)\}$, $\xi_n^- = (|\xi_n| - \xi_n)/2$, yields the inequality $E(\zeta_-^\alpha) \le \alpha/(\alpha - 1)E(\xi_n^-)^\alpha$. We obtain conclusion 3 upon noting that $\max |\xi_k| = \max(\zeta, \zeta_-)$ and that inequality 2 also holds with ζ replaced by ζ_- and ξ_n^+ replaced by ξ_n^-. Finally conclusion 4 follows if we note further that $|\xi_n|^\alpha = (\xi_n^+)^\alpha + (\xi_n^-)^\alpha$. □

Returning to Eq. (3.3.6), we see that the random variables $\xi_k^n = \int_0^{s_{nk}} f_s\, dW_s$ are $\mathscr{F}_{s_{nj}}$-measurable for $k \le j$, so that $E(\xi_n^n - \xi_j^n|\xi_1^n, \ldots, \xi_j^n) = 0$. Thus the random variables ξ_k^n satisfy the conditions of the preceding lemma, and we have

$$P(Z_n > a) \le \frac{1}{a^2}\int_0^T E(f_s^2)\, ds, \tag{3.3.7}$$

$$E(Z_n^2) \le 4\int_0^T E(f_s^2)\, ds. \tag{3.3.8}$$

Let the partitions $\Pi_n = \{s_{nj}\}$ be so chosen as to make $\Pi_n \subset \Pi_{n+1}$ and $\bigcup_{n=1}^{\infty} \Pi_n$ a dense subset of $[0,T]$. By the continuity of the process $\int_0^t f_s\, dW_s$, making $n \to \infty$ in expressions (3.3.7) and (3.3.8), it follows that

$$P\left[\sup_{0 \leq t \leq T} \left| \int_0^t f_s\, dW_s \right| > a \right] \leq \frac{1}{a^2} \int_0^T E(f_s^2)\, ds \tag{3.3.9}$$

and

$$E\left[\sup_{0 \leq t \leq T} \left| \int_0^t f_s\, dW_s \right|^2 \right] \leq 4 \int_0^T E(f_s^2)\, ds. \tag{3.3.10}$$

Now let f be measurable, $(\mathscr{F}_t)$-adapted, and let $\int_0^T E(f_s^2)\, ds < \infty$. Then there is a sequence (f^n) of simple processes such that $\lim_{n \to \infty} \int_0^T E(f_s - f_s^n)^2\, ds = 0$. Choose a subsequence (n_k) such that $\int_0^T E(f_s - f_s^{n_k})^2\, ds \leq 1/2^k$. Since $\int_0^T E[f_s^{n_{k+1}} - f_s^{n_k}]^2\, ds \leq 3/2^k$, we obtain the following inequality from Equation (3.3.9) applied to the simple process $f^{n_{k+1}} - f^{n_k}$:

$$P\left[\sup_{0 \leq t \leq T} \left| \int_0^t f_s^{n_{k+1}}\, dW_s - \int_0^t f_s^{n_k}\, dW_s \right| > \frac{1}{k^2} \right] \leq \frac{3k^4}{2^k}.$$

It then follows from the Borel-Cantelli lemma that the series $\int_0^t f_s^{n_1}\, dW_s + \sum_{k=1}^{\infty} [\int_0^t f_s^{n_{k+1}}\, dW_s - \int_0^t f_s^{n_k}\, dW_s]$ converges uniformly with respect to t in $[0,T]$ almost surely. It is easy to see that the sum of this series, say, (Y_t), is a version of the stochastic integral $\int_0^t f_s\, dW_s$ and, moreover, that the process (Y_t) is sample continuous. Combining what has just been proved with property 4(a) and (b) of Section 3.2, we arrive at the following result.

Theorem 3.3.1. *The stochastic process* $(\int_0^t f_s\, dW_s)$, $0 \leq t \leq T$, *is a continuous square-integrable* $(\mathscr{F}_t, P)$*-martingale for all* $f \in \mathscr{M}_2^W[0,T]$.

Remark 3.3.2. In establishing this result we have not used the concept of separability of a process due to Doob. The argument we have given above can actually be used to show that a separable version of $(\int_0^t f_s\, dW_s)$ is continuous.

Remark 3.3.3. Note that the inequality (3.3.10) which is now satisfied by the martingale $\int_0^t f_s\, dW_s$ is a special case of the inequality of 7, Section 2.2.

Remark 3.3.4. The variation process associated with the L^2-martingale $M_t = \int_0^t f_s\, dW_s$ on $[0,T]$ is given by $A_t = \int_0^t f_s^2\, ds$.

Proof of Remark 4. Let $A_t = \int_0^t f_s^2\, ds$ for $0 \leq t \leq T$. This defines $A_t(\omega)$ for a.a. ω. Let $A_t(\omega) = 0$ otherwise. We take a version of (f_s) which is progressively measurable, so that A_t is $\mathscr{F}_t$-adapted, nonnegative, $A_0 = 0$, and $A_t \geq A_s$ for $t \geq s$, (a.s.). The requirement that $f \in \mathscr{M}_2^W[0,T]$ implies that

$$EA_t \leq EA_T < \infty.$$

Also $M_t^2 - A_t$ is a martingale with respect to $(\mathscr{F}_t)$ since

$$E(M_t^2 - A_t|\mathscr{F}_s) = E\left(\left(\int_0^t f_u\,dW_u\right)^2\middle|\mathscr{F}_s\right) - E\left(\int_0^t f_u^2\,du|\mathscr{F}_s\right)$$

$$= E\left(\left(\int_0^s f_u\,dW_u\right)^2\middle|\mathscr{F}_s\right) - E\left(\left(\int_0^s f_u^2\,du\right)\middle|\mathscr{F}_s\right)$$

$$+ E\left(\left(\int_s^t f_u\,dW_u\right)^2\middle|\mathscr{F}_s\right) - E\left(\left(\int_s^t f_u^2\,du\right)\middle|\mathscr{F}_s\right)$$

$$= E(M_s^2 - A_s|\mathscr{F}_s) \qquad s \le t.$$

□

Remark 3.3.4 also follows directly from property 9 of Section 3.2.

Finally, since $A_t(\omega)$ is continuous and increasing in t for all ω, A_t is a predictable increasing process.

The Ito integral for $\mathscr{L}_2^W[0,T]$. Denote by $\mathscr{L}_2^W[0,T]$ the class of measurable processes f which are $(\mathscr{F}_t)$-adapted and which satisfy the condition

$$P\left[\omega\colon \int_0^T f_s^2(\omega)\,ds < \infty\right] = 1. \tag{3.3.11}$$

In order to extend the definition of the Ito integral to $f \in \mathscr{L}_2^W[0,T]$, we need the following two results.

Lemma 3.3.2. *Let f be a simple process adapted to $(\mathscr{F}_t)$. Then for every $N > 0$ and $c > 0$,*

$$P\left[\left|\int_0^T f_s\,dW_s\right| > c\right] \le \frac{N}{c^2} + P\left[\int_0^T f_s^2\,ds > N\right]. \tag{3.3.12}$$

PROOF. Let $f_t = \sum f_{t_j} I_{[t_j,t_{j+1})}(t)$ and define $f_t^N(\omega) = f_t(\omega)$ if $t_i \le t < t_{i+1}$ and $\int_0^{t_{i+1}} f_t^2(\omega)\,dt \le N$, and $=0$ if $t_i \le t < t_{i+1}$ and $\int_0^{t_{i+1}} f_t^2(\omega)\,dt > N$. Since $\int_0^{t_{i+1}} f_t^2\,dt$ is $\mathscr{F}_{t_i}$-measurable, f^N is a simple, $(\mathscr{F}_t)$-adapted process. Furthermore, $\int_0^T (f_t^N)^2\,dt \le N$ so that $E\int_0^T (f_t^N)^2\,dt \le N$ and $P\{\sup_t |f_t^N - f_t| > 0\} = P\{\int_0^T |f_t|^2\,dt > N\}$. From this relation it is easy to see that

$$P\left\{\left|\int_0^T f_t\,dW_t\right| > c\right\} \le P\left\{\left|\int_0^T f_t^N\,dW_t\right| > c\right\} + P\left\{\sup_t |f_t^N - f_t| > 0\right\}$$

$$\le \frac{E\left[\int_0^T f_t^N\,dW_t\right]^2}{c^2} + P\left\{\int_0^T f_t^2\,dt > N\right\}$$

$$\le \frac{N}{c^2} + P\left\{\int_0^T f_t^2\,dt > N\right\}$$

which is (3.3.12). □

Lemma 3.3.3. *Let $f \in \mathscr{L}_2^W[0,T]$. Then there exists a sequence (f^n) of simple processes in $\mathscr{L}_2^W[0,T]$ such that*

$$\int_0^T [f_t - f_t^n]^2 \, dt \to 0 \quad \textit{in probability.}$$

PROOF. Let

$$\rho(t) = \begin{cases} c \exp\left[-\dfrac{1}{1-t^2} \right] & \text{if } |t| < 1, \\ 0 & \text{if } |t| \geq 1, \end{cases}$$

where $c > 0$ is such that $\int_{-\infty}^{\infty} \rho(t) \, dt = 1$. Define $f_t(\omega) = 0$ for all ω if $t < 0$ and let

$$(\theta_\varepsilon f)_t = \frac{1}{\varepsilon} \int_{-1}^T \rho\left(\frac{t - s - \varepsilon}{\varepsilon} \right) f_s \, ds \qquad (0 < \varepsilon < \tfrac{1}{2}).$$

Then $\theta_\varepsilon f$ is continuous and

$$(\theta_\varepsilon f)_t = \frac{1}{\varepsilon} \int_{t-2\varepsilon}^t \rho\left(\frac{t - s - \varepsilon}{\varepsilon} \right) f_s \, ds = \int_{-1}^1 \rho(u) f_{t - \varepsilon u - \varepsilon} \, du.$$

Since f is measurable and adapted to $(\mathscr{F}_t)$ we may work with a progressively measurable modification. This is justified by Proposition 1.1.4. It follows that $(\theta_\varepsilon f)_t$ is $\mathscr{F}_t$-measurable. Furthermore by Schwarz's inequality,

$$\begin{aligned} \int_0^T (\theta_\varepsilon f)_t^2 \, dt &\leq \int_0^T \left\{ \int_{-1}^1 \rho(u) \, du \cdot \int_{-1}^1 \rho(u) f_{t-\varepsilon u - \varepsilon}^2 \, du \right\} dt \\ &\leq \int_{-1}^1 \rho(u) \left\{ \int_{-1}^T f_t^2 \, dt \right\} du. \end{aligned}$$

Hence

$$\int_0^T (\theta_\varepsilon f)_t^2 \, dt \leq \int_0^T f_t^2(t) \, dt. \tag{3.3.13}$$

Fix ω such that $\int_0^T f_t^2(\omega) \, dt < \infty$. Let g_t^n be continuous functions of t (defined to be 0 for $t < 0$) such that

$$\int_0^T |g_t^n - f_t(\omega)|^2 \, dt \to 0 \quad \text{as } n \to \infty. \tag{3.3.14}$$

Since g^n is continuous, it is easy to see that $(\theta_\varepsilon g^n)_t \to g_t^n$ uniformly in $t \in [0,T]$ as $\varepsilon \to 0$.

By the inequality

$$\begin{aligned} \int_0^T [(\theta_\varepsilon f)_t(\omega) - f_t(\omega)]^2 \, dt \leq 3 \bigg[&\int_0^T \{(\theta_\varepsilon f)_t(\omega) - \theta_\varepsilon g_t^n\}^2 \, dt \\ &+ \int_0^T (\theta_\varepsilon g_t^n - g_t^n)^2 \, dt + \int_0^T (g_t^n - f_t(\omega))^2 \, dt \bigg] \end{aligned}$$

using expression (3.3.13) with f replaced by $f - g^n$ and making $\varepsilon \to 0$, we have

$$\limsup_{\varepsilon \downarrow 0} \int_0^T [(\theta_\varepsilon f)_t(\omega) - f_t(\omega)]^2 \, dt \le 6 \int_0^T [g_t^n - f_t(\omega)]^2 \, dt.$$

Now making $n \to \infty$ it follows from expression (3.3.14) that

$$\lim_{\varepsilon \downarrow 0} \int_0^T [(\theta_\varepsilon f)_t(\omega) - f_t(\omega)]^2 \, dt = 0 \qquad \text{(a.s.)}.$$

Let us set $\phi^n = \theta_{1/n} f$ and $h_t^{n,m} = \phi_{k/m}^n$ if

$$\frac{k}{m} \le t < \frac{k+1}{m} \qquad (0 \le k \le [mT]).$$

Then clearly for each n,

$$\lim_{m \to \infty} \int_0^T [h_t^{n,m} - \phi_t^n]^2 \, dt = 0 \qquad \text{(a.s.)}.$$

For any $\delta > 0$, there exists n_0 such that

$$P\left\{\int_0^T [f_t - \phi_t^n]^2 \, dt > \frac{1}{4}\delta\right\} < \frac{\delta}{2} \quad \text{for all } n \ge n_0.$$

Taking $n = n_0$, we have

$$P\left\{\int_0^T [\phi_t^{n_0} - h_t^{n_0,m}]^2 \, dt > \frac{1}{4}\delta\right\} < \frac{\delta}{2} \quad \text{for all } m \ge m_0$$

(where n_0 and m_0 depend on δ). Hence

$$P\left\{\int_0^T [f_t - h_t^{n_0,m_0}]^2 \, dt > \delta\right\} < \delta.$$

Letting $\delta = 1/j$ and denoting the corresponding h^{n_0,m_0} by f^j, we have $f^j \in \mathscr{L}_2^W[0,T]$ and

$$\int_0^T [f_t - f_t^j]^2 \, dt \to 0 \quad \text{in probability as } j \to \infty.$$

This completes the proof of Lemma 3.3.3. □

Now suppose $f \in \mathscr{L}_2^W[0,T]$ and let $f^n \in \mathscr{L}_2^W[0,T]$ be simple processes such that $\int_0^T [f_t - f_t^n]^2 \, dt \to 0$ in probability. Then $\int_0^T [f_t^m - f_t^n]^2 \, dt \to 0$ in probability as $m,n \to \infty$. Hence for $\varepsilon > 0$ and $\delta > 0$, expression (3.3.12) yields the inequality

$$\begin{aligned}\limsup_{m,n \to \infty} P\left\{\left|\int_0^T f_t^n \, dW_t - \int_0^T f_t^m \, dW_t\right| > \delta\right\} &\le \frac{\varepsilon}{\delta^2} + \limsup_{m,n \to \infty} P\left\{\int_0^T [f_t^n - f_t^m]^2 \, dt > \varepsilon\right\} \\ &= \frac{\varepsilon}{\delta^2}.\end{aligned}$$

Since $\varepsilon > 0$ is arbitrary, we have

$$\lim_{m,n\to\infty} P\left\{\left|\int_0^T f_t^n\,dW_t - \int_0^T f_t^m\,dW_t\right| > \delta\right\} = 0.$$

Thus the sequence $\{\int_0^T f_t^n\,dW_t\}$ converges in probability to a limit which can be shown to be independent of the choice of the approximating sequence $\{f^n\}$. We define the stochastic integral

$$\int_0^T f_t\,dW_t = p\text{-lim} \int_0^T f_t^n\,dW_t. \tag{3.3.15}$$

We now show that if $f \in \mathscr{L}_2^W[0,T]$, the indefinite Ito integral,

$$\int_0^t f_s\,dW_s \qquad (0 \le t \le T)$$

has a continuous version. For $n \ge 1$ define the stopping time sequence

$$\tau_n(\omega) = \begin{cases} \inf\left\{t: t \le T, \int_0^t f_s^2(\omega)\,ds > n\right\} & \\ T & \text{if the set in the braces is empty.} \end{cases}$$

It is easy to see that each τ_n is a stopping time with respect to the family $(\mathscr{F}_t)$. Letting $f_s^n(\omega) = f_s(\omega) I_{[0,\tau_n(\omega)]}(s)$, we see that $f^n \in \mathscr{M}_2^W[0,T]$, so that a continuous version of the process $(\int_0^t f_s^n\,dW_s)$ exists by Theorem 3.3.1, which we denote by $I(f^n;t)$. Let $A_n = \{\omega: \int_0^T f_s^2(\omega)\,ds \le n\}$. Clearly $A_n \uparrow A$, where $A = \bigcup_1^\infty A_n$. If $\omega \in A_n$ then $f_t^m(\omega) = f_t^n(\omega)$ for all $m \ge n$. Taking $M = W$ in property 12 of Section 3.2, it follows that for a.a. $\omega \in A_n$,

$$I(f^m;t)(\omega) = I(f^n;t)(\omega) \quad \text{all } t \text{ and all } m \ge n.$$

Hence for $\omega \in A$, that is, for a.a. ω, $\lim_{m\to\infty} I(f^m;t)(\omega)$ exists. Define (for all t)

$$I(f;t)(\omega) = \begin{cases} \lim\limits_{m\to\infty} I(f^m;t)(\omega) & \text{if } \omega \in A, \\ 0 & \text{if } \omega \in A^c. \end{cases}$$

Now $I(f;t)(\omega)$ is continuous for every ω since for $\omega \in A$, $\lim_{m\to\infty} I(f^m;t)$ is a continuous function of t. Moreover,

$$P\left\{\int_0^t [f_s - f_s^m]^2\,ds > 0\right\} = P\left\{\int_0^t f_s^2\,ds > m\right\} \to 0$$

as $m \to \infty$. Hence by Lemma 3.3.12, extended for all $f \in L_2^W[0,T]$,

$$\int_0^t f_s\,dW_s = \underset{m\to\infty}{p\text{-lim}}\, I(f^m;t).$$

Therefore

$$\int_0^t f_s\,dW_s = I(f;t) \text{ (a.s.) for each } t,$$

proving the existence of a continuous version. Henceforth we shall always refer to the continuous version of the process $(\int_0^t f_s\,dW_s)$.

It is easy to see (using Remark 2.6.4) that we have the following analog of Theorem 3.3.1.

Theorem 3.3.2. *Let $f \in \mathscr{L}_2^W[0,T]$. Then $(\int_0^t f_s\,dW_s, \mathscr{F}_t, P)$ is a continuous local martingale with associated increasing process (A_t) given by $A_t = \int_0^t f_s^2\,ds$.*

The classes $\mathscr{L}_*^W$ and $\mathscr{L}_{2,\infty}^W$. So far we have considered only integrands belonging to the class $\mathscr{L}_2^W[0,T]$, where $T > 0$ is fixed.

If $0 < T_1 < T_2$ and $f \in \mathscr{L}_2^W[0,T_2]$, then its restriction f^R to $[0,T_1] \times \Omega$ obviously belongs to $\mathscr{L}_2^W[0,T_1]$. The notation $\int_0^{T_1} f_s\,dW_s$ can then be interpreted either as $\int_0^{T_1} f_s^R\,dW_s$ or as

$$\int_0^{T_2} f_s I_{[0,T_1]}(s)\,dW_s,$$

where $f_s I_{[0,T_1]}(s) \in \mathscr{L}_2^W[0,T_2]$. It can be easily seen, however, that these two random variables are the same a.s. and hence yield the same stochastic integral.

It is necessary, in many situations, to consider integrands $f_s(\omega)$ defined for (s,ω) in $\mathbf{R}_+ \times \Omega$.

Define $\mathscr{L}_*^W$ to be the *class of all real-valued functions on* $\mathbf{R}_+ \times \Omega$ *with the following property*:

(i) For every $T > 0$, the restriction of f to $[0,T] \times \Omega$ belongs to $\mathscr{L}_2^W[0,T]$.

For $f \in \mathscr{L}_*^W$, the stochastic integral $\int_0^t f_s\,dW_s$ is defined for all $t \geq 0$ as follows. Fix $t \geq 0$ and let $T \geq t$. Then with f restricted to $[0,T] \times \Omega$, consider the random variable $\int_0^T f_s I_{[0,t]}(s)\,dW_s$, where $f_s I_{[0,t]}(s)$ is regarded as a member of $\mathscr{L}_2^W[0,T]$. The remarks made in the preceding paragraph show that this integral does not depend on T as long as $T \geq t$. We denote it by $\int_0^t f_s\,dW_s$. This definition, valid for all $t \geq 0$ when $f \in \mathscr{L}_*^W$, coincides with the previous definition given for each finite interval. It follows, in particular, that a continuous version of the indefinite integral $\int_0^t f_s\,dW_s$, $t \geq 0$, can always be found on each $[0,T]$ and hence on $\mathbf{R}_+$. Other properties carry over in a similar fashion; for example, if τ is a finite stopping time for $(\mathscr{F}_t)$, then for $f \in \mathscr{L}_*^W$ we have $f_s I_{[0,\tau]}(s) \in \mathscr{L}_*^W$ and

$$\int_0^{t \wedge \tau} f_s\,dW_s = \int_0^t f_s I_{[0,\tau]}(s)\,dW_s \quad \text{for all } t \in \mathbf{R}_+.$$

Next consider the subclass $\mathscr{L}_{2,\infty}^W \subseteq \mathscr{L}_*^W$ of processes $f = (f_s)$ satisfying the stronger requirement $\int_0^\infty f_s^2\,dW_s < \infty$ (a.s.). In this case

$$\int_{t_1}^{t_2} f_s^2\,ds \to 0 \qquad \text{(a.s.) as } t_1,t_2 \to \infty\ (t_1 < t_2)$$

and hence from the extended Lemma 3.3.2,

$$\limsup_{\substack{t_1,t_2\to\infty\\ t_1<t_2}} P\left\{\left|\int_{t_1}^{t_2} f_s\,dW_s\right| > \varepsilon\right\} \le \frac{\eta}{\varepsilon^2} + \limsup_{\substack{t_1,t_2\to\infty\\ t_1<t_2}} P\left\{\int_{t_1}^{t_2} f_s^2\,ds > \eta\right\}$$

$$= \frac{\eta}{\varepsilon^2},$$

where ε and η are arbitrary positive numbers. Hence

$$\lim_{t_1,t_2\to\infty} P\left\{\left|\int_0^{t_1} f_s\,dW_s - \int_0^{t_2} f_s\,dW_s\right| > \varepsilon\right\} = 0.$$

The stochastic integral $\int_0^\infty f_s\,dW_s$ for $f \in \mathscr{L}^W_{2,\infty}$ is now defined as the limit in probability as $t \to \infty$ of $\int_0^t f_s\,dW_s$.

Remark 3.3.5. The Ito stochastic integral may also be defined in a similar manner for integrand processes $f = (f_s)$ satisfying the condition $E(\int_0^t f_s^2\,ds) < \infty$ for every $t \in \mathbf{R}_+$. This class, which we denote by $\mathscr{M}^W_*$, contains the class $\mathscr{M}^W_{2,\infty}$ for which the integral has already been defined at the beginning of this section. The properties of the stochastic integral for $f \in \mathscr{M}^W_*$ can be established without difficulty.

3.4 The Stochastic Integral with Respect to Continuous Local Martingales

The aim of this section is to define the stochastic integral for a wider class of predictable integrands and, at the same time, to relax the restrictions on M to include all continuous, local martingales. We begin with a lemma and an observation based on it.

Lemma 3.4.1. *Let M be a continuous L^2-martingale and $f \in L^2(M)$. If σ is any stopping time relative to $(\mathscr{F}_t)$, then $f \in L^2(M^\sigma)$ and*

$$\int_0^t f_s\,dM_s^\sigma = \int_0^{t\wedge\sigma} f_s\,dM_s, \tag{3.4.1}$$

where $M_t^\sigma = M_{t\wedge\sigma}$.

Proof. We have already seen that M^σ is a continuous L^2-martingale. Since $\langle M^\sigma\rangle_t = \langle M\rangle_{t\wedge\sigma}$, it follows that $L^2(M) \subset L^2(M^\sigma)$. From properties 7 to 10 of Section 3.2 and recalling that we have denoted the right-hand-side integral in Equation (3.4.1) by Y_t^σ, we obtain

$$E\left[Y_t^\sigma - \int_0^t f_s\,dM_s^\sigma\right]^2 = E(Y_t^\sigma)^2 - 2E\left[Y_t^\sigma \cdot \int_0^t f_s\,dM_s^\sigma\right]$$

$$+ E\left[\int_0^t f_s\,dM_s^\sigma\right]^2$$

$$= E\int_0^t f_s^2 I_{[0,\sigma]}\, d\langle M\rangle_s - 2E\int_0^t f_s\, d\langle M^\sigma, Y^\sigma\rangle_s$$

$$+ E\int_0^t f_s^2\, d\langle M^\sigma\rangle_s = 0. \qquad \square$$

Remark 3.4.1. Let (σ_n) be any sequence of stopping times such that $\sigma_n \uparrow T$ (a.s.). Then as $n \to \infty$, for each t,

$$\int_0^t f_s\, dM_{s\wedge\sigma_n} = \int_0^{t\wedge\sigma_n} f_s\, dM_s \to \int_0^t f_s\, dM_s \qquad \text{(a.s.)}.$$

Throughout the rest of this section we shall take M to be a continuous local martingale relative to $(\mathscr{F}_t)$. *Let $L^2_{\text{loc}}(M)$ be the class of all predictable processes f for which*

$$\int_0^T f_s^2\, d\langle M\rangle_s < \infty \qquad \text{(a.s.)}.$$

With each $f \in L^2_{\text{loc}}(M)$ is associated the family of $(\mathscr{F}_t)$ stopping times given by

$$\sigma_n = \begin{cases} \inf\left\{t\colon |M_t| \ge n \text{ or. } \int_0^t f_s^2\, d\langle M\rangle_s \ge n\right\} \\ T & \text{if the above set is empty.} \end{cases}$$

Note that $\sigma_n \le \sigma_{n+1}$ and that by the continuity of M and $\langle M\rangle$, if $A_n = [\omega\colon \sigma_n(\omega) = T]$, then $P(\bigcup_1^\infty A_n) = 1$.

For each n, let $M_t^n = M_{t\wedge\sigma_n}$. Each M^n is a continuous L^2-martingale (in fact, bounded) and $\langle M^n\rangle_t = \langle M\rangle_{t\wedge\sigma_n}$. Thus $f \in L^2(M^n)$, and we fix a continuous version,

$$Y_t^n = \int_0^t f_s\, dM_s^n.$$

From Lemma 3.4.1, for $m \ge n$,

$$Y^m_{\tau\wedge\sigma_n}(\omega) = Y_t^n(\omega) \quad \text{on } [0,T], \text{(a.s.)}. \tag{3.4.2}$$

Suppose $A = \{\omega\colon t \to Y_t^n(\omega)$ is continuous for all n and Equation (3.4.2) holds for all n,m $(n \le m)\}$. Set $B = (\bigcup_1^\infty A_n) \cap A$. Then $P(B) = 1$; for all $\omega \in B$, there exists n such that $\omega \in A_n$. For this n, $\sigma_n(\omega) = T$ and if $m \ge n$,

$$Y_t^m(\omega) = Y_t^n(\omega) \quad \text{on } [0,T]. \tag{3.4.3}$$

Definition 3.4.1. For each $t \in [0,T]$, define

$$\int_0^t f_s\, dM_s = \lim_{n\to\infty} \int_0^t f_s\, dM_s^n \qquad \text{(a.s.)}.$$

This limit exists since Equation (3.4.3) shows Y_t^n converges (a.s.) and the integral so defined has a continuous version. In fact, we may let

$$Y_t(\omega) = \begin{cases} \lim Y_t^n(\omega) & \text{if this limit exists,} \\ 0 & \text{otherwise.} \end{cases}$$

Then Equation (3.4.3) says that for $\omega \in B$, there exists n such that

$$Y_t(\omega) = Y_t^n(\omega) \quad \text{on } [0,T].$$

Therefore, Y is a continuous $\mathscr{F}_t$-adapted process and a version $\int_0^t f_s \, dM_s$.

Remark 3.4.2. In the case of a continuous L^2-martingale we have $L^2(M) \subset L^2_{\text{loc}}(M)$. For $f \in L^2(M)$ there are thus two notions at $\int_0^t f_s \, dM_s$. However, Remark 3.4.1 shows that they coincide.

Remark 3.4.3. The value of the integral defined above does not depend on the sequence σ_n used. We shall make this precise later; but for now it suffices to see that if $\tau_n \leq \sigma_n$ is a second sequence such that $\tau_n \uparrow T$ (a.s.), then for each t,

$$\int_0^t f_s \, dM_{s \wedge \tau_n} \to \int_0^t f_s \, dM_s \qquad \text{(a.s.)}.$$

PROOF. For a given n and all $m \geq n$, $\tau_n = \sigma_m \wedge \tau_n$ so that by Lemma 3.4.1,

$$\int_0^t f_s \, dM_{s \wedge \tau_n} = \int_0^{t \wedge \tau_n} f_s \, dM_{s \wedge \sigma_m} = Y^m_{t \wedge \tau_n}.$$

From the properties of the processes Y^m, for a.a. ω, there exists $N(\omega)$ such that $m \geq N(\omega)$ implies $Y_t^m(\omega) = Y_t(\omega)$ on $[0,T]$. So $Y^m_{t \wedge \tau_n} \to Y_{t \wedge \tau_n}$ (a.s.) as $m \to \infty$. Hence $\int_0^t f_s \, dM_{s \wedge \tau_n} = Y_{t \wedge \tau_n}$. But since Y is a continuous process and $\tau_n \uparrow T$ (a.s.), we must have $Y_{t \wedge \tau_n} \to Y_t$ as $n \to \infty$. Hence $\int_0^t f_s \, dM_{s \wedge \tau_n} \to \int_0^t f_s \, dM_s$ (a.s.) as $n \to \infty$. □

Properties of the Integral on $L^2_{\text{loc}}(M)$

a. Linearity

PROOF. Clearly $L^2_{\text{loc}}(M)$ is a linear space of processes. Given f and g in $L^2_{\text{loc}}(M)$, their integrals are defined by means of two sequences σ_n and σ'_n. The integral of $(f + g)$ is defined by means of a third sequence σ''_n. By the remark above, we may replace each sequence by $\tau_n = \sigma_n \wedge \sigma'_n \wedge \sigma''_n$, to obtain, (a.s.),

$$\int_0^t f_s \, dM_{s \wedge \tau_n} \to \int_0^t f_s \, dM_s,$$

$$\int_0^t g_s \, dM_{s \wedge \tau_n} \to \int_0^t g_s \, dM_s,$$

$$\int_0^t (f_s + g_s) \, dM_{s \wedge \tau_n} \to \int_0^t (f_s + g_s) \, dM_s.$$

The linearity of the stochastic integral on $L^2(M_{t \wedge \tau_n})$ then yields $\int_0^t f_s \, dM_s + \int_0^t g_s \, dM_s = \int_0^t (f_s + g_s) \, dM_s$ (a.s.). A similar argument works for scalar multiplication. □

b. For any $f \in L^2_{\text{loc}}(M)$, and $\varepsilon > 0, \eta > 0$,

$$P\left(\sup_{0 \leq t \leq T} \left| \int_0^t f_s \, dM_s \right| \geq \varepsilon \right) \leq \frac{4\eta}{\varepsilon^2} + P\left(\int_0^T f_s^2 \, d\langle M \rangle_s \geq \eta \right).$$

PROOF. Fix f, ε, and η. For any integer $m > \eta$, let

$$\tau_m = \begin{cases} \inf\{t: |M_t| \geq m \text{ or } \int_0^t f_s^2 \, d\langle M \rangle_s \geq \eta\} \\ T & \text{if the above set is empty.} \end{cases}$$

Then τ_m is an $\mathscr{F}_t$-stopping time and $\tau_m \leq \sigma_m$, where $\{\sigma_m\}$ is the stopping time sequence associated with f in the definition of $Y_t = \int_0^t f_s \, dM_s$ above. For each m, $M_{t \wedge \tau_m}$ is a bounded continuous martingale and $f \in L^2(M_{t \wedge \tau_m})$. By Lemma 3.4.1, we may fix a continuous version of the L^2-martingale

$$Z_t^m = \int_0^t f_s \, dM_{s \wedge \tau_m}.$$

We have

$$P\left(\sup_{0 \leq t \leq T} |Y_t| \geq \varepsilon\right) \leq P\left(\sup_{0 \leq t \leq T} |Z_t^m| \geq \varepsilon\right) + P\left(\sup_{0 \leq t \leq T} |Y_t - Z_t^m| \neq 0\right). \quad (3.4.4)$$

The first term on the right-hand side may be estimated by the martingale inequality:

$$\begin{aligned} P\left(\sup_{0 \leq t \leq T} |Z_t^m| \geq \varepsilon\right) &\leq \frac{1}{\varepsilon^2} E\left(\sup_{0 \leq t \leq T} \left|\int_0^t f_s \, dM_{s \wedge \tau_m}\right|^2\right) \\ &\leq \frac{4}{\varepsilon^2} E \int_0^T f_s^2 \, d\langle M \rangle_{s \wedge \tau_m} \\ &\leq \frac{4\eta}{\varepsilon^2}. \end{aligned} \quad (3.4.5)$$

To deal with the second term, first note that by Lemma 3.4.1,

$$\int_0^t f_s \, dM_{s \wedge \tau_m} = \int_0^{t \wedge \tau_m} f_s \, dM_{s \wedge \sigma_m},$$

so that

$$Z_t^m = Y_{t \wedge \tau_m}^m \quad \text{for } t \in [0,T] \qquad \text{(a.s.)}.$$

Let $B_m = \{\omega: \tau_m(\omega) = T\}$ and $A_m = \{\omega: \sigma_m(\omega) = T\}$. Then $B_m \subset A_m$. Recall that for a.a. $\omega \in A_m$,

$$Y_t^m(\omega) = Y_t(\omega) \quad \text{on } [0,T].$$

Thus, for a.a. $\omega \in B_m$,

$$Z_t^m(\omega) = Y_{t \wedge \tau_m}^m(\omega) = Y_t^m(\omega) = Y_t(\omega) \quad \text{on } [0,T].$$

Therefore,

$$\begin{aligned} P\left(\sup_{0 \leq t \leq T} |Z_t^m - Y_t| \neq 0\right) &\leq P(B_m^c) \\ &\leq P\left(\sup_{0 \leq t \leq T} |M_t| \geq m\right) + P\left(\int_0^T f_s^2 \, d\langle M \rangle_s \geq \eta\right). \end{aligned}$$

Combining this estimate with expressions (3.4.4) and (3.4.5) yields

$$P\left(\sup_{0\le t\le T}|Y_t|\ge\varepsilon\right)\le\frac{4\eta}{\varepsilon^2}+P\left(\sup_{0\le t\le T}|M_t|\ge m\right)+P\left(\int_0^T f_s^2\,d\langle M\rangle_s\ge\eta\right).$$

Since m is arbitrary here and M is continuous (a.s.) on $[0,T]$, the term $P(\sup_{0\le t\le T}|M_t|\ge m)$ may be made arbitrarily small, proving part b. □

c. Suppose f^n and f belong to $L^2_{\text{loc}}(M)$, and

$$\int_0^T (f_s^n - f_s)^2\,d\langle M\rangle_s \to 0 \quad \text{in probability.}$$

Then $\sup_{0\le t\le T}|\int_0^t f_s^n\,dM_s - \int_0^t f_s\,dM_s| \to 0$ in probability. In particular, for each t,

$$\int_0^t f_s^n\,dM_s \to \int_0^t f_s\,dM_s.$$

PROOF. The assertion follows immediately from part b. □

d. Let $f\in L^2_{\text{loc}}(M)$ and τ be any stopping time. Then

$$\int_0^{t\wedge\tau} f_s\,dM_s = \int_0^t f_s\,dM_{s\wedge\tau}.$$

PROOF. Notice that $M_{t\wedge\tau}$ is a continuous local martingale and $f\in L^2_{\text{loc}}(M_{t\wedge\tau})$, so that the right side of the equation is defined. As usual we take continuous versions Y_t and Z_t of $\int_0^t f_s\,dM_s$ and $\int_0^t f_s\,dM_{s\wedge\tau}$, respectively.

These integrals are defined by the stopping times

$$\sigma_n = \begin{cases} \inf\left\{t: |M_t|\ge n \text{ or } \int_0^t f_s^2\,d\langle M\rangle_s \ge n\right\} & \\ T & \text{otherwise} \end{cases}$$

and

$$\sigma_n' = \begin{cases} \inf\left\{t: |M_{t\wedge\tau}|\ge n \text{ or } \int_0^t f_s^2\,d\langle M\rangle_{t\wedge\tau} \ge n\right\} & \\ T & \text{otherwise.} \end{cases}$$

Since $\sigma_n\le\sigma_n'$, Remark 3.4.3 shows that for each t,

$$\int_0^t f_s\,dM_{t\wedge\tau\wedge\sigma_n} \to \int_0^t f_s\,dM_{t\wedge\tau} = Z_t \qquad \text{(a.s.).}$$

For a.a. ω, there exists $N(\omega)$ such that $n > N(\omega)$ implies $Y_t^n(\omega) = Y_t(\omega)$ for all $t\in[0,T]$. For such ω and $n\ge N(\omega)$, $Y_{t\wedge\tau}^n = Y_{t\wedge\tau}$. Hence $Y_{t\wedge\tau}^n \to Y_{t\wedge\tau}$ (a.s.). By Lemma 3.4.1,

$$Y_{t\wedge\tau}^n = \int_0^{t\wedge\tau} f_s\,dM_{s\wedge\sigma_n} = \int_0^t f_s\,dM_{s\wedge\tau\wedge\sigma_n}.$$

Hence $Y_{t \wedge \tau} = Z_t$ (a.s.) for each t. As these are continuous processes, part d holds on $[0,T]$, (a.s.). □

e. The process $Y_t = \int_0^t f_s \, dM_s$ is a continuous local martingale with quadratic variation given by

$$\langle Y \rangle_t = \int_0^t f_s^2 \, d\langle M \rangle_s.$$

PROOF. In fact, part d shows that

$$Y_{t \wedge \sigma_n} = \int_0^{t \wedge \sigma_n} f_s \, dM_s = \int_0^t f_s \, dM_{s \wedge \sigma_n}$$
$$= Y_t^n.$$

Each Y^n is a continuous L^2-martingale with variation

$$\langle Y^n \rangle_t = \int_0^t f_s^2 \, d\langle M^n \rangle_s = \int_0^{t \wedge \sigma_n} f_s^2 \, d\langle M \rangle_s.$$ □

Remark 3.4.4. We also see from part d that the integral $Y_t = \int_0^t f_s \, dM_s$ does not depend on the choice of stopping times $\{\sigma_n\}$. For if $f \in L^2_{\text{loc}}(M)$ and τ_n is any sequence of stopping times for which $\tau_n \uparrow T$ (a.s.), then (a.s.)

$$\int_0^t f_s \, dM_{s \wedge \tau_n} = \int_0^{t \wedge \tau_n} f_s \, dM_s = Y_{t \wedge \tau_n} \quad \text{on } [0,T].$$

By continuity of Y, $\sup_{0 \le t \le T} |Y_{t \wedge \tau_n} - Y_t| \to 0$ (a.s.). Thus the processes $\int_0^t f_s \, dM_{s \wedge \tau_n}$ converge to Y in the sense that

$$\sup_{0 \le t \le T} \left| \int_0^t f_s \, dM_{s \wedge \tau_n} - Y_t \right| \to 0 \qquad \text{(a.s.)}.$$

Remark 3.4.5. For simple left-continuous elements of $L^2_{\text{loc}}(M)$ our integral is given by the usual formula. In fact, if

$$f = \sum_{j=0}^{m-1} f_j I_{(t_j, t_{j+1}]},$$

where $0 = t_0 < t_1 < \cdots < t_m = T$, take

$$\int_0^t f_s \, dM_{s \wedge \sigma_n} = \sum_{j=0}^{m-1} f_j (M_{(t_{j+1} \wedge \sigma_n) \wedge t} - M_{(t_j \wedge \sigma_n) \wedge t})$$
$$\to \sum_{j=0}^{m-1} f_j (M_{t_{j+1} \wedge t} - M_{t_j \wedge t}) \qquad \text{(a.s.)}.$$

Hence

$$\int_0^t f_s \, dM_s = \sum_{j=0}^{m-1} f_j (M_{t_{j+1} \wedge t} - M_{t_j \wedge t}) \quad \text{for all } t, \text{(a.s.)}.$$

Remark 3.4.6. Suppose $M = (M_t)$ with $M_0 = 0$ is a right-continuous, square-integrable martingale over $\mathbf{R}_+$. Then the stochastic integral $\int_0^\infty f_t \, dM_t$

can be defined for predictable integrand processes $f = (f_t)$ satisfying the condition $E \int_0^\infty f_t^2 \, d\langle M \rangle_t < \infty$. Denote this class of integrands by $L_{2,\infty}(M)$. The definition follows mutatis mutandis the procedure adopted in Section 3.2. The integral is first defined in the usual way for a simple process f in $L_{2,\infty}(M)$,

$$f_t(\omega) = \sum_{j=0}^{n-1} f_j(\omega) I_{(t_j, t_{j+1}]}(t) \quad \text{for all } t \in \mathbf{R}_+,$$

where $0 \leq t_0 < \cdots < t_n < \infty$. Note that Lemmas 3.1.2 and 3.1.3 are valid if $[0,T]$ is replaced by $\mathbf{R}_+$. The extension of the definition to all of $L_{2,\infty}(M)$ is then carried out as in Section 3.2.

The Ito Formula 4

4.1 Vector-Valued Processes

A process $M_t = (M_t^1, \ldots, M_t^d)$ taking values in $\mathbf{R}^d$ is a martingale with respect to the increasing σ-field family $(\mathscr{F}_t)$ if $(M_t^i, \mathscr{F}_t)$ is a martingale for each $i = 1, \ldots, d$, or equivalently, if (θ, M_t) is a real-valued martingale with respect to $(\mathscr{F}_t)$ for every $\theta \in \mathbf{R}^d$. Here we use $(\, , \,)$ to denote inner product in $\mathbf{R}^d$. $(M_t, \mathscr{F}_t)$ with $M_0 = 0$ (a.s.) is a d-dimensional, continuous L^2-martingale if for every $\theta \in \mathbf{R}^d$, (θ, M_t) is a continuous L^2-martingale with respect to $\mathscr{F}_t$. It is then easy to verify the existence of a unique $d \times d$-matrix–valued process $A_t = (A_t^{ij})$ with the following properties:

a. Each A_t^{ij} is $\mathscr{F}_t$-measurable.
b. $A_0 = 0$ and $A_t(\omega)$ is continuous in t for almost all ω.
c. For $\theta \in \mathbf{R}^d$, $(A_t\theta, \theta)$ is the (continuous) increasing process associated with (θ, M_t).

It is easy to see that $A_t^{ii} = \langle M^i \rangle_t$ and $A_t^{ij} = \langle M^i, M^j \rangle_t$. As in the real-valued case, we adopt the notation $\langle M \rangle$ for A and call it the *increasing process* (or *quadratic variation process*) of M.

We define $(M_t, \mathscr{F}_t)$ to be a d-dimensional local martingale if, for each $\theta \in \mathbf{R}^d$, $((\theta, M_t), \mathscr{F}_t)$ is a local martingale. If $(M_t, \mathscr{F}_t)$ with $M_0 = 0$ is a continuous local martingale, there exists a unique $d \times d$-matrix–valued process $A_t = (A_t^{ij})$ with the following properties:

a′. Each A_t^{ij} is $\mathscr{F}_t$-measurable.
b′. $A_0 = 0$ and $A_t(\omega)$ is continuous in t for almost every ω.
c′. For $\theta \in \mathbf{R}^d$, $(A_t\theta, \theta)$ is the (continuous) *locally* increasing process associated with the local martingale (θ, M_t).

In this case $A_t^{ii} = \langle M^i \rangle_t$, where the latter symbol has the meaning attached to it in Chapter 2; viz., $\langle M^i \rangle_t$ is the locally increasing process of the local martingale M_t^i. Also, $A_t^{ij} = \langle M^i, M^j \rangle_t$, where the latter is defined to be $\frac{1}{4}\{\langle M^i + M^j \rangle_t - \langle M^i - M^j \rangle_t\}$. The process A_t will be called the *locally increasing process*, (or the *quadratic variation process*) of the local martingale M_t and denoted by $\langle M \rangle_t$. The following result is a useful extension to the vector-valued case of Lemma 2.6.3. The proof is quite similar.

Lemma 4.1.1. *Let $(M_t, \mathscr{F}_t)_{t \in [0,T]}$ $(M_0 = 0)$ be a d-dimensional, continuous local martingale with associated locally increasing process A_t. If $E\langle A_T\theta, \theta\rangle < \infty$ for every $\theta \in \mathbf{R}^d$, then M_t is an L^2-martingale with A_t for its quadratic variation process.*

Denote by $\mathscr{B}$ the class of d-dimensional processes (B_t), $B_t = (B_t^1, \ldots, B_t^d)$, satisfying the following conditions: for each i,

(a) $B_0^i = 0$ (a.s.).
(b) The function $t \to B_t^i(\omega)$ is, for almost all ω, continuous and of bounded variation in every finite interval.
(c) $E|B^i|_t < \infty$ for every t, where $|B^i|_t$ is the total variation of B_s^i in the interval $[0,t]$.

By $\mathscr{B}_{\text{loc}}$ we shall mean the class of d-dimensional processes (B_t) satisfying the conditions (a) and (b) stated above.

Definition 4.1.1. (a) $X = (X_t)$, where X_t is a d-dimensional measurable, $(\mathscr{F}_t)$-adapted process is a *continuous semimartingale* if X_t is continuous and has the form

$$X_t = X_0 + M_t + B_t \tag{4.1.1}$$

for all t (a.s.), where $E|X_0| < \infty$, (i) $M = (M_t)$ is a continuous L^2-$(\mathscr{F}_t)$-martingale with $M_0 = 0$ (a.s.) and (ii) $(B_t) \in \mathscr{B}$.

(b). If in the decomposition (4.1.1), (M_t) is a continuous *local* martingale and (B_t) belongs to $\mathscr{B}_{\text{loc}}$, then (X_t) will be called a *continuous local semimartingale*.

4.2 The Ito Formula

This chapter is devoted to the derivation of Ito's formula, or as it is sometimes called, the change of variables formula for the stochastic integral. Because it is one of the most important tools of stochastic calculus, we shall present a complete treatment here, proving different versions of the formula for convenience in applications. We first consider the case of continuous semimartingales.

Theorem 4.2.1. *Let*

$$X_t = X_0 + M_t + B_t$$

be a d-dimensional, continuous semimartingale. Let $F \in C_b^2(\mathbf{R}^d)$, *that is, let* $F: \mathbf{R}^d \to \mathbf{R}$ *be bounded and continuous and have bounded, continuous derivatives of orders* 1 *and* 2. *Then*

$$F(X_t) = F(X_0) + \sum_{i=1}^{d} \int_0^t \frac{\partial F}{\partial x^i}(X_s)\, dM_s^i + \sum_{i=1}^{d} \int_0^t \frac{\partial F}{\partial x^i}(X_s)\, dB_s^i + \frac{1}{2} \sum_{i,j=1}^{d} \int_0^t \frac{\partial^2 F}{\partial x^i \partial x^j}(X_s)\, d\langle M^i, M^j \rangle_s \tag{4.2.1}$$

Proof. We shall give the proof for $d = 1$. As a first step in the proof we show the following:

First step: Suppose there exists a sequence of continuous semimartingales (X^N), $X_t^N = X_0^N + M_t^N + B_t^N$, such that

1. $E|X_0^N - X_0| \to 0$, $E(M_T^N - M_T)^2 \to 0$, and $E \int_0^T d|B^N - B|_t \to 0$ as $N \to \infty$.
2. The formula (4.2.1) holds with X_t replaced by X_t^N.

   Then the Ito formula (4.2.1) is valid for the process (X_t).

From item 2 above,

$$F(X_t^N) = F(X_0^N) + \int_0^t F'(X_s^N)\, dM_s^N + \int_0^t F'(X_s^N)\, dB_s^N + \tfrac{1}{2} \int_0^t F''(X_s^N)\, d\langle M^N \rangle_s.$$

Clearly, $F(X_t^N) \to F(X_t)$ and $F'(X_t^N) \to F'(X_t)$ in probability. Also from $E(M_T^N - M_T)^2 \to 0$ and Doob's martingale inequality it follows that $\sup_s |M_s^N - M_s| \to 0$ in probability. We shall use this fact and the third condition in item 1 above in the argument which follows. Let C be a number bounding $|F'|$ and $|F''|$, and let $I_1 = \int_0^t F'(X_s^N)\, dM_s^N - \int_0^t F'(X_s)\, dM_s = \int_0^t [F'(X_s^N) - F'(X_s)]\, dM_s + \int_0^t F'(X_s^N)\, d(M_s^N - M_s) = I_2 + I_3$, say. $E(I_2^2) = E \int_0^t [F'(X_s^N) - F'(X_s)]^2\, d\langle M \rangle_s \to 0$ by the dominated convergence theorem, since $\int_0^t [F'(X_s^N) - F'(X_s)]^2\, d\langle M \rangle_s \leq 4C^2 \langle M \rangle_T$ and $E\langle M \rangle_T < \infty$. $E(I_3^2) \leq C^2 E(M_T^N - M_T)^2 \to 0$. Hence $I_1 \to 0$ in probability. Next, write $I_4 = \int_0^t F'(X_s^N)\, dB_s^N - \int_0^t F'(X_s)\, dB_s = \int_0^t [F'(X_s^N) - F'(X_s)]\, dB_s + \int_0^t F'(X_s^N)\, d(B_s^N - B_s) = I_5 + I_6$, say. $E|I_5| \leq E \int_0^t |F'(X_s^N) - F'(X_s)|\, d|B|_s \to 0$ by the dominated convergence theorem. $E|I_6| \leq CE \int_0^T d|B^N - B|_s \to 0$. Hence $I_4 \to 0$ in probability. Let

$$I_7 = \int_0^t F''(X_s^N)\, d\langle M^N \rangle_s - \int_0^t F''(X_s)\, d\langle M \rangle_s = I_8 + I_9,$$

where

$$I_8 = \int_0^t F''(X_s^N)\, d[\langle M^N \rangle_s - \langle M \rangle_s]$$

and

$$I_9 = \int_0^t [F''(X_s^N) - F''(X_s)]\, d\langle M \rangle_s.$$

Now $|I_8| \le C\int_0^t d|\langle M^N + M, M^N - M\rangle|_s \le C\{\langle M^N + M\rangle_t \langle M^N - M\rangle_t\}^{\frac{1}{2}}$. So

$$E|I_8| \le C[E\langle M^N + M\rangle_t E\langle M^N - M\rangle_t]^{\frac{1}{2}} \to 0,$$

since $E\langle M^N - M\rangle_t = E(M_t^N - M_t)^2 \to 0$ and $E\langle M^N + M\rangle_t \le 2[E(M_t^N)^2 + E(M_t)^2] \to 4E(M_t)^2$. Hence $I_8 \to 0$ in probability. Arguing as in the case of I_5, it follows that $I_9 \to 0$ in probability. Collecting all these results, we see that the assertion in the first step is proved.

Second step: Given the semimartingale (X_t) as in the theorem, we show that a sequence (X_t^N) of semimartingales can be chosen to satisfy condition 1, and furthermore that X_0^N, M_t^N, and B_t^N can be chosen to be *bounded* in (t,ω) for each N. Define

$$\sigma_N(\omega) = \begin{cases} \inf\{t\colon |M_t(\omega)| \ge N \text{ or } |B|_t(\omega) \ge N\} & \\ T & \text{if the above set is empty.} \end{cases}$$

(Here $|B|_t$ denotes the total variation process corresponding to B_t.) Write $X_0^N = X_0 I_{[|X_0| \le N]}$, $M_t^N = M_{t\wedge\sigma_N}$, $B_t^N = B_{t\wedge\sigma_N}$, and $X_t^N = X_0^N + M_t^N + B_t^N$. Clearly, $E|X_0^N - X_0| \to 0$. Since $M - M^N$ is a continuous L^2-martingale with $M_0 - M_0^N = 0$, we have for each t, $E[M_t - M_t^N]^2 = E\langle M - M^N\rangle_t = E\langle M\rangle_t - 2E\langle M,M^N\rangle_t + E\langle M^N\rangle_t$. Since $\langle M,M^N\rangle_t = \langle M\rangle_{t\wedge\sigma_N} = \langle M^N\rangle_t$ by Lemma 2.6.4, $E[M_t - M_t^N]^2 = E\langle M\rangle_t - E\langle M^N\rangle_t$. Noting that $\langle M^N\rangle_t(\omega) \to \langle M\rangle_t(\omega)$ (a.s.) and $\langle M^N\rangle_t(\omega) \le \langle M\rangle_T(\omega)$ (a.s.) for all N, it follows that $E\langle M^N\rangle_t \to E\langle M\rangle_t$ by the dominated convergence theorem. Hence

$$E[M_t - M_t^N]^2 \to 0.$$

On the other hand,

$$\int_0^t d|B - B^N|_s \le \int_{\sigma_N(\omega)}^T d|B|_s \to 0 \text{ (a.s.)}$$

and

$$\int_{\sigma_N(\omega)}^T d|B|_s \le \int_0^T d|B|_s.$$

From $E\int_0^T d|B|_s < \infty$ and dominated convergence, it follows that

$$E\int_0^t d|B - B^N|_s \to 0.$$

Thus property 1 is verified for the sequence (X_t^N).

Third step: In view of what has been shown above, it only remains to establish the theorem for the case of a bounded semimartingale (X_t), that is, where $X_t = X_0 + M_t + B_t$, $|X_0| \le K$, M_t is a continuous martingale with $|M_t| \le K$ for all t and with $\int_0^T d|B|_s \le K$, (B_t) being a continuous process of bounded variation in $[0,T]$.

The process (X_t) takes its values in $[-3K,3K]$. Let C denote a constant majorizing $|F'(x)|$, $|F''(x)|$ for $x \in [-3K,3K]$. From Taylor's formula, if a and b are in the interval $[-3K,3K]$,

$$F(b) - F(a) = (b - a)F'(a) + \tfrac{1}{2}(b - a)^2 F''(a) + r(a,b),$$

where since F'' is uniformly continuous over the interval, we have $|r(a,b)| \le \varepsilon(|b-a|)(b-a)^2$, $\varepsilon(h)$ being an increasing function of h, tending to 0 as $h \downarrow 0$. We now introduce the following stochastic subdivision of $[0,t]$. Fix $a > 0$, $t_0 = 0$, and

$$t_{i+1} = t \wedge (t_i + a) \wedge \inf\{s > t_i : |M_s - M_{t_i}| > a \text{ or } |B_s - B_{t_i}| > a\}.$$

As $a \to 0$, $\sup_i (t_{i+1} - t_i) \le a \to 0$ and $\sup_i |M_{t_{i+1}} - M_{t_i}| \le a \to 0$. Write

$$\begin{aligned} F(X_t) - F(X_0) &= \sum_i [F(X_{t_{i+1}}) - F(X_{t_i})] \\ &= \sum_i F'(X_{t_i})(X_{t_{i+1}} - X_{t_i}) + \tfrac{1}{2}\sum_i F''(X_{t_i})(X_{t_{i+1}} - X_{t_i})^2 \\ &\quad + \sum_i r(X_{t_i}, X_{t_{i+1}}) \\ &= S_1 + \tfrac{1}{2}S_2 + S_3, \text{ say.} \end{aligned}$$

Consider the three sums in succession. Write $S_1 = S_1' + S_1''$, where

$$S_1' = \sum_i F'(X_{t_i})(M_{t_{i+1}} - M_{t_i}) \quad \text{and} \quad S_1'' = \sum_i F'(X_{t_i})(B_{t_{i+1}} - B_{t_i}).$$

$$\begin{aligned} E\left[S_1' - \int_0^t F'(X_s)\,dM_s\right]^2 &= E\left[\sum_i \int_{t_i}^{t_{i+1}} (F'(X_s) - F'(X_{t_i}))^2\, d\langle M\rangle_s\right] \\ &\le E\left[\left\{\sup_i \sup_{s \in [t_i, t_{i+1}]} (F'(X_s) - F'(X_{t_i}))^2\right\}\langle M\rangle_t\right] \end{aligned}$$

$\to 0$ since $\langle M\rangle_t$ is integrable and the sup in braces converges uniformly to 0. Now

$$\begin{aligned} \left|S_1'' - \int_0^t F'(X_s)\,dB_s\right| &\le \sum_i \int_{t_i}^{t_{i+1}} |F'(X_s) - F'(X_{t_i})|\, d|B|_s \\ &\le \left\{\sup_i \sup_{s \in [t_i, t_{i+1}]} |F'(X_s) - F'(X_{t_i})|\right\} \int_0^t d|B|_s. \end{aligned}$$

Hence $E|S_1'' - \int_0^t F'(X_s)\,dB_s| \to 0$. In the above calculations we have made use of properties of the stochastic integral given in Section 3.2. Let us now write S_2 as the sum of the following three terms:

$$\begin{aligned} S_2' &= \sum_i F''(X_{t_i})(M_{t_{i+1}} - M_{t_i})^2, \\ S_2'' &= 2\sum_i F''(X_{t_i})(B_{t_{i+1}} - B_{t_i})(M_{t_{i+1}} - M_{t_i}), \\ S_2''' &= \sum_i F''(X_{t_i})(B_{t_{i+1}} - B_{t_i})^2. \end{aligned}$$

$S_2''' \to 0$ in L^1 since $|S_2'''| \le C \sup_i |B_{t_{i+1}} - B_{t_i}| \int_0^t d|B|_s$, $\sup_i |B_{t_{i+1}} - B_{t_i}| \le a$. Similarly $S_2'' \to 0$ in L_1. Before passing on to S_2', we note that $E\langle M\rangle_t^2 < \infty$ for every $t \in [0,T]$. This can be seen as follows. First of all we have $E\langle M\rangle_t = E(M_t^2) \le K^2$ for $t \in [0,T]$. Now applying Theorem 2.4.7 to the continuous potential $Z_t = E(M_T^2|\mathscr{F}_t) - M_t^2$ (here the index set is $[0,T]$ instead of $\mathbf{R}_+$) with natural increasing process $\langle M\rangle_t$, we obtain

$$E\langle M\rangle_T^2 = 2E\left[\int_0^T Z_t\,d\langle M\rangle_t\right] \le 2K^2 E\langle M\rangle_T < \infty.$$

Next write

$$S_2^* = \sum_i F''(X_{t_i})[\langle M\rangle_{t_{i+1}} - \langle M\rangle_{t_i}].$$

It is easy to verify that

$$E\left|S_2^* - \int_0^t F''(X_s)\,d\langle M\rangle_s\right| \to 0.$$

Let us now show that $E(S_2^* - S_2')^2 \to 0$. Since $E[(\langle M\rangle_{t_{i+1}} - \langle M\rangle_{t_i}) - (M_{t_{i+1}} - M_{t_i})^2|\mathscr{F}_{t_i}] = 0$, we obtain

$$\begin{aligned} E(S_2^* - S_2')^2 &= \sum_i E[(F''(X_{t_i}))^2\{(\langle M\rangle_{t_{i+1}} - \langle M\rangle_{t_i}) - (M_{t_{i+1}} - M_{t_i})^2\}^2] \\ &\le 2C^2\left(E\left[\sum_i(\langle M\rangle_{t_{i+1}} - \langle M\rangle_{t_i})^2\right] + E\left[\sum_i (M_{t_{i+1}} - M_{t_i})^4\right]\right). \end{aligned}$$

It is enough to show that each of the terms on the right-hand side tends to 0. The first expected value is dominated by $E[(\sup_i(\langle M\rangle_{t_{i+1}} - \langle M\rangle_{t_i}))\cdot\langle M\rangle_t]$. The sup $\to 0$ by the uniform continuity in $[0,t]$ of $\langle M\rangle_s$ and is also dominated by $\langle M\rangle_t$ and $E\langle M\rangle_t^2 < \infty$, as we have just seen. Hence the first term $\to 0$ by the dominated convergence theorem. The second term is majorized by

$$\begin{aligned} E\left[\left\{\sup_i(M_{t_{i+1}} - M_{t_i})^2\right\}\sum_i(M_{t_{i+1}} - M_{t_i})^2\right] &\le a^2E\left[\sum_i(M_{t_{i+1}} - M_{t_i})^2\right] \\ &= a^2E(M_t^2) \to 0 \quad \text{as } a \to 0. \end{aligned}$$

[Observe that the fact that $\{t_i\}$ is a stochastic partition of $[0,t]$ as defined makes $|M_{t_{i+1}}(\omega) - M_{t_i}(\omega)| \le a$ for every i and all ω.]

Finally, S_3 is majorized by

$$\sum_i(X_{t_{i+1}} - X_{t_i})^2\varepsilon(|X_{t_{i+1}} - X_{t_i}|) \le 2\varepsilon(2a)\sum_i[(B_{t_{i+1}} - B_{t_i})^2 + (M_{t_{i+1}} - M_{t_i})^2].$$

Now $E[\sum_i(M_{t_{i+1}} - M_{t_i})^2] = E(M_t^2)$,

$$E\left[\sum_i(B_{t_{i+1}} - B_{t_i})^2\right] \le aE\left[\sum_i|B_{t_{i+1}} - B_{t_i}|\right] \le aE\int_0^t d|B|_s,$$

and $\varepsilon(2a) \to 0$ as $a \to 0$. Hence $E|S_3| \to 0$. Collecting all the facts proved above, we see that as $a \to 0$,

$$S_1 + \tfrac{1}{2}S_2 + S_3 \to \int_0^t F'(X_s)\,dM_s + \int_0^t F'(X_s)\,dB_s + \tfrac{1}{2}\int_0^t F''(X_s)\,d\langle M\rangle_s$$

in probability. The proof of the theorem is complete. □

The following result is a slightly different version of the Ito formula and is proved exactly as the theorem established above.

Theorem 4.2.2. *Suppose given the following:*

(a) $M_t = (M_t^i)$ *is a continuous,* $\mathbf{R}^d$*-valued* L^2*-martingale with respect to* $(\mathscr{F}_t)$.
(b) (V_t) *is a continuous,* $(\mathscr{F}_t)$*-adapted, real-valued process almost all of whose sample paths are of bounded variation in every finite interval of* $\mathbf{R}_+$ *and such that* $E|V|_t < \infty$.
(c) $F: \mathbf{R} \times \mathbf{R}^d \to \mathbf{R}$ *is a bounded, continuous function of* $(v,x^1, \ldots, x^d)$ *with bounded, continuous first partial derivative with respect to* v *and bounded, continuous first and second partial derivatives with respect to* $(x^1, \ldots, x^d)$.

Then

$$F(V_t,M_t) = F(V_0,M_0) + \int_0^t \frac{\partial F}{\partial v}(V_s,M_s)\,dV_s + \sum_{i=1}^d \int_0^t \frac{\partial F}{\partial x^i}(V_s,M_s)\,dM_s^i + \frac{1}{2}\sum_{i,j=1}^d \int_0^t \frac{\partial^2 F}{\partial x^i\,\partial x^j}(V_s,M_s)\,d\langle M^i,M^j\rangle_s. \tag{4.2.2}$$

We shall now derive an Ito formula involving continuous local semimartingales. In the proof the index set is taken to be the finite interval $[0,T]$.

Theorem 4.2.3. *Let* $X_t = X_0 + M_t + B_t$ *be a d-dimensional, continuous local semimartingale. Let* $F: \mathbf{R}^d \to \mathbf{R}$ *be a continuous function with continuous partial derivatives of the first two orders. Then*

$$F(X_t) = F(X_0) + \sum_{i=1}^d \int_0^t \frac{\partial}{\partial x^i} F(X_s)\,dM_s^i + \sum_{i=1}^d \int_0^t \frac{\partial}{\partial x^i} F(X_s)\,dB_s^i + \frac{1}{2}\sum_{i,j} \int_0^t \frac{\partial^2 F}{\partial x^i\,\partial x^j}(X_s)\,d\langle M^i,M^j\rangle_s. \tag{4.2.3}$$

PROOF. We give the proof for $d = 1$, the general case being quite similar. Since $F(X_s)$, $F'(X_s)$, and $F''(X_s)$ are continuous, each of the integrals $\int_0^T |F'(X_s)|\,d|B|_s$, $\int_0^T |F''(X_s)|\,d\langle M\rangle_s$, and $\int_0^T [F'(X_s)]^2\,d\langle M\rangle_s$ is finite (a.s.). Thus the right-hand side of Equation (4.2.3) has a meaning. Define

$$\sigma_n = \begin{cases} \inf\left\{t \le T: |M_t| \ge n \text{ or } \int_0^t [F'(X_s)]^2\,d\langle M\rangle_s \ge n \text{ or } |B|_t \ge n\right\} & \\ T & \text{if the above set is empty.} \end{cases}$$

Let $\sigma_n'(\omega) = T$ if $|X_0(\omega)| \le n$, and $= 0$ if $|X_0(\omega)| > n$. Setting $\tau_n = \sigma_n \wedge \sigma_n'$, $X_0^n = X_0 I_{[|X_0| \le n]}$, $M_t^n = M_{t \wedge \tau_n}$, and $B_t^n = B_{t \wedge \tau_n}$, we define $X_t^n = X_0^n + M_t^n + B_t^n$. Then X_t^n is a continuous semimartingale (where M_t^n is a continuous

L^2-martingale and B_t^n is an integrable bounded variation process). Furthermore, $X_t^n(\omega) \in [-3n,3n]$. Let U and V be bounded open intervals such that $[-3n,3n] \subset V \subset \bar{V} \subset U$. Let ϕ be a $C_b^2(\mathbf{R})$-function such that $\phi(x) = 1$ for $x \in \bar{V}$, and $= 0$ for $x \in U^c$. Defining $H(x) = F(x)\phi(x)$, it is easy to see that $H \in C_b^2(\mathbf{R})$. Furthermore, for $x \in [-3n,3n]$, $H(x)$, $H'(x)$, and $H''(x)$ coincide with $F(x)$, $F'(x)$, and $F''(x)$, respectively. We now apply the Ito formula established in Theorem 4.2.1 to $H(x)$ and the semimartingale X_t^n.

$$F(X_t^n) = F(X_0^n) + \int_0^t F'(X_s^n)\,dM_s^n + \int_0^t F'(X_s^n)\,dB_s^n + \tfrac{1}{2}\int_0^t F''(X_s^n)\,d\langle M^n\rangle_s. \tag{4.2.4}$$

Since $X_t^n \to X_t$ and $X_0^n \to X_0$ (a.s.), we have $F(X_t^n) \to F(X_t)$ and $F(X_0^n) \to F(X_0)$. The stochastic integral

$$\int_0^t F'(X_s^n)\,dM_s^n = \int_0^{t\wedge\tau_n} F'(X_s)\,dM_s \to \int_0^t F'(X_s)\,dM_s$$

in probability as $n \to \infty$ by the definition given in Chapter 3 of the stochastic integral with respect to a continuous local martingale. For a.a. ω, there is some $N(\omega)$ such that for all $n \geq N(\omega)$,

$$\int_0^t F'(X_s^n)\,dB_s^n = \int_0^t F'(X_s)\,dB_s.$$

Therefore,

$$\int_0^t F'(X_s^n)\,dB_s^n \to \int_0^t F'(X_s)\,dB_s \qquad \text{(a.s.)}.$$

An identical argument yields

$$\int_0^t F''(X_s^n)\,d\langle M^n\rangle_s \to \int_0^t F''(X_s)\,d\langle M\rangle_s \qquad \text{(a.s.)} \quad \text{as } n \to \infty.$$

Hence making $n \to \infty$ in Equation (4.2.4), we obtain the desired formula

$$F(X_s) = F(X_0) + \int_0^t F'(X_s)\,dM_s + \int_0^t F'(X_s)\,dB_s + \tfrac{1}{2}\int_0^t F''(X_s)\,d\langle M\rangle_s. \qquad \square$$

4.3 Ito Formula (General Version)

We turn now to the task of obtaining a version of the Ito formula where $F(x)$ is replaced by a time-dependent function $G(t,x)$. Throughout this section

$$X_t = X_0 + M_t + B_t$$

will denote a continuous local semimartingale taking values in $\mathbf{R}^d$.

For the purposes of proving our theorem we introduce the temporary notation (used only in this section)

$$\int_0^t f_s\,dX_s$$

to stand for the vector-valued process

$$\left(\int_0^t f_s\,dM_s^j + \int_0^t f_s\,dB_s^j\right)_{j=1}^d,$$

where (f_s) is any continuous, $(\mathscr{F}_t)$-adapted real-valued process.

Lemma 4.3.1.

$$\int_0^t f_s\,dX_s = \underset{|\Pi_n|\to 0}{p\text{-lim}} \sum_{\Pi_n} f_{t_k^n}[X_{t_{k+1}^n} - X_{t_k^n}].$$

PROOF. Here Π_n is a partition of $[0,t]$,

$$\Pi_n = \{0 = t_0^n < t_1^n < \cdots < t_{m_n}^n = t\},$$
$$|\Pi_n| = \max_k |t_{k+1}^n - t_k^n|.$$

Now

$$\sum_{\Pi_n} f_{t_k^n}[X_{t_{k+1}^n} - X_{t_k^n}] = \sum_{\Pi_n} f_{t_k^n}[M_{t_{k+1}^n} - M_{t_k^n}] + \sum_{\Pi_n} f_{t_k^n}[B_{t_{k+1}^n} - B_{t_k^n}].$$

Let $f_s^n = \sum_{k=0}^{m_n-1} f_{t_k^n} \cdot I_{(t_k^n, t_{k+1}^n]}(s) + f_0 \cdot I_{\{0\}}(s)$. By uniform continuity of f_s on $[0,t]$ and since $B_0 = 0$, we have

$$\sum_{\Pi_n} f_{t_k^n}[B_{t_{k+1}^n} - B_{t_k^n}] = \left(\int_0^t f_s^n\,dB_s^j\right)_{j=1}^d \to \left(\int_0^t f_s\,dB_s^j\right)_{j=1}^d \qquad \text{(a.s.)}$$

as $|\Pi_n| \to 0$. On the other hand, for each $1 \le j \le d$, $f, f^n \in L_{\text{loc}}^2(M^j)$ for each continuous local martingale M^j, and

$$\int_0^t (f_s - f_s^n)^2\,d\langle M^j\rangle_s \to 0 \qquad \text{(a.s.)}$$

as $|\Pi_n| \to 0$. Hence by properties of the stochastic integral proved in Section 3.4, we obtain

$$\sum_{\Pi_n} f_{t_k^n}(M_{t_{k+1}^n}^j - M_{t_k^n}^j) = \int_0^t f_s^n\,dM_s^j \to \int_0^t f_s\,dM_s^j$$

in probability as $|\Pi_n| \to 0$. The statement in Lemma 4.3.1 now follows. □

Now let h_t be a continuously differentiable (nonrandom) real-valued function on $\mathbf{R}_+$ and $F\colon \mathbf{R}^d \to \mathbf{R}$ be continuous together with its partial derivatives $\partial F/\partial x^j$, $\partial^2 F/\partial x^i\,\partial x^j$. From Ito's formula for $F(X_t)$, we see that $F(X_t)$ is again a continuous local semimartingale. Hence $\int_0^t h_s\,dF(X_s)$ is defined above.

Lemma 4.3.2.

$$h_t F(X_t) - h_0 F(X_0) = \int_0^t h_s\,dF(X_s) + \int_0^t F(X_s)\,dh_s. \tag{4.3.1}$$

PROOF. Take a sequence of partitions

$$\Pi_n = \{0 = t_0^n < \cdots < t_{m_n}^n = t\}$$

for which $|\Pi_n| \to 0$ as $n \to \infty$. Using the previous lemma applied to $F(X_s)$, we see that the right-hand side of Equation (4.3.1) is the limit in probability of the expression

$$\sum_{k=0}^{m_n-1} h_{t_k^n}[F(X_{t_{k+1}^n}) - F(X_{t_k^n})] + \sum_{k=0}^{m_n-1} F(X_{t_{k+1}^n})[h_{t_{k+1}^n} - h_{t_k^n}]$$
$$= F(X_{t_{m_n}^n})h_{t_{m_n}^n} - F(X_{t_0^n})h_{t_0^n}$$
$$= h_t F(X_t) - h_0 F(X_0).$$

This establishes Equation (4.3.1). □

The Ito formula of Theorem 4.2.3 shows that $F(X_t)$ is a continuous local semimartingale with the decomposition

$$F(X_t) = F(X_0) + M'_t + B'_t,$$

where

$$M'_t = \sum_{i=1}^{d} \int_0^t \frac{\partial}{\partial x^i} F(X_s)\, dM_s^i$$

and

$$B' = \sum_{i=1}^{d} \int_0^t \frac{\partial}{\partial x^i} F(X_s)\, dB_s^i + \frac{1}{2} \sum_{i,j=1}^{d} \int^t \frac{\partial^2}{\partial x^i\, \partial x^j} F(X_s)\, d\langle M^i, M^j\rangle_s.$$

By our definition,

$$\int_0^t h_s\, dF(X_s) = \int_0^t h_s\, dM'_s + \int_0^t h_s\, dB'_s$$

Hence we obtain

$$\int_0^t h_s\, dF(X_s) = \sum_i \int_0^t h_s \frac{\partial F}{\partial x^i}(X_s)\, dM_s^i + \sum_i \int_0^t h_s \frac{\partial F}{\partial x^i}(X_s)\, dB_s^i + \frac{1}{2}\sum_{i,j} \int_0^t h_s \frac{\partial^2 F}{\partial x^i\, \partial x^j}(X_s)\, d\langle M^i, M^j\rangle_s. \tag{4.3.2}$$

If we let

$$G(t,x) = h_t F(x)$$

for $t \in \mathbf{R}_+$ and $x \in \mathbf{R}^d$, we may write Equation (4.3.1) in the form

$$G(t,X_t) - G(0,X_0) = \int_0^t \frac{\partial G}{\partial s}(s,X_s)\, ds + \sum_i \int_0^t \frac{\partial G}{\partial x^i}(s,X_s)\, dM_s^i + \sum_i \int_0^t \frac{\partial G}{\partial x^i}(s,X_s)\, dB_s^i + \frac{1}{2}\sum_{i,j} \int_0^t \frac{\partial^2 G}{\partial x^i\, \partial x^j}(s,X_s)\, d\langle M^i, M^j\rangle_s. \tag{4.3.3}$$

By linearity we will then have Equation (4.3.3) for functions of the form

$$G(t,x) = \sum_{k=1}^{m} h^k(t)F^k(x). \tag{4.3.4}$$

In general, suppose G is any function

$$G: \mathbf{R}_+ \times \mathbf{R}^d \to \mathbf{R}$$

which is continuous together with its partials

$$\frac{\partial G}{\partial t}, \quad \frac{\partial G}{\partial x^i}, \quad \frac{\partial^2 G}{\partial x^i \partial x^j}.$$

Then there exists a sequence G_m, each a function of the form (4.3.4) with

$$G_m \to G, \qquad \frac{\partial G_m}{\partial x^i} \to \frac{\partial G}{\partial x^i}, \qquad \frac{\partial^2 G_m}{\partial x^i \partial x^j} \to \frac{\partial^2 G}{\partial x^i \partial x^j}$$

uniformly on compact sets of $\mathbf{R}_+ \times \mathbf{R}^d$. (See [13], problem 11, p. 95.) Define $I^N = \{x \in \mathbf{R}^d \colon |x^j| \le N \text{ for all } 1 \le j \le d\}$. Let g represent any one of the functions G, $\partial G/\partial x^j$, or $\partial G/\partial x^i \partial x^j$, and let g_m denote the corresponding function with G replaced by G_m.

Fix $T > 0$. For each N,

$$\sup_{[0,T] \times I^N} |g_m(t,x) - g(t,x)| \to 0 \quad \text{as } m \to \infty.$$

Write

$$\|X^j(\omega)\| = \sup_{[0,T]} |X_t^j(\omega)|.$$

Then $\sup |g_m(t,X_t(\omega)) - g(t,X_t(\omega)| \to 0$ as $m \to \infty$, where the sup is taken over the set $\{(t,\omega) | (t,X_t(\omega)) \in [0,T] \times I^N\}$. Hence

$$\begin{aligned} P\Bigg[\omega\colon \lim_{m\to\infty} \Bigg(&\sum_{i,j=1}^{d} \int_0^T \{g_m(t,X_t(\omega)) - g(t,X_t(\omega))\}^2 \, d\langle M^i,M^j\rangle_t(\omega) \\ &+ \sum_{j=1}^{d} \int_0^T \{g_m(t,X_t(\omega) - g(t,X_t(\omega)\}^2 \, dB_t^j(\omega) \\ &+ \int_0^T \{g_m(t,X_t(\omega)) - g(t,X_t(\omega))\}^2 \, dt \Bigg) = 0 \Bigg] \\ \ge P[\omega\colon &(\|X^1(\omega)\|, \ldots, \|X^d(\omega)\|) \in I^N]. \end{aligned} \tag{4.3.5}$$

Since N is arbitrary, we see that the probability on the left-hand side of (4.3.5) equals 1.

We have formula (4.3.3) for each G_m, and we may now pass to the limit under each integral to obtain the following theorem.

Theorem 4.3.1. *Let $X_t = X_0 + M_t + B_t$ be a continuous local semimartingale. Let $G(t,x)$, $(t,x) \in \mathbf{R}_+ \times \mathbf{R}^d$, be a real-valued function, continuous together with* *its partial derivatives* $\partial G/\partial t$, $\partial G/\partial x^j$, $\partial^2 G/\partial x^i \partial x^j$. *Then the process* $G(t,X_t)$ *is again a continuous local semimartingale for which formula* (4.3.3) *holds.*

4.4 Applications of the Ito Formula

We will derive the following characterization of the Wiener process: Note that in the theorem to be proved below we do not assume right-continuity of $(\mathscr{F}_t)$.

Theorem 4.4.1. *Let* $(M_t,\mathscr{F}_t)$, $(t \in \mathbf{R}_+)$ *be a continuous local martingale in* $\mathbf{R}^d$ *such that* $M_0 = 0$ *and* $\langle M \rangle_t = tI$ *(I being the* $d \times d$ *identity matrix). Then*

(i) (M_t) *is a d-dimensional Wiener process.*
(ii) *For all* $s < t$, $\sigma[M_v - M_u : s \le u < v \le t] \perp\!\!\!\perp \mathscr{F}_s$.

PROOF. First assume that $(\mathscr{F}_t)$ is right-continuous. From Lemma 2.6.3, we can conclude that on each interval $[0,T]$, $(M_t,\mathscr{F}_t)$ is a continuous L^2-martingale. Fix s, $0 \le s \le T$, and $u = (u_1, \ldots, u_d) \in \mathbf{R}^d$. For t, $s \le t \le T$, define

$$g(t) = \int_A e^{i(u,M_t - M_s)}\, dP,$$

where $A \in \mathscr{F}_s$ and $(\cdot,\cdot)$ denotes the inner product of $\mathbf{R}^d$.

Applying Ito's formula to $F(x) = e^{i(u,x)}$, we obtain

$$e^{i(u,M_t)} = 1 + i\sum_j u_j \int_0^t e^{i(u,M_v)}\, dM_v^j - \tfrac{1}{2}|u|^2 \int_0^t e^{i(u,M_v)}\, dv,$$

where $|u|^2 = \sum_j u_j^2$. Hence

$$e^{i(u,M_t - M_s)} = 1 + i\sum_j u_j \int_s^t e^{i(u,M_v - M_s)}\, dM_v^j - \tfrac{1}{2}|u|^2 \int_s^t e^{i(u,M_v - M_s)}\, dv.$$

Keeping s fixed and integrating both sides with respect to P over A and noting that

$$E\left\{\int_s^t e^{i(u,M_v - M_s)}\, dM_v^j \middle| \mathscr{F}_s\right\} = 0 \qquad \text{(a.s.)},$$

we see that

$$g(t) = P(A) - \tfrac{1}{2}|u|^2 \int_s^t g(v)\, dv.$$

Differentiating with respect to t, we get

$$g'(t) = -\tfrac{1}{2}|u|^2 g(t) \quad \text{on } t \in [s,T].$$

Hence $g(t) = e^{-\frac{1}{2}|u|^2(t-s)}P(A)$. This yields $\int_A e^{i(u,M_t - M_s)}\, dP = \int_A e^{-\frac{1}{2}|u|^2(t-s)}\, dP$ for all $A \in \mathscr{F}_s$. Hence since T is arbitrary, we have for all $s < t$,

$$E\{e^{i(u,M_t - M_s)} | \mathscr{F}_s\} = e^{-\frac{1}{2}|u|^2(t-s)} \qquad \text{(a.s.)} \tag{4.4.1}$$

which proves (i).

Conclusion (ii) follows almost immediately. Take $s = t_1 < t_2 < \cdots < t_{n+1} = t$, and $(u^k)_{k=1}^n$, vectors in $\mathbf{R}^d$. Then

$$E\left[\prod_{k=1}^{n} e^{i(u^k, M_{t_{k+1}} - M_{t_k})} \Big| \mathscr{F}_s\right]$$

$$= E\left\{\prod_{k=1}^{n-1} e^{i(u^k, M_{t_{k+1}} - M_{t_k})} \cdot E\left[e^{i(u^n, M_{t_{n+1}} - M_{t_n})} \big| \mathscr{F}_{t_n}\right] \Big| \mathscr{F}_s\right\}$$

$$= E\left[e^{i(u^n, M_{t_{n+1}} - M_{t_n})}\right] \cdot E\left[\prod_{k=1}^{n-1} e^{i(u^k, M_{t_{k+1}} - M_{t_k})} \Big| \mathscr{F}_s\right]$$

$$= \prod_{k=1}^{n} E e^{i(u^k, M_{t_{k+1}} - M_{t_k})}$$

by induction and Equation (4.4.1). Hence $\sigma(M_{t_{k+1}} - M_{t_k}, 1 \le k \le n) \perp\!\!\!\perp \mathscr{F}_s$, and (ii) is proved.

If $(\mathscr{F}_t)$ is not right-continuous, the result just proved holds if in the hypothesis we replace $(\mathscr{F}_t)$ by $(\mathscr{F}_{t+})$. This is justified because if $(M_t, \mathscr{F}_t)$ is a continuous local martingale, so is $(M_t, \mathscr{F}_{t+})$ since M_t is continuous. Conclusion (ii) then becomes $\sigma[M_v - M_u : s \le u < v \le t] \perp\!\!\!\perp \mathscr{F}_{s+}$. But this implies that the σ-field in question is $\perp\!\!\!\perp \mathscr{F}_s$. The proof of the theorem is complete. □

Remark 4.4.1. Note that although the statement of the theorem is made for a continuous local martingale on $\mathbf{R}_+$, the proof proceeds for each fixed T by obtaining Equation (4.4.1) for all s and t such that $0 \le s \le t \le T$. Hence the same conclusions follow if the local martingale M is given initially on some interval $[0,T]$.

Stochastic Integral with Respect to a Wiener Process in $\mathbf{R}^n$

Let $(W_t, \mathscr{F}_t)$ be an n-dimensional Wiener martingale, $W = (W_t^i)$, $i = 1,2,\ldots,n$. A matrix-valued process $(f_t = (f_t^{ij}), (i = 1,\ldots,m;\ j = 1,\ldots,n)$ will be said to belong to $\mathscr{L}_2^W[0,T]$ if each f_t^{ij} is measurable, $(\mathscr{F}_t)$-adapted, and satisfies (for each i,j)

$$\int_0^T (f_t^{ij})^2\, dt < \infty \qquad \text{(a.s.)}.$$

The family $(\mathscr{F}_t)$ is not assumed to be right-continuous. We shall say that $f_t = (f_t^{ij})$ belongs to $\mathscr{M}_2^W[0,T]$ if this last condition is replaced by

$$E\int_0^T (f_t^{ij})^2\, dt < \infty.$$

We then define the m-vector–valued stochastic integral $\int_0^T f_t\, dW_t$ as the vector

$$\left(\sum_{j=1}^{n} \int_0^T f_t^{ij}\, dW_t^j\right)_{i=1,\ldots,m}.$$

Let $\alpha_t = (\alpha_t^i)$ be a vector-valued process. We say that $(\alpha_t) \in \mathscr{L}_1^W[0,T]$ if each α_t^i is measurable, $(\mathscr{F}_t)$-adapted, and

$$\int_0^T |\alpha_t^i|\, dt < \infty \qquad \text{(a.s.) for each } i.$$

The process (α_t) belongs to $\mathscr{M}_1^W[0,T]$ if the stronger condition $E\int_0^T |\alpha_t^i| dt < \infty$ holds for each i.

Ito Differential

Definition 4.4.1. A continuous process $X = (X_t)$, $X_t \in \mathbf{R}^m$, has a *stochastic* or *Ito differential* on $[0,T]$ with respect to an n-dimensional Wiener martingale $W = (W_t)$ if there exists

(i) An m-dimensional process $\alpha \in \mathscr{L}_1^W[0,T]$
(ii) An $m \times n$-matrix–valued process $\beta \in \mathscr{L}_2^W[0,T]$ such that for all $0 \le t \le T$,

$$X_t - X_0 = \int_0^t \alpha_u\, du + \int_0^t \beta_u\, dW_u \qquad \text{(a.s.).} \tag{4.4.2}$$

This statement is commonly written in differential notation as

$$dX_t = \alpha_t\, dt + \beta_t\, dW_t. \tag{4.4.3}$$

Note now that if X has a stochastic differential, then (4.4.2) shows that X is a continuous semimartingale. Hence we may apply the Ito formula.

Let $\psi(t,x)$, $(t,x) \in \mathbf{R}_+ \times \mathbf{R}^m$, be continuous together with its derivatives $\partial\psi/\partial t$, $\partial\psi/\partial x^i$, and $\partial^2\psi/\partial x^i\, \partial x^j$. In Ito's formula let us set

$$B_t^i = \int_0^t \alpha_s^i\, ds \quad \text{and} \quad M_t^i = \sum_{k=1}^n \int_0^t \beta_s^{ik}\, dW_s^k.$$

We then have

$$\begin{aligned} \langle M^i, M^j \rangle_t &= \left\langle \sum_k \int_0^{(\cdot)} \beta_s^{ik}\, dW_s^k, \sum_l \int_0^{(\cdot)} \beta_s^{jl}\, dW_s^l \right\rangle_t \\ &= \int_0^t \left(\sum_k \beta_s^{ik} \beta_s^{jk} \right) ds. \end{aligned}$$

Let $\gamma_s = \beta_s \beta_s^*$ (where $*$ denotes transpose) and define the operator

$$L_s = \sum_i \alpha_s^i \frac{\partial}{\partial x^i} + \frac{1}{2} \sum_{i,j} \gamma_s^{ij} \frac{\partial^2}{\partial x^i \partial x^j} \tag{4.4.4}$$

The Ito formula of Theorem 4.3.1 becomes

$$\psi(t,X_t) - \psi(0,X_0) = \int_0^t \left(\frac{\partial}{\partial s} + L_s \right) \psi(s,X_s)\, ds + \sum_{i,k} \int_0^t \frac{\partial\psi(s,X_s)}{\partial x^i} \beta_s^{ik}\, dW_s^k. \tag{4.4.5}$$

From Equation (4.4.5) we obtain the following result.

Theorem 4.4.2. *Let $\psi(t,x)$, $(t,x) \in \mathbf{R}_+ \times \mathbf{R}^d$, be continuous with continuous partial derivatives $\partial\psi/\partial t$, $\partial\psi/\partial x^i$, and $\partial^2\psi/\partial x^i \partial x^j$. Fix $T > 0$. Suppose X has the stochastic differential* (4.4.2) *and that*

$$\left(\frac{\partial\psi}{\partial s}(s,X_s) + L_s\psi(s,X_s)\right) \in \mathscr{M}_1^W[0,T]$$

and

$$\left(\frac{\partial\psi(s,X_s)}{\partial x^i}\beta_s^{ik}\right) \in \mathscr{M}_2^W[0,T] \quad \text{for all } i,k,$$

where L_s is given by (4.4.4). Then, if $\psi(0,X_0)$ is integrable,

$$E\psi(t,X_t) - E\psi(0,X_0) = E\left[\int_0^t \left(\frac{\partial}{\partial s} + L_s\right)\psi(s,X_s)ds\right]. \tag{4.4.6}$$

Let τ be a stopping time for $(\mathscr{F}_t)$, $0 \leq \tau \leq T$, (a.s.). Set $Y_t = \psi(t,X_t) - \psi(0,X_0) - \int_0^t (\partial/\partial s + L_s)\psi(s,X_s)\,ds$. The stochastic integrals in Equation (4.4.5) are continuous L^2-martingales because of the assumption

$$\frac{\partial\psi(s,X_s)}{\partial x^i}\beta_s^{ik} \in \mathscr{M}_2^W[0,T].$$

It follows that $(Y_t,\mathscr{F}_t)$ is a continuous L^2-martingale and hence a continuous, uniformly integrable martingale. Hence by result 10 in Section 2.2, $(Y_{t\wedge\tau})$, $0 \leq t \leq T$, is also a martingale, and we get

$$E\psi(t\wedge\tau,X_{t\wedge\tau}) - E\psi(0,X_0) = E\left\{\int_0^{t\wedge\tau}\left(\frac{\partial}{\partial s} + L_s\right)\psi(s,X_s)\,ds\right\}. \tag{4.4.7}$$

If ψ is taken to be only a function of x, Equation (4.4.7) becomes

$$E\psi(X_{t\wedge\tau}) - E\psi(X_0) = E\left[\int_0^{t\wedge\tau}(L_s\psi)(X_s)\,ds\right] \tag{4.4.8}$$

Proposition 4.4.1. *Assume the conditions of Theorem 4.4.2 on $\psi(t,x)$, $(t,x) \in \mathbf{R}_+ \times \mathbf{R}^d$, and let the conditions on the stochastic differential for X imposed in Theorem 4.4.2 hold for every $T > 0$. Suppose that σ is a stopping time such that*

$$E(\sigma) < \infty. \tag{4.4.9}$$

Let us further assume that ψ does not involve t and

$$\sup_{\substack{x\in\mathbf{R}^n\\ s\geq 0}} [|\psi(x)| + |L_s\psi|(x)] \leq C < \infty \qquad \text{(a.s.)}. \tag{4.4.10}$$

Then

$$E\psi(X_\sigma) - E\psi(X_0) = E\left\{\int_0^\sigma (L_s\psi)(X_s)\,ds\right\}. \tag{4.4.11}$$

Proof. From Equation (4.4.8) (putting $t = n$, $\tau = \sigma \wedge n$), we have

$$E\psi(X_{\sigma\wedge n}) - E\psi(X_0) = E\int_0^{\sigma\wedge n}(L_s\psi)(s,X_s)ds.$$

From expression (4.4.10), clearly $E\psi(X_{\sigma\wedge n}) \to E\psi(X_\sigma)$ as $n \to \infty$ and $|\int_0^{\sigma\wedge n} L_s\psi(s,X_s)\,ds| \le C\sigma$. By the Lebesgue dominated convergence it follows that

$$E\int_0^{\sigma\wedge n}(L_s\psi)(s,X_s)\,ds \to E\int_0^{\sigma}(L_s\psi)(s,X_s)\,ds.$$

and Equation (4.4.11) is proved. □

EXAMPLE 4.4.1. Consider the following special case of the above proposition.

Take $X_t = x + W_t$, where $x \in \mathbf{R}^n$ is fixed and (W_t) is a Wiener process in $\mathbf{R}^n$. Let $\psi(x)$ be bounded and continuous together with its derivatives $\partial\psi/\partial x^i$ and $\partial^2\psi/\partial x^i\,\partial x^j$. In this case, $L\psi = \frac{1}{2}\Delta\psi$ ($\Delta = \sum_i \partial^2/\partial x^{i^2}$), and Equation (4.4.11) becomes

$$E\psi(X_\sigma) = \psi(x) + \tfrac{1}{2}E\int_0^\sigma (\Delta\psi)(X_s)\,ds, \tag{4.4.12}$$

where σ is a stopping time for W such that $E\sigma < \infty$. As a corollary we derive the following well-known result.

Proposition 4.4.2. *Let D be a bounded domain in $\mathbf{R}^n$ and τ_D the exit time from D for the Wiener process starting at x (that is, the hitting time to $\mathbf{R}^d - D$). Then $E(\tau_D) < \infty$.*

PROOF. If $x \notin D$, $P(\tau_D = 0) = 1$ and the conclusion is trivially true. So assume $x \in D$. Choosing $\psi(x) = -x_1^2$, where $x = (x_1, \ldots, x_n)$, we have $\Delta\psi(x) = -2$. Since D is bounded, $|\psi(x)| \le K$, a constant, for $x \in D$. From Equation (4.4.12) we then obtain

$$\begin{aligned} E\psi(X_{\tau_D\wedge n}) - \psi(x) &= \tfrac{1}{2}E\int_0^{\tau_D\wedge n}\Delta\psi(X_s)\,ds \\ &= -E(\tau_D \wedge n). \end{aligned}$$

Hence

$$E(\tau_D \wedge n) = [-E\psi(X_{\tau_D\wedge n}) + \psi(x)] \le 2K.$$

Making $n \uparrow \infty$, the result follows. □

4.5 A Vector-Valued Version of Ito's Formula

Let (X_t) be an m-dimensional process having an Ito differential given by (4.4.2) or (4.4.3). Let $G(t,x)$, $(t,x) \in \mathbf{R}_+ \times \mathbf{R}^m$, be an $\mathbf{R}^m$-valued function which is continuous together with its derivatives $\partial G/\partial t$, $\partial G/\partial x^i$, and $\partial^2 G/\partial x^i\,\partial x^j$. Note that the derivatives are m-vector–valued functions. Letting G_k ($k = 1, \ldots, m$) denote the components of G, we note that

$$\frac{\partial G}{\partial t} = \left(\frac{\partial G_k}{\partial t}\right), \frac{\partial G}{\partial x^i} = \left(\frac{\partial G_k}{\partial x^i}\right), \quad \text{and} \quad \frac{\partial^2 G}{\partial x^i\,\partial x^j} = \left(\frac{\partial^2 G_k}{\partial x^i\,\partial x^j}\right).$$

We now derive an Ito differential for $G(t,X_t)$ using the differential formula (4.4.5) for a real-valued process $\psi(t,X_t)$.

For any $\theta = (\theta_1, \ldots, \theta_m) \in \mathbf{R}^m$, set $\psi(t,x) = (\theta, G(t,x))$. Applying (4.4.5) to $\psi(t,x)$, we obtain

$$(\theta, G(t,X_t)) - (\theta, G(0,X_0)) = \int_0^t \left(\frac{\partial}{\partial s} + L_s\right) \{(\theta, G(s,X_s))\}\, ds$$

$$+ \sum_{i,k} \int_0^t \frac{\partial}{\partial x^i} (\theta, G(s,X_s)) \beta_s^{ik}\, dW_s^k,$$

where L_s is the differential operator given by (4.4.4). The first term on the right-hand side of the above relation becomes

$$\int_0^t \left(\theta, \frac{\partial G(s,X_s)}{\partial s} + L_s G(s,X_s)\right) ds,$$

where $L_s G$ is the vector with components

$$(L_s G)_k = \sum_{i=1}^m \alpha_s^i \frac{\partial G_k}{\partial x^i} + \frac{1}{2} \sum_{i,j=1}^m \gamma_s^{ij} \frac{\partial^2 G_k}{\partial x^i\, \partial x^j}.$$

The second term becomes

$$\int_0^t \sum_{i,k} \left(\theta, \frac{\partial G(s,X_s)}{\partial x^i}\right) \beta_s^{ik}\, dW_s^k = \sum_{l=1}^m \theta_l \int_0^t \left[\left(\sum_{i,k} \beta_s^{ik} \frac{\partial G_l(s,X_s)}{\partial x^i}\right)\right] dW_s^k$$

$$= \left(\theta, \int_0^t g_s\, dW_s\right),$$

where $g_s = (g_s^{lk})$ is the $m \times n$ matrix with $g_s^{lk} = \sum_i \beta_s^{ik}\, \partial G_l(s,X_s)/\partial x^i$ and $\int_0^t g_s\, dW_s$ is the m-vector–valued Ito integral $(\sum_k \int_0^t g_s^{lk}\, dW_s^k)_{l=1,\ldots,m}$. Since $\theta \in \mathbf{R}^m$ is arbitrary, we obtain the formula

$$G(t,X_t) - G(0,X_0) = \int_0^t \left[\frac{\partial G(s,X_s)}{\partial s} + L_s G(s,X_s)\right] ds$$

$$+ \int_0^t g_s\, dW_s. \tag{4.5.1}$$

Formula (4.5.1) can be written componentwise as follows:

$$G_l(t,X_t) - G_l(0,X_0) = \int_0^t \left[\frac{\partial G_l(s,X_s)}{\partial s} + \sum_i \alpha_s^i \frac{\partial G_l(s,X_s)}{\partial x^i}\right.$$

$$\left. + \frac{1}{2} \sum_{i,j} \gamma_s^{ij} \frac{\partial^2 G_l(s,X_s)}{\partial x^i\, \partial x^j}\right] ds$$

$$+ \int_0^t \sum_{i,k} \beta_s^{ik} \frac{\partial G_l(s,X_s)}{\partial x^i}\, dW_s^k \qquad (l = 1, \ldots, m). \tag{4.5.2}$$

5 Stochastic Differential Equations

5.1 Existence and Uniqueness of Solutions

Denote by $C_d = C([0,T], \mathbf{R}^d)$, $d \geq 1$, the space of continuous functions on $[0,T]$ taking values in $\mathbf{R}^d$. Let S be a complete, separable metric space and $D = D([0,T], S)$, the space of right-continuous functions from $[0,T]$ to S having left-hand limits. Let $\mathscr{B}_t(C_d)$ be the minimal σ-field with respect to which the coordinate functions $f(s)(0 \leq s \leq t, f \in C_d)$ are measurable. The σ-field $\mathscr{B}_t(D)$ is similarly defined. We write $\mathscr{B}(C_d) = \mathscr{B}_T(C_d)$ and $\mathscr{B}(D) = \mathscr{B}_T(D)$.

Definition 5.1.1. The functional $\gamma\colon [0,T] \times D \times C_d \to \mathbf{R}$ is *causal* if the following conditions are satisfied:

(i) γ is $\mathscr{B}[0,T] \times \mathscr{B}(D) \times \mathscr{B}(C_d)$-measurable.
(ii) For each t, $\gamma(t,\cdot,\cdot)$ is measurable with respect to $\mathscr{B}_{t+}(D) \times \mathscr{B}_{t+}(C_d)$.

A d-dimensional functional $b = (b_1, \ldots, b_d)$ is causal if each component b_i is causal in the sense defined above. Similarly, the matrix-valued functional $\sigma = (\sigma^{ij})$, $(i = 1, \ldots, d, j = 1, \ldots, n)$ is causal if each σ^{ij} is causal.

Causal functionals are also referred to as *nonanticipative* functionals.

Certain L^2-estimates of solutions of stochastic differential equations will be obtained with the aid of the following result, known as Gronwall's inequality.

Proposition 5.1.1. *Let $\alpha(t)$ and $\beta(t)$ be Lebesgue integrable functions on $[a,b]$, and suppose H is a constant such that*

$$\alpha(t) \leq \beta(t) + H \int_a^t \alpha(s)\,ds \quad \textit{for } t \in [a,b].$$

Then

$$\alpha(t) \le \beta(t) + H \int_a^t e^{H(t-s)} \beta(s)\,ds.$$

We will use this result only when $\beta(t) \equiv B$, a constant, in which case the conclusion becomes

$$\alpha(t) \le B e^{H(t-a)}.$$

PROOF. Let $A(t) = \int_a^t \alpha(s)\,ds$. The function $g(t) = A(t)e^{-Ht}$ is absolutely continuous on $[a,b]$ and

$$g'(t) \equiv \frac{d}{dt} g(t) = \alpha(t)e^{-Ht} - HA(t)e^{-Ht} \qquad \text{(a.e.)}$$

Since

$$g'(t) \le \beta(t)e^{-Ht} \qquad \text{(a.e.) on } [a,b],$$

$$g(t) - g(a) = \int_a^t g'(s)\,ds \le \int_a^t \beta(s)e^{-Hs}\,ds,\ t \in [a,b].$$

Hence

$$A(t) \le e^{Ht} \int_a^t \beta(s)e^{-Hs}\,ds$$

and thus

$$\alpha(t) \le \beta(t) + HA(t) \le \beta(t) + H \int_a^t \beta(s)e^{H(t-s)}\,ds. \qquad \square$$

Definition 5.1.2. Let $(\Omega,\mathscr{A},P)$ be a complete probability space on which are defined the following:

(i) $X = (X_t)$, $t \in [0,T]$, where $X_t(\omega)$ is a measurable, S-valued process with $X(\omega) \in D$ for almost all ω.
(ii) $W = (W_t)$, $t \in [0,T]$, an n-dimensional Wiener process.
(iii) η, a d-dimensional random vector.

$\{\mathscr{F}_t\}$ is an increasing family of sub-σ-fields of $\mathscr{A}$, $\mathscr{F}_0$ containing all P-null sets in $\mathscr{A}$, such that for each t in $[0,T]$,

(A) $\mathscr{F}_t^{\eta,X,W} \subseteq \mathscr{F}_t$ and $\mathscr{F}_t \perp\!\!\!\perp \sigma[W_v - W_u, t \le u \le v \le T]$,

where $\mathscr{F}_t^{\eta,X,W} = \sigma[\eta, X_s, W_s, 0 \le s \le t] \vee [\text{all } P\text{-null sets in } \mathscr{A}]$.

Suppose that b is a d-dimensional causal functional and σ a $d \times n$-matrix–valued causal functional.

Then a d-dimensional process $\xi = (\xi_t)$, $t \in [0,T]$, defined on $(\Omega,\mathscr{A},P)$ is called a *solution* or *strong solution* of the stochastic differential equation

$$d\xi_t = b(t,X,\xi)\,dt + \sigma(t,X,\xi)\,dW_t \tag{5.1.1}$$

with initial condition $\xi_0 = \eta$ if the following assertions are true:

(a) (ξ_t) is continuous and $(\mathscr{F}_t)$-adapted.
(b) $\int_0^T |b(t,X,\xi)|\,dt + \int_0^T |\sigma(t,X,\xi)|^2\,dt < \infty$ (a.s.).

(c) For each $t \in [0,T]$,

$$\xi_t = \eta + \int_0^t b(s,X,\xi)\,ds + \int_0^t \sigma(s,X,\xi)\,dW_s \qquad \text{(a.s.)}.$$

In (b), the symbol $|\cdot|$ denotes the length of a d-dimensional vector as well as the norm of a matrix, that is,

$$|\sigma| = \left[\sum_{i,j} (\sigma^{ij})^2\right]^{\frac{1}{2}}.$$

Observe that the causality of b and σ together with the assumption that X and ξ are $(\mathscr{F}_t)$-adapted implies that the processes $b(t,X(\omega),\xi(\omega))$ and $\sigma(t,X(\omega),\xi(\omega))$ are measurable and $(\mathscr{F}_{t+})$-adapted. Furthermore, (A) implies that $(W_t,\mathscr{F}_t)$ and hence $(W_t,\mathscr{F}_{t+})$ is a Wiener martingale. In view of condition (b), the stochastic integration in (c) is legitimate.

Definition 5.1.3. The stochastic differential Equation (5.1.1) with initial condition $\xi_0 = \eta$ has a *unique* strong solution if for any two strong solutions $\xi = (\xi_t)$ and $\tilde{\xi} = (\tilde{\xi}_t)$ in the sense defined above, we have

$$P[\omega\colon \xi_t(\omega) = \tilde{\xi}_t(\omega) \quad \text{for all } t \in [0,T]] = 1.$$

Remark 5.1.1. The definition of a strong solution presupposes that a setup $\{(\Omega,\mathscr{A},P),(\mathscr{F}_t),X,W,\eta,b,\sigma\}$ is given beforehand.

For our main result on the existence and uniqueness of strong solutions of Equation *(5.1.1)* we need to impose the following additional restrictions on b, σ, and the process X: for all $t \in [0,T]$, $g \in D$, and $f, \tilde{f} \in C_d$,

(B) $|b(t,g,f) - b(t,g,\tilde{f})|^2 + |\sigma(t,g,f) - \sigma(t,g,\tilde{f})|^2 \le K\|f - \tilde{f}\|_t^2$

and

(C) $|b(t,g,f)|^2 + |\sigma(t,g,f)|^2 \le K[1 + \|f\|_t^2 + |L(t,g)|^2]$,

where K is a positive constant independent of t, f, and g, $\|f\|_t = \sup_{0\le s\le t}|f(s)|$ and $L\colon[0,T] \times D \to \mathbf{R}$ is a $\mathscr{B}[0,T] \times \mathscr{B}(D)$-measurable function such that $E\int_0^T |L(t,X)|^2\,dt < \infty$.

Two alternative sets of conditions are also of interest. We state them here for use later in this section.

(B$_1$) $|b(t,g,f) - b(t,g,\tilde{f})|^2 + |\sigma(t,g,f) - \sigma(t,g,\tilde{f})|^2 \le K\|f - \tilde{f}\|_T^2$

and

(C$_1$) $|b(t,g,f)|^2 + |\sigma(t,g,f)|^2 \le K[1 + \|f\|_T^2 + |L(t,g)|^2]$.

Let (B$_2$) be the condition obtained by replacing the right-hand side of (B$_1$) by $\int_0^T |f(s) - \tilde{f}(s)|^2\Gamma(ds)$, and let (C$_2$) be obtained by replacing $\|f\|_T^2$ in the right-hand side of (C_1) by $\int_0^T |f(s)|^2 \Gamma(ds)$, where Γ is a finite Borel measure on $[0,T]$ whose total mass is denoted by $\|\Gamma\|$. Then we note here for later use that (B) and (C) imply (B_1) and (C_1). Clearly, also (B_2) and (C_2) imply (B_1) and (C_1).

Theorem 5.1.1. *Suppose that* $(\Omega,\mathscr{A},P)$, X, W, η, $\{\mathscr{F}_t\}$, b, *and* σ *are given as in the definition above. Let conditions* (B) *and* (C) *be satisfied. Then the stochastic differential Equation* (5.1.1) *with initial condition* η *possesses a unique strong solution* $\xi = (\xi_t)$ *such that for each* t,

$$\mathscr{F}_t^{\xi} \subset \mathscr{F}_t^{\eta,X,W},$$

where, as usual, $\mathscr{F}_t^{\xi} = \sigma[\xi_s, 0 \le s \le t] \vee [P\text{-null sets of } \mathscr{A}]$.

Proof. *Existence.* We first prove existence under the assumption $E|\eta|^2 < \infty$. The proof proceeds by the well-known method of successive approximations. For all t and ω, define

$$\xi_t^0(\omega) = \eta(\omega)$$

and define the continuous process

$$\xi_t^1 = \eta + \int_0^t b(s,X,\xi^0)\,ds + \int_0^t \sigma(s,X,\xi^0)\,dW_s.$$

Then (ξ_t^1) is $\mathscr{F}_t^{\eta,X,W}$-adapted, and it follows from condition (C) that

$$E \sup_{0 \le t \le T} |\xi_t^1|^2 < \infty.$$

Suppose that the processes (ξ_t^k), $(0 \le k \le m)$ have been defined to be continuous, $\mathscr{F}_t^{\eta,X,W}$-adapted and to have the following properties:

$$E \sup_{0 \le t \le T} |\xi_t^k|^2 < \infty$$

and for $0 < k \le m$,

$$\xi_t^k = \eta + \int_0^t b(s,X,\xi^{k-1})\,ds + \int_0^t \sigma(s,X,\xi^{k-1})\,dW_s. \tag{5.1.2}$$

Now let

$$\xi_t^{m+1} = \eta + \int_0^t b(s,X,\xi^m)\,ds + \int_0^t \sigma(s,X,\xi^m)\,dW_s. \tag{5.1.3}$$

By the inductive assumption on ξ^m, the stochastic integrals on the right-hand side of Equation (5.1.3) have meaning, and (ξ_t^{m+1}) is $\mathscr{F}_t^{\eta,X,W}$-adapted. From condition (C)

$$E \int_0^T |b(s,X,\xi^m)|^2\,ds + E \int_0^T |\sigma(s,X,\xi^m)|^2\,ds$$

$$\le K\left[T + TE\left(\sup_{0 \le t \le T} |\xi_t^m|^2\right) + E\int_0^T |L(s,X)|^2\,ds\right] < \infty. \tag{5.1.4}$$

Using the martingale inequality and condition (C) once again, we have

$$
\begin{aligned}
E \sup_{0 \le t \le T} |\xi_t^{m+1}|^2 &\le 2E|\eta|^2 + 4TE \int_0^T |b(s,X,\xi^m)|^2 \, ds \\
&\quad + 4E \sup_{0 \le t \le T} \left| \int_0^t \sigma(s,X,\xi^m) \, dW_s \right|^2 \\
&\le 2E|\eta|^2 + 4KT\left[T + TE \sup_{0 \le t \le T} |\xi_t^m|^2 + \int_0^T E|L(t,X)|^2 \, dt \right] \\
&\quad + 16K\left[T + TE \sup_{0 \le t \le T} |\xi_t^m|^2 + \int_0^T E|L(t,X)|^2 \, dt \right] < \infty.
\end{aligned}
$$

Hence Equation (5.1.2) is satisfied for $k = m + 1$, and the induction is complete. We thus have a sequence of continuous, $\mathscr{F}_t^{\eta,X,W}$-adapted processes (ξ_t^k), $k \ge 1$, satisfying Equation (5.1.2). Let us now write

$$
\rho_1(t) = E \sup_{0 \le s \le t} |\xi_s^1 - \eta|^2
$$

and

$$
\rho_{m+1}(t) = E \sup_{0 \le s \le t} |\xi_s^{m+1} - \xi_s^m|^2 \quad \text{for } m \ge 1.
$$

Then

$$
\begin{aligned}
\rho_1(t) &\le 2tE \int_0^t |b(s,X,\eta)|^2 \, ds + 8E \int_0^t |\sigma(s,X,\eta)|^2 \, ds \\
&\le 2(T + 4)\left[T + TE|\eta|^2 + E \int_0^T |L(s,X)|^2 \, ds \right] \\
&= A, \text{ say.}
\end{aligned}
$$

Similarly, applying condition (B) in the form (B$_1$), we obtain

$$
\begin{aligned}
\rho_{m+1}(t) &\le 2tE \int_0^t |b(s,X,\xi^m) - b(s,X,\xi^{m-1})|^2 \, ds \\
&\quad + 8E \int_0^t |\sigma(s,X,\xi^m) - \sigma(s,X,\xi^{m-1})|^2 \, ds \\
&\le 2(T + 4)K \int_0^t E \sup_{0 \le u \le s} |\xi_u^m - \xi_u^{m-1}|^2 \, ds.
\end{aligned}
$$

Hence

$$
\rho_{m+1}(t) = 2(T + 4)K \int_0^t \rho_m(s) \, ds. \tag{5.1.5}
$$

Denote the constant $2(T + 4)K$ by C. Then it follows by induction from Equation (5.1.5) that

$$
\rho_{m+1}(t) \le \frac{C^m t^m}{m!} \rho_1(t) \le \frac{AC^m t^m}{m!}
$$

for each $m \geq 0$ and t in $[0,T]$. We then have

$$\sum_{m=1}^{\infty} P\left[\sup_{0 \leq t \leq T} |\xi_t^{m+1} - \xi_t^m| > \frac{1}{m^2}\right] \leq \sum_{m=1}^{\infty} m^4 E\left[\sup_{0 \leq t \leq T} |\xi_t^{m+1} - \xi_t^m|^2\right]$$
$$\leq A \sum_{m=1}^{\infty} \frac{m^4 (CT)^m}{m!}.$$

By the Borel-Cantelli lemma, the series

$$\xi_t^0(\omega) + \sum_{m=1}^{\infty} [\xi_t^m(\omega) - \xi_t^{m-1}(\omega)]$$

converges uniformly on $[0,T]$ (P-a.s.) to a continuous, $\mathscr{F}_t^{n,X,W}$-adapted process which we denote by $\xi = (\xi_t)$. It remains to verify properties (b) and (c) of the Definition 5.1.2. For $p > m$, we have

$$|\xi_t^p - \xi_t^m|^2 = \left|\sum_{j=m+1}^{p} (\xi_t^j - \xi_t^{j-1})\right|^2$$
$$\leq \left(\sum_{j=m+1}^{p} 2^{-j}\right)\left(\sum_{j=m+1}^{p} 2^j |\xi_t^j - \xi_t^{j-1}|^2\right)$$
$$\leq 2^{-m} \sum_{j=1}^{\infty} 2^j |\xi_t^j - \xi_t^{j-1}|^2,$$

and so

$$\sup_{0 \leq t \leq T} |\xi_t^p - \xi_t^m|^2 \leq 2^{-m} \sum_{j=1}^{\infty} 2^j \sup_{0 \leq t \leq T} |\xi_t^j - \xi_t^{j-1}|^2. \tag{5.1.6}$$

For each fixed m,

$$\sup_{0 \leq t \leq T} |\xi_t^p - \xi_t^m|^2 \to \sup_{0 \leq t \leq T} |\xi_t - \xi_t^m|^2 \qquad \text{(a.s.)}$$

as $p \to \infty$ because $\xi_t^p \to \xi_t$ uniformly (a.s.). Since

$$2^{-m} E \sum_{j=1}^{\infty} 2^j \sup_{0 \leq t \leq T} |\xi_t^j - \xi_t^{j-1}|^2 \leq 2^{-m} \sum_{1}^{\infty} 2^j \rho_j(T)$$
$$\leq 2^{-m} A \sum_{j=0}^{\infty} \frac{2^{j+1}(CT)^j}{j!},$$

we conclude from (5.1.6) and the dominated convergence theorem that

$$E \sup_{[0,T]} |\xi_t - \xi_t^m|^2 \leq 2^{-m} R, \tag{5.1.7}$$

where $R = A \sum_{j=0}^{\infty} 2^{j+1}(CT)^j/j!$ depends only on K, T, and A. In particular, $E \sup_{0 \leq t \leq T} |\xi_t|^2 < \infty$. By the computation in (5.1.4) with ξ^m replaced by ξ we obtain

$$E \int_0^T |b(s,X,\xi)|^2 \, ds + E \int_0^T |\sigma(s,X,\xi)|^2 \, ds < \infty.$$

Fix $t \in [0,T]$ and observe that

$$E\left[\int_0^t |b(s,X,\xi) - b(s,X,\xi^{m-1})|^2\,ds + \int_0^t |\sigma(s,X,\xi) - \sigma(s,X,\xi^{m-1})|^2\,ds\right]$$
$$\leq K\int_0^t E \sup_{0\leq u\leq s} |\xi_u - \xi_u^{m-1}|^2\,ds$$
$$\leq KTE \sup_{0\leq t\leq T} |\xi_t - \xi_t^{m-1}|^2.$$

Using (5.1.7) again it follows that

$$\int_0^t b(s,X,\xi^{m-1})\,ds \to \int_0^t b(s,X,\xi)\,ds,$$
$$\int_0^t \sigma(s,X,\xi^{m-1})\,dW_s \to \int_0^t \sigma(s,X,\xi)\,dW_s$$

and

$$\xi_t^m \to \xi_t$$

in the L^2 sense as $m \to \infty$. Now, from Equation (5.1.2) (with k replaced by m) we obtain, in the limit,

$$\xi_t = \eta + \int_0^t b(s,X,\xi)\,ds + \int_0^t \sigma(s,X,\xi)\,dW_s \qquad \text{(a.s.)}$$

The process (ξ_t) constructed above is therefore a strong solution in the case $E|\eta|^2 < \infty$.

The assumption $E|\eta|^2 < \infty$ will now be removed. For $N \geq 1$, let $\eta_N = \eta\psi_N$, where $\psi_N(\omega) = I_{[|\eta(\omega)|\leq N]}(\omega)$. Then η_N is $\mathscr{F}_0^{\eta,X,W}$-measurable so that condition (A) holds for this new random vector η_N. Since $E|\eta_N|^2 < \infty$, the above special case furnishes a continuous, $\mathscr{F}_t^{\eta,X,W}$-adapted process (ξ_t^N) which is a strong solution of Equation (5.1.1) with initial condition $\xi_0^N = \eta_N$. For $M > N$,

$$\xi_t^M - \xi_t^N = [\eta_M - \eta_N] + \int_0^t [b(s,X,\xi^M) - b(s,X,\xi^N)]\,ds$$
$$+ \int_0^t [\sigma(s,X,\xi^M) - \sigma(s,X,\xi^N)]\,dW_s.$$

Noting that $(\eta_M - \eta_N)\psi_N = \eta\psi_M\psi_N - \eta\psi_N = 0$, and that ψ_N is $\mathscr{F}_0$-measurable, we have

$$\sup_{0\leq u\leq t} (|\xi_u^M - \xi_u^N|^2\psi_N) \leq 2 \sup_{0\leq u\leq t} \left|\psi_N \int_0^u [b(s,X,\xi^M) - b(s,X,\xi^N)]\,ds\right|^2$$
$$+ 2 \sup_{0\leq u\leq t} \left|\psi_N \int_0^u [\sigma(s,X,\xi^M) - \sigma(s,X,\xi^N)]\,dW_s\right|^2$$
$$\leq 2T \int_0^t \psi_N |b(s,X,\xi^M) - b(s,X,\xi^N)|^2\,ds$$
$$+ 2 \sup_{0\leq u\leq t} \left|\int_0^u \psi_N[\sigma(s,X,\xi^M) - \sigma(s,X,\xi^N)]\,dW_s\right|^2.$$

Hence from condition (B)

$$E \sup_{[0,t]} (|\xi_u^M - \xi_u^N|^2 \psi_N) \leq 2(T + 4)KE \int_0^t \psi_N \sup_{[0,s]} |\xi_u^M - \xi_u^N|^2 \, ds.$$

Writing $R^{M,N}(t) = E \sup_{[0,t]} (|\xi_u^M - \xi_u^N|^2 \psi_N)$, we get

$$R^{M,N}(t) \leq 2(T + 4)K \int_0^t R^{M,N}(s) \, ds.$$

From Proposition 5.1.1,

$$R^{M,N}(t) = 0, \qquad M \geq N, t \in [0,T];$$

that is,

$$E \sup_{[0,T]} (|\xi_t^M - \xi_t^N|^2 \psi_N) = 0 \quad \text{for all } M,N \ (M \geq N).$$

It is now easy to verify that for any positive ε,

$$P\left[\sup_{0 \leq t \leq T} |\xi_t^M - \xi_t^N| > \varepsilon \right] \leq P[|\eta| > N] \to 0$$

as $N,M \to \infty$. Hence $\sup_{0 \leq t \leq T} |\xi_t^M - \xi_t^N| \to 0$ in probability, and there exists a continuous, $\mathscr{F}_t^{\eta,X,W}$-adapted process (ξ_t) such that $\sup_{0 \leq t \leq T} |\xi_t^N - \xi_t| \to 0$ in probability as $N \to \infty$. To verify that (ξ_t) is the desired solution first observe that

$$\int_0^T |b(s,X,\xi)|^2 \, ds + \int_0^T |\sigma(s,X,\xi)|^2 \, ds$$
$$\leq KT + KT \sup_{0 \leq t \leq T} |\xi_t|^2 + K \int_0^T |L(s,X)|^2 \, ds < \infty \qquad \text{(a.s.)}.$$

Condition (B) gives

$$\int_0^T |b(s,X,\xi) - b(s,X,\xi^N)|^2 \, ds + \int_0^T |\sigma(s,X,\xi) - \sigma(s,X,\xi^N)|^2 \, ds$$
$$\leq K \int_0^T \sup_{0 \leq u \leq s} |\xi_u - \xi_u^N|^2 \, ds \leq KT \sup_{0 \leq t \leq T} |\xi_t - \xi_t^N|^2 \to 0$$

in probability as $N \to \infty$. Therefore, for each fixed t in $[0,T]$ we have

$$\int_0^t b(s,X,\xi^N) \, ds \to \int_0^t b(s,X,\xi) \, ds$$

and

$$\int_0^t \sigma(s,X,\xi^N) \, dW_s \to \int_0^t \sigma(s,X,\xi) \, dW_s$$

in probability. Finally, passing to the limit in each term of

$$\xi_t^N = \eta_N + \int_0^t b(s,X,\xi^N) \, ds + \int_0^t \sigma(s,X,\xi^N) \, dW_s \qquad \text{(a.s.)},$$

we obtain

$$\xi_t = \eta + \int_0^t b(s,X,\xi) \, ds + \int_0^t \sigma(s,X,\xi) \, dW_s \qquad \text{(a.s.)}.$$

The existence assertion of the theorem is thus established.

Uniqueness. Suppose ξ and $\tilde{\xi}$ are two strong solutions [with respect to $\{(\Omega,\mathscr{A},P), (\mathscr{F}_t),X,W,\eta,b,\sigma\}$]. Then

$$\xi_t - \tilde{\xi}_t = \int_0^t [b(s,X,\xi) - b(s,X,\tilde{\xi})]\,ds + \int_0^t [\sigma(s,X,\xi) - \sigma(s,X,\tilde{\xi})]\,dW_s.$$

For each $N \geq 1$, define

$$\tau_N = \begin{cases} \inf\{t: |\xi_t|^2 + |\tilde{\xi}_t|^2 \geq N\} & \\ T & \text{if this set is empty.} \end{cases}$$

Then each τ_N is a stopping time with respect to $(\mathscr{F}_t)$, and if we let $I_t^N = I_{[0,\tau_N]}(t)$, then

$$\begin{aligned} \xi_{t\wedge\tau_N} - \tilde{\xi}_{t\wedge\tau_N} &= \int_0^{t\wedge\tau_N} [b(s,X,\xi) - b(s,X,\tilde{\xi})]\,ds \\ &\quad + \int_0^{t\wedge\tau_N} [\sigma(s,X,\xi) - \sigma(s,X,\tilde{\xi})]\,dW_s \\ &= \int_0^t I_{[0,\tau_N]}(s)[b(s,X,\xi) - b(s,X,\tilde{\xi})]\,ds \\ &\quad + \int_0^t I_{[0,\tau_N]}(s)[\sigma(s,X,\xi) - \sigma(s,X,\tilde{\xi})]\,dW_s, \end{aligned}$$

so that

$$\begin{aligned} I_t^N(\xi_t - \tilde{\xi}_t) &= I_t^N(\xi_{t\wedge\tau_N} - \tilde{\xi}_{t\wedge\tau_N}) \\ &= I_t^N \int_0^t I_s^N[b(s,X,\xi) - b(s,X,\tilde{\xi})]\,ds \\ &\quad + I_t^N \int_0^t I_s^N[\sigma(s,X,\xi) - \sigma(s,X,\tilde{\xi})]\,dW_s. \end{aligned}$$

Noting that $I_t^N \cdot I_u^N = I_t^N$ whenever $u \leq t$, we obtain

$$\begin{aligned} E\left(I_t^N \sup_{0\leq u\leq t} |\xi_u - \tilde{\xi}_u|^2\right) &\leq E\left(\sup_{0\leq u\leq t} I_u^N|\xi_u - \tilde{\xi}_u|^2\right) \\ &\leq 2E \sup_{0\leq u\leq t} I_u^N \left|\int_0^u I_s^N[b(s,X,\xi) - b(s,X,\tilde{\xi})]\,ds\right|^2 \\ &\quad + 2E \sup_{0\leq u\leq t} I_u^N \left|\int_0^u I_s^N[\sigma(s,X,\xi) - \sigma(s,X,\tilde{\xi})]\,dW_s\right|^2 \\ &\leq 2TE\int_0^t I_s^N|b(s,X,\xi) - b(s,X,\tilde{\xi})|^2\,ds \\ &\quad + 8E\int_0^t I_s^N|\sigma(s,X,\xi) - \sigma(s,X,\tilde{\xi})|^2\,ds \\ &\leq 2(T+4)K\int_0^t E\left(I_s^N \sup_{0\leq u\leq s} |\xi_u - \tilde{\xi}_u|^2\right)ds. \end{aligned}$$

Writing $\phi(t) = E(I_t^N \sup_{0\leq u\leq t}|\xi_u - \tilde{\xi}_u|^2)$, we have by definition of I_t^N that $\phi(t) < \infty$ and, by the above inequality, $\phi(t) \leq 2(T+4)K\int_0^t \phi(s)\,ds$. By

Proposition 5.1.1 this implies $\phi(t) = 0$ for all t. In particular, for each $N \geq 1$, we have

$$E\left(I_T^N \sup_{0 \leq t \leq T} |\xi_t - \tilde{\xi}|^2\right) = 0$$

and hence

$$I_T^N \sup_{0 \leq t \leq T} |\xi_t - \tilde{\xi}_t|^2 = 0 \qquad \text{(a.s.)}.$$

Denoting this ω set by Λ, we easily see that

$$\left[\omega\colon \sup_{0 \leq t \leq T} (|\xi_t(\omega)|^2 + |\tilde{\xi}_t(\omega)|^2) < N\right] \cap \Lambda \subseteq \left[\omega\colon \sup_{0 \leq t \leq T} |\xi_t(\omega) - \tilde{\xi}_t(\omega)|^2 = 0\right].$$

Hence

$$P\left[\omega\colon \sup_{0 \leq t \leq T} |\xi_t(\omega) - \tilde{\xi}_t(\omega)|^2 > 0\right] \leq P\left[\omega\colon \sup_{0 \leq t \leq T} (|\xi_t(\omega)|^2 + |\tilde{\xi}_t(\omega)|^2) \geq N\right].$$

Since N is arbitrary, making $N \to \infty$, we obtain

$$P\left[\omega\colon \sup_{0 \leq t \leq T} |\xi_t(\omega) - \tilde{\xi}_t(\omega)|^2 > 0\right] = 0$$

and uniqueness is proved. This completes the proof of Theorem 5.1.1. □

Suppose that the causal functionals b and σ satisfy the stronger requirement:

(ii)$_1$. For each $t, b_i(t,\cdot,\cdot)$ and $\sigma^{ij}(t,\cdot,\cdot)$ are $\mathscr{B}_t(D) \times \mathscr{B}_t(C_d)$-measurable $(i = 1, \ldots, d; j = 1, \ldots, n)$.

Then a different version of Theorem 5.1.1. is available.

Theorem 5.1.1a. *Assume all the conditions of Theorem 5.1.1 except that the causal functionals b and σ satisfy* (ii)$_1$ *and conditions* (B) *and* (C) *are replaced by* (B$_1$) *and* (C$_1$). *Then the conclusions of Theorem 5.1.1 hold.*

To prove Theorem 5.1.1a, it is enough to note that the following proposition implies the equivalence of the two sets of conditions (B), (C) and (B$_1$), (C$_1$) if condition (ii)$_1$ is satisfied. The proof of Proposition 5.1.2 is left to the reader.

Proposition 5.1.2. *Let γ be a causal functional satisfying* (ii)$_1$, *$f_i \in C$ and $g_i \in D$ $(i = 1,2)$, such that for each t, $f_1(s) = f_2(s)$ and $g_1(s) = g_2(s)$ for $s \in [0,t]$. Then $\gamma(t,g_1,f_1) = \gamma(t,g_2,f_2)$.*

Let C_d^- be the space of all continuous functions from $[-T,0]$ to $\mathbf{R}^d$ with the usual uniform topology, and let D^- be the space of all right-continuous functions with left-hand limits from $[-T,0]$ to S.

The assumptions of the result just proved can be recast in a somewhat different form as is done, for example, in the work of Fujisaki, Kallianpur, and Kunita [14]. Let $\hat{b} = (\hat{b}_i)$ and $\hat{\sigma} = (\hat{\sigma}^{ij})$ $(i = 1, \ldots, d; j = 1, \ldots, n)$, where $\hat{b}_i$ and $\hat{\sigma}^{ij}$ are real-valued functions on $[0,T] \times D^- \times C_d^-$, measurable with respect to $\mathscr{B}[0,T] \times \mathscr{B}(D^-) \times \mathscr{B}(C_d^-)$. Conditions (B_2) and (C_2) are replaced by the following: there exists a finite measure Γ on $[-T,0]$ such that

(B_2') $|\hat{b}(t,g,f) - \hat{b}(t,g,\tilde{f})|^2 + |\hat{\sigma}(t,g,f) - \hat{\sigma}(t,g,\tilde{f})|^2 \le \int_{-T}^{0} |f(s) - \tilde{f}(s)|^2 \Gamma(ds).$

(C_2') There exists a Borel measurable, real function $L(t,g)$ on $[0,T] \times D^-$ such that

$$|\hat{b}(t,g,f)|^2 + |\hat{\sigma}(t,g,f)|^2 \le K\left[1 + \int_{-T}^{0} |f(s)|^2 \Gamma(ds) + |L(t,g)|^2\right],$$

where K is a positive constant and

$$\int_0^T E|L(t,\pi_t X)|^2\, dt < \infty.$$

Here π_t $(0 \le t \le T)$ is the map from $D \to D^-$ given by

$$(\pi_t g)(s) = \begin{cases} g(s + t) & \text{if } -t \le s \le 0 \\ g(0) & \text{if } -T \le s \le -t. \end{cases}$$

We shall also denote by the same symbol the map, similarly defined, from $C_d \to C_d^-$.

Theorem 5.1.1b. *Let conditions* (A), (B_2'), *and* (C_2') *be satisfied. Then there exists a unique solution* (ξ_t) *to the stochastic differential equation*

$$d\xi_t = \hat{b}(t,\pi_t X,\pi_t \xi)\, dt + \hat{\sigma}(t,\pi_t X,\pi_t \xi)\, dW_t,$$

$$\xi_0 = \eta,$$

such that $\mathscr{F}_t^{\eta,X,W} \subseteq \mathscr{F}_t$.

The proof is identical to that of Theorem 5.1.1 if we note that the functionals b and σ defined by

$$b(t,g,f) \equiv \hat{b}(t,\pi_t g,\pi_t f)$$

and

$$\sigma(t,g,f) \equiv \hat{\sigma}(t,\pi_t g,\pi_t f) \qquad (g \in D,\ f \in C_d)$$

are causal.

Theorem 5.1.1 is of importance in the theory of nonlinear stochastic filtering. An important special case arises when the coefficients $b(t,g,f)$ and $\sigma(t,g,f)$ do not depend on g. Condition (C_2) is then replaced by

(C_2'') $|b(t,f)|^2 + |\sigma(t,f)|^2 \le K\left[1 + \int_0^T |f(s)|^2 \Gamma(ds)\right]$

and since the process (X_t) does not figure in the result, condition (A) simply becomes

(A″) $\mathscr{F}_t^{\eta,W} \subseteq \mathscr{F}_t$ and $\mathscr{F}_t \perp\!\!\!\perp \sigma[W_v - W_u, t \le u < v \le T]$.

Theorem 5.1.2. *Under conditions* (A″), (B$_2$), *and* (C$_2''$), *the stochastic differential equation*

$$d\xi_t = b(t,\xi)\,dt + \sigma(t,\xi)\,dW_t, \qquad \xi_0 = \eta \tag{5.1.8}$$

has a unique solution (ξ_t) *which is continuous and for which*

$$\mathscr{F}_t^{\xi} \subseteq \mathscr{F}_t^{\eta,W}.$$

Let us specialize further and suppose that $b(t,f)$ and $\sigma(t,f)$ depend on f only through its initial value, that is, $b(t,f) = b(t,f(0))$, $\sigma(t,f) = \sigma(t,f(0))$. It is natural to write $b(t,y)$ and $\sigma(t,y)$, where $y \in \mathbf{R}^d$.

Theorem 5.1.3. *The stochastic differential equation*

$$d\xi_t = b(t,\xi_t)\,dt + \sigma(t,\xi_t)\,dW_t, \qquad \xi_0 = \eta \tag{5.1.9}$$

has a unique solution provided the coefficients satisfy the Lipschitz and growth conditions given below. There is a positive constant K such that for all $t \in [0,T]$ *and* $y, \tilde{y} \in \mathbf{R}^d$.

$$\begin{aligned} |b(t,y) - b(t,\tilde{y})|^2 + |\sigma(t,y) - \sigma(t,\tilde{y})^2 &\le K|y - \tilde{y}|^2, \\ |b(t,y)|^2 + |\sigma(t,y)|^2 &\le K[1 + |y|^2]. \end{aligned} \tag{5.1.10}$$

The proof follows immediately from Theorem 5.1.1 if, in (B$_2'$) and (C$_2'$), we take $L = 0$ and Γ to be a point measure which concentrates a mass equal to K at 0.

We shall return to a more detailed study of the process of Theorem 5.1.3 later in this chapter.

5.2 Strong and Weak Solutions

The definition of a solution or strong solution of a stochastic differential equation has been given in the previous section, There is another concept of solution which has proved useful, especially in applications to stochastic control theory. We give the definition with reference to Equation (5.1.8)

Definition 5.2.1. Let causal functionals b and σ be given as above, and let F_η be a distribution function on $\mathbf{R}^d$.

Suppose that on some probability space $(\Omega,\mathscr{A},P)$ we can define an increasing family $(\mathscr{F}_t)$ of sub-σ-fields of $\mathscr{A}$, a d-dimensional random variable

η with the given distribution function F_η and continuous processes $\xi = (\xi_t)$, $W = (W_t)$, such that

(i) $P[\omega: \int_0^T |b(t,\xi(\omega))|\, dt < \infty] = 1$.
(ii) $P[\omega: \int_0^T |\sigma(t,\xi(\omega))|^2\, dt < \infty] = 1$.
(iii) $(W_t,\mathscr{F}_t,P)$ is a Wiener martingale and (ξ_t) is $\mathscr{F}_t$-adapted.
(iv) $\xi_t(\omega) = \eta(\omega) + \int_0^t b(s,\xi(\omega))\, ds + \int_0^t \sigma(s,\xi(\omega))\, dW_s(\omega)$ (a.s.) for all t.

Then the family $(\Omega,\mathscr{A},(\mathscr{F}_t),\eta,(W_t),(\xi_t))$ is called a *weak solution* of Equation (5.1.8). It is convenient to refer to (ξ_t) as the weak solution associated with the model $(\Omega,\mathscr{A},(\mathscr{F}_t),\eta,(W_t))$. Equation (5.1.8) is said to have a *unique* weak solution if, for any two weak solutions (ξ_t^i) associated with $(\Omega^i,\mathscr{A}^i,\mathscr{F}_t^i,\eta^i,(W_t^i))$ $(i = 1,2)$, (ξ_t^1) and (ξ_t^2) induce the same measure in C_d, that is, if

$$P^1(\xi^1)^{-1} = P^2(\xi^2)^{-1}.$$

It is clear from the definition that a weak solution is essentially the measure $\mu = P\xi^{-1}$ induced by (ξ_t). The following remarks clarify the notions of weak and strong solutions:

1. A strong solution obviously gives a weak solution. Under condition (5.1.10), the strong solution of Equation (5.1.8) constructed in Section 5.1 always satisfies

$$\mathscr{F}_t^\xi \subseteq \mathscr{F}_t^{\eta,W}.$$

In this sense the solution ξ can be regarded as a nonanticipative functional of the given Wiener process W whenever the initial condition $\eta = x$ is non random.

2. A weak solution ξ is nonanticipative in the following sense: for every t,

$$\mathscr{F}_t^\xi \perp\!\!\!\perp \sigma(W_v - W_u,\, v > u \geq t).$$

This follows from the fact that $(W_t,\mathscr{F}_t)$ is a Wiener martingale and $\mathscr{F}_t^\xi \subseteq \mathscr{F}_t$.

3. Let (ξ_t) be a weak solution of Equation (5.1.8) associated with $(\Omega,\mathscr{A},P,W_t,\mathscr{F}_t)$. Then the process

$$M_t \equiv \xi_t - \eta - \int_0^t b(s,\xi)\, ds$$

has a continuous version which is a d-dimensional local martingale relative to $(\mathscr{F}_t)$. We shall now assume that the $d \times d$ matrix $a(t,f) = \sigma(t,f)\sigma^*(t,f)$ is positive definite for all (t,f) (σ^* being the transpose of σ).

Let $\tau_N \uparrow T$ (a.s) be a sequence of stopping times with respect to $\mathscr{F}_t^{\eta,\xi}$ such that $M_t^N \equiv M_{t\wedge\tau_N}$ is a continuous L^2-martingale. Its quadratic variation process

$$\langle M^N\rangle_t = \langle M\rangle_{t\wedge\tau_N} = \int_0^{t\wedge\tau_N} \sigma_s\sigma_s^*\, ds = \int_0^t a_s I_{[0,\tau_N]}(s)\, ds.$$

Define

$$\tilde{W}_t^N = \int_0^t f_s\, dM_s^N,$$

where

$$f_s(\omega) = a^{-\frac{1}{2}}(s,\xi(\omega)).$$

(Here $a^{\frac{1}{2}}$ denotes the unique positive definite square root of a.) The ith component of $\tilde{W}_t^N$ is given by

$$\tilde{W}_t^{N,i} = \sum_k \int_0^t f_s^{ik}\, dM_s^{N,k}.$$

It is easy to verify that $\tilde{W}_t^N$ is a d-dimensional L^2-martingale. Furthermore,

$$\langle \tilde{W}^{N,i}, \tilde{W}^{N,j} \rangle_t = \sum_{k,1} \int_0^{t \wedge \tau_N} f_s^{ik} f_s^{j1} a_s^{k,1}\, ds.$$

Since

$$f_s^{ik}(\omega) = (a_s^{-\frac{1}{2}}(\omega))^{ik},$$

$(a^{-\frac{1}{2}})^{ik}$ being the (i,k)th element of $a^{-\frac{1}{2}}$ it follows that (a.s.)

$$\langle \tilde{W}^{N,i}, \tilde{W}^{N,j} \rangle_t = (t \wedge \tau_N)\delta_{ij} \quad \text{for all } t.$$

Hence

$$\langle \tilde{W}^N \rangle_t = (t \wedge \tau_N) \cdot I,$$

I denoting the identity matrix. Thus, the stochastic integral

$$\int_0^t f_s\, dM_s \quad \text{or} \quad \int_0^t a_s^{-\frac{1}{2}}\, dM_s$$

exists and $\tilde{W}_t$ defined by $\int_0^t a_s^{-\frac{1}{2}}\, dM_s$ is a d-dimensional Wiener process which, moreover, is $(\mathscr{F}_t^{\eta,\xi})$-adapted. We then have

$$M_t = \int_0^t a_s^{\frac{1}{2}}\, d\tilde{W}_s.$$

In the case when $n = d$, the above argument is valid with a_s replaced by σ_s, and we have the following result.

Proposition 5.2.1. *Let $(\Omega,\mathscr{A},(\mathscr{F}_t),\eta,(W_t),(\xi_t))$ be a weak solution of Equation (5.1.8). Assume further that $\sigma(s,x)$ is positive definite for every $(s,x) \in [0,T] \times C_d$. Then*

$$\mathscr{F}_t^W \subseteq \mathscr{F}_t^{\eta,\xi}$$

for each t.

Proof. The Wiener process $\tilde{W}_t$ (see 3. above) is $\mathscr{F}_t^{\eta,\xi}$ measurable and

$$\tilde{W}_t = \int_0^t \sigma_s^{-1}\, dM_s.$$

The stochastic integral on the right side equals W_t since, by hypothesis,

$$M_t = \int_0^t \sigma_s\, dW_s.$$

Hence $P[W_t = \tilde{W}_t \text{ for all } t] = 1$ and the conclusion follows. □

5.3 Linear Stochastic Differential Equations

The stochastic differential equation $d\xi_t = b(t,\xi_t)\,dt + \sigma(t,\xi_t)\,dW_t$ with initial condition ξ_0 is called *linear* if the coefficients $b(t,y)$ and $\sigma(t,y)$ are linear in y, that is,

$$b(t,y) = A_t + B_t y, \qquad \sigma(t,y) = C_t + D_t y,$$

where A_t, B_t, C_t, and D_t are nonrandom continuous functions of t. We consider only the one-dimensional case. The linear equation has a unique solution since the conditions (5.1.10) are obviously satisfied. In this section we shall solve the equation

$$d\xi_t = (A_t + B_t\xi_t)\,dt + (C_t + D_t\xi_t)\,dW_t \tag{5.3.1}$$

with initial value ξ_0 ($\xi_0 \perp\!\!\!\perp \mathscr{F}_T^W$). This is done in several steps.

EXAMPLE 5.3.1. Let $D_t \equiv 0$. It is instructive to work this out by the method of successive approximations. Let $\xi_t^0 = \xi_0$ for all t and

$$\xi_t^n = \xi_0 + \int_0^t (A_s + B_s\xi_s^{n-1})\,ds + \int_0^t C_s\,dW_s \qquad (n \geq 1).$$

For convenience, set

$$\alpha_t = \int_0^t A_s\,ds, \qquad \beta_s = \int_0^t B_s\,ds, \qquad J_t = \int_0^t C_s\,dW_s.$$

We shall also define the Volterra operator L on $L^2[0,T]$ by

$$(Lh)(t) = \int_0^t B_s h_s\,ds.$$

Then

$$\begin{aligned}
\xi_t^1 &= \xi_0(1 + \beta_t) + \alpha_t + J_t \quad \text{and for any } n \geq 1,\\
\xi_t^{n+1} &= \xi_0[1 + \beta_t + \cdots + L^n\beta_t] + [\alpha_t + L\alpha_t + \cdots + L^n\alpha_t]\\
&\quad + \int_0^t \left[1 + (\beta_t - \beta_s) + \frac{1}{2!}(\beta_t - \beta_s)^2 + \cdots + \frac{1}{n!}(\beta_t - \beta_s)^n\right] C_s\,dW_s.
\end{aligned} \tag{5.3.2}$$

Since L is a Volterra operator,

$$\alpha_t + L\alpha_t + \cdots + L^n\alpha_t \to (I - L)^{-1}\alpha_t.$$

It is easy to verify that

$$(I - L)^{-1}\alpha_t = e^{\int_0^t B_s\,ds} \int_0^t e^{-\int_0^s B_u\,du} A_s\,ds. \tag{5.3.3}$$

In a similar fashion

$$[1 + \beta_t + \cdots + L^n\beta_t] \to 1 + (I - L)^{-1}\beta_t = 1 + e^{\beta_t}\int_0^t e^{-\beta_s} B_s\,ds = e^{\beta_t}. \tag{5.3.4}$$

For $s < t$, write $L_n(t,s)$ for the series in the integrand of the stochastic integral on the right-hand side of Equation (5.3.2). Then

$$E\left[\int_0^t C_s L_n(t,s)\,dW_s - \int_0^t C_s e^{\beta_t - \beta_s}\,dW_s\right]^2 \to 0 \quad \text{as } n \to \infty.$$

Hence the unique solution ξ_t is given by

$$\xi_t = e^{\int_0^t B_s\,ds}\left[\xi_0 + \int_0^t e^{-\int_0^s B_u\,du} A_s\,ds + \int_0^t e^{-\int_0^s B_u\,du} C_s\,dW_s\right]. \tag{5.3.5}$$

Remark 5.3.1. It will be seen that we obtain Equation (5.3.5) if we treat Equation (5.3.1) with $D_t = 0$ *formally* as an ordinary differential equation

$$\frac{d\xi_t}{dt} = A_t + B_t\xi_t + C_t\frac{dW_t}{dt}$$

and regard the white noise dW_t/dt as a known function of t. Its formal solution can easily be written down and coincides with Equation (5.3.5) if we identify the integral

$$\int_0^t e^{-\int_0^s B_u\,du} C_s\left(\frac{dW_s}{ds}\right) ds$$

as the stochastic integral $\int_0^t (-)\,dW_s$. This method of guessing at the solution of a stochastic differential equation by solving (formally) an ordinary differential equation involving white noise breaks down when $D_t \neq 0$ even in the simplest case, as is shown by the following well-known example. Consider

$$dX_t = X_t\,dW_t. \tag{5.3.6}$$

If we solve $X_t' = X_t W_t'$ (primes indicating differentiation), then we get

$$X_t = X_0 e^{W_t}. \tag{5.3.7}$$

However, the Ito differential of Equation (5.3.7) is

$$dX_t = X_t[dW_t + \tfrac{1}{2}dt]$$

which is *not* Equation (5.3.6).

Since Equation (5.3.6) has a unique solution, it is enough to find a process which satisfies (5.3.6). Let us write as a trial solution

$$X_t = X_0 e^{W_t + f(t)},$$

where $f(t)$ is a differentiable nonrandom function of t to be chosen suitably. The Ito differential is

$$dX_t = X_t\,dW_t + \tfrac{1}{2}X_t\,dt + X_t f'(t)\,dt = X_t\,dW_t$$

if we take $f(t) = -\tfrac{1}{2}t$. Hence the solution is

$$X_t = X_0 e^{W_t - \frac{1}{2}t}.$$

An obvious modification of this example gives

EXAMPLE 5.3.2. The unique solution of $d\eta_t' = D_t\eta_t'\,dW_t$ with initial value η_0' is given by

$$\eta_t' = \eta_0' \exp\left[\int_0^t D_s\,dW_s - \tfrac{1}{2}\int_0^t D_s^2\,ds\right]. \tag{5.3.8}$$

EXAMPLE 5.3.3. Let us now consider

$$d\eta_t = B_t\eta_t\,dt + D_t\eta_t\,dW_t \tag{5.3.9}$$

with initial value η_0.

Let us seek a solution of the form $\eta_t = \eta_t'\eta_t''$, where η_t' is the solution (5.3.8) and η_t'' has the Ito differential

$$d\eta_t'' = b_t''\,dt + \sigma_t''\,dW_t \quad \text{and initial value } \eta_0''.$$

Then

$$\eta_0 = \eta_0'\eta_0''.$$

Since the initial value η_0 is given, we are at liberty to choose $\eta_0' = \eta_0$ and $\eta_0'' = 1$ (a.s.). Writing $d\eta_t' = b_t'\,dt + \sigma_t'\,dW_t$, we have $b_t' \equiv 0$ and $\sigma_t' = D_t\eta_t'$. By the Ito formula it is easy to verify that

$$\begin{aligned} d\eta_t &= d(\eta_t'\eta_t'') = \eta_t' d\eta_t'' + \eta_t''\,d\eta_t' + \sigma_t'\sigma_t''\,dt \\ &= \eta_t'\,d\eta_t'' + \eta_t''\,D_t\eta_t'\,dW_t + \sigma_t'\sigma_t''\,dt \\ &= D_t\eta_t\,dW_t + (\eta_t'\,d\eta_t'' + \sigma_t'\sigma_t''\,dt). \end{aligned} \tag{5.3.10}$$

Let us choose η_t'' such that

$$\eta_t'\,d\eta_t'' + \sigma_t'\sigma_t''\,dt = B_t\eta_t'\eta_t''\,dt.$$

For this, take $\sigma_t'' \equiv 0$ and $b_t'' = B_t\eta_t''$. Now Equation (5.3.10) becomes

$$d\eta_t = B_t\eta_t\,dt + D_t\eta_t\,dW_t,$$

where $\eta_t = \eta_t'\eta_t''$, and (since $\eta_0'' = 1$) $\eta_t'' = e^{\int_0^t B_s\,ds}$ by Example 5.3.1. Hence using the expression (5.3.8), we obtain the solution of (5.3.9) to be

$$\eta_t = \eta_0 \exp\left[\int_0^t (B_s - \tfrac{1}{2}D_s^2)\,ds + \int_0^t D_s\,dW_s\right]. \tag{5.3.11}$$

EXAMPLE 5.3.4. We now consider the most general linear equation

$$d\xi_t = (A_t + B_t\xi_t)\,dt + (C_t + D_t\xi_t)\,dW_t \tag{5.3.12}$$

with prescribed initial value ξ_0.

Let η_t be the process of Example 5.3.3 with initial value $\eta_0 = 1$ (a.s.). We look for a process ζ_t satisfying $d\zeta_t = b_t\,dt + \sigma_t\,dW_t$, $\zeta_0 = \xi_0$, such that $\xi_t = \eta_t\zeta_t$ is a solution of (5.3.12). Proceeding as in the preceding example and using Equation (5.3.9),

$$\begin{aligned} d\xi_t = d(\eta_t \zeta_t) &= \zeta_t \, d\eta_t + \eta_t \, d\zeta_t + (D_t \eta_t)\sigma_t \, dt \\ &= B_t \xi_t \, dt + D_t \xi_t \, dW_t + \eta_t(b_t \, dt + \sigma_t \, dW_t) + D_t \eta_t \sigma_t \, dt. \end{aligned}$$

Choose $b_t = (A_t - D_t C_t)\eta_t^{-1}$ and $\sigma_t = C_t \eta_t^{-1}$. Note that since we have specified $\eta_0 = 1$, Equation (5.3.11) implies that

$$P[\eta_t > 0 \text{ for all } t] = 1.$$

Hence the solution of (5.3.12) turns out to be

$$\begin{aligned} \xi_t = \exp&\left[\int_0^t (B_s - \tfrac{1}{2}D_s^2)\, ds + \int_0^t D_s \, dW_s\right] \\ &\times \left\{\xi_0 + \int_0^t \exp\left[-\int_0^s (B_u - \tfrac{1}{2}D_u^2)\, du - \int_0^s D_u \, dW_u\right](A_s - D_s C_s)\, ds\right. \\ &\left. + \int_0^t \exp\left[-\int_0^s (B_u - \tfrac{1}{2}D_u^2)\, du - \int_0^s D_u \, dW_u\right] C_s \, dW_s\right\}. \end{aligned} \tag{5.3.13}$$

Remark 5.3.2. (i) It will follow from a general result which will be proved in the next section that (ξ_t) given by (5.3.13) is a Markov process. This fact can also be directly verified from the explicit expression for (ξ_t) given in (5.3.13).

(ii) If $D_t \equiv 0$ and ξ_0 is constant, (ξ_t) is obviously a Gaussian process. Hence, in this case, the solution of (5.3.12) is a Gaussian, Markov process.

The remainder of this chapter is devoted to studying more deeply, the properties of the solution of the stochastic differential equation (5.1.9) (under the Lipschitz and growth conditions on the coefficients stated above). For this, we first need the definition of a Markov process and some general concepts associated with it.

5.4 Markov Processes

An $\mathbf{R}^d$-valued process (X_t) defined on $(\Omega,\mathscr{A},P)$ and adapted to the family $(\mathscr{F}_t)$ is said to be a *Markov process* with respect to $(\mathscr{F}_t)$ [or Markovian with respect to $(\mathscr{F}_t)$] if the following property is satisfied. For all s and t $(t \geq s)$,

(M_1) $E[f(X_t)|\mathscr{F}_s] = E[f(X_t)|X_s]$ (a.s.) for every bounded real-valued Borel function f on $\mathbf{R}^d$.

Note that (M_1) may be replaced by the condition

(M_1') $P[X_t \in A|\mathscr{F}_s] = P[X_t \in A|X_s]$ (a.s.) for $A \in \mathscr{B}(\mathbf{R}^d)$.

Proposition 5.4.1. *The following condition is equivalent to* (M_1):

(M_2) $E(Y|\mathscr{F}_s) = E(Y|X_s)$ (a.s.) *for every bounded,* $\sigma[X_u, u \geq s]$*-measurable, real random variable* Y.

PROOF. It is obvious that (M_2) implies (M_1). To show the converse, let $s \leq t_1 < \cdots < t_n$ and $f_1, \ldots, f_n$ be bounded Borel functions. We first show by induction that

$$E[f_1(X_{t_1}) \cdots f_n(X_{t_n})|\mathscr{F}_s] = E[f_1(X_{t_1}) \cdots f_n(X_{t_n})|X_s] \qquad \text{(a.s.)}.$$

For $n = 1$, the assertion is (M_1) itself. Suppose it to be true for $n = m - 1$. Then the left-hand side with $n = m$ equals

$$\begin{aligned} &E[E\{f_1(X_{t_1}) \cdots f_m(X_{t_m})|\mathscr{F}_{t_{m-1}}\}|\mathscr{F}_s] \\ &\qquad = E[f_1(X_{t_1}) \cdots f_{m-1}(X_{t_{m-1}})E\{f_m(X_{t_m})|X_{t_{m-1}}\}|\mathscr{F}_s] \\ &\qquad = E[f_1(X_{t_1}) \cdots f_{m-1}(X_{t_{m-1}})\tilde{f}(X_{t_{m-1}})|\mathscr{F}_s], \end{aligned}$$

where $\tilde{f}$ is a Borel function such that $\tilde{f}(X_{t_{m-1}}) = E(f_m(X_{t_m})|X_{t_{m-1}})$. Hence setting $\tilde{f} \cdot f_{m-1} = f'_{m-1}$, the above

$$\begin{aligned} &= E[f_1(X_{t_1}) \cdots f_{m-2}(X_{t_{m-2}})f'_{m-1}(X_{t_{m-1}})|\mathscr{F}_s] \\ &= E[f_1(X_{t_1}) \cdots f_{m-2}(X_{t_{m-2}})f'_{m-1}(X_{t_{m-1}})|X_s] \end{aligned}$$

by the inductive hypothesis. But this

$$\begin{aligned} &= E[f_1(X_{t_1}) \cdots f_{m-2}(X_{t_{m-2}})f_{m-1}(X_{t_{m-1}})E\{f_m(X_{t_m})|X_{t_{m-1}}\}|X_s] \\ &= E[f_1(X_{t_1}) \cdots f_m(X_{t_m})|X_s]. \end{aligned}$$

Let $\mathscr{S}$ denote the class of all bounded, $\sigma[X_u, u \geq s]$-measurable sets A of Ω for which (M_2) holds for $Y = I_A$. By what has been just shown above, $\mathscr{S}$ contains all sets $[X_{t_1} \in B_1, \ldots, X_{t_n} \in B_n]$, where $t_i \geq s$ and $B_i \in \mathscr{B}(\mathbf{R}^d)$. Hence $\mathscr{S}$ contains the field generated by $\{X_u, u \geq s\}$. Since $\mathscr{S}$ is easily seen to be a monotone class, it follows that $\mathscr{S}$ contains $\sigma[X_u, u \geq s]$. The assertion now follows immediately. □

We say that (X_t) is a Markov process if it is a Markov process with respect to $(\mathscr{F}_t^X)$.

Proposition 5.4.2. (X_t) *is a Markov process if and only if condition* (M_3) *holds*:

(M_3) *For* $0 \leq s_1 \leq s_2 \leq \cdots \leq s_n \leq s \leq t$ *and for every real, bounded Borel function* f *on* $\mathbf{R}^d$,

$$E[f(X_t)|X_{s_1}, X_{s_2}, \ldots, X_{s_n}, X_s] = E[f(X_t)|X_s] \qquad \text{(a.s.)}.$$

PROOF. $(\mathrm{M}_1) \Rightarrow (\mathrm{M}_3)$. Since $\sigma(X_{s_1}, \ldots, X_{s_n}, X_s) \subset \mathscr{F}_s^X$,

$$\begin{aligned} E[f(X_t)|X_{s_1}, \ldots, X_{s_n}, X_s] &= E[E\{f(X_t)|\mathscr{F}_s^X\}|X_{s_1}, \ldots, X_{s_n}, X_s] \\ &= E[E\{f(X_t)|X_s\}|X_{s_1}, \ldots, X_{s_n}, X_s] \\ &= E[f(X_t)|X_s] \qquad \text{(a.s.)}. \end{aligned}$$

(M_3) implies (M_1): Let h be a bounded, $\mathscr{F}_s^X$-measurable random variable. Suppose we have for each such h,

$$E[f(X_t)h] = E[E[f(X_t)|X_s]h]. \tag{5.4.1}$$

Then for $A \in \mathscr{F}_s^X$,

$$\int_A E[f(X_t)|\mathscr{F}_s^X]\,dP = \int_A f(X_t)\,dP = E[f(X_t)I_A].$$

With $h = I_A$ in (5.4.1) the right-hand side becomes

$$E\{E[f(X_t)|X_s]I_A\} = \int_A E[f(X_t)|X_s]\,dP.$$

Hence

$$E[f(X_t)|\mathscr{F}_s^X] = E[f(X_t)|X_s] \qquad \text{(a.s.)}.$$

Thus it is enough to show (5.4.1).

Let $h = I_A$, where $A \in \sigma(X_{s_1},X_{s_2},\ldots,X_{s_n},X_s)$. Then

$$\begin{aligned} E[f(X_t)I_A] &= \int_A E[f(X_t)|X_{s_1},\ldots,X_{s_n},X_s]\,dP \\ &= \int_A E[f(X_t)|X_s]\,dP \quad \text{by } (M_3) \\ &= E\{E[f(X_t)|X_s]I_A\}, \quad \text{which is Equation (5.4.1).} \end{aligned}$$

Since $s_1 < s_2 < \cdots < s_n$ ($\leq s$) is an arbitrary choice of indices, it follows that (5.4.1) holds for $h = I_A$, $A \in \mathscr{F}_s^X$. Hence (5.4.1) holds for simple, $\mathscr{F}_s^X$-measurable functions and, by monotone convergence, for any, bounded, nonnegative, $\mathscr{F}_s^X$-measurable function h and thus for any bounded, $\mathscr{F}_s^X$-measurable function h. □

In the above definition of a Markov process we may take the state space (that is, the space from which X_t takes values) to be a complete, separable metric space S. We then take $\mathscr{B}(S)$ to be the topological Borel σ-field of S.

Let (X_t) be a Markov process given on a complete probability space $(\Omega,\mathscr{A},P)$. Since S, the state space of X_t, is a separable, complete metric space, a result of Doob (Ref. 8, Theorem 9.4, Chapter II, p. 29) on the existence of wide-sense conditional distributions applies in this case, and it follows that there exists a function $P(s,x,t,A)$ ($s < t$, $x \in S$, $A \in \mathscr{B}(S)$) with the following properties:

(a) For all (s,x,t), $P(s,x,t,\cdot)$ is a probability measure on $\mathscr{B}(S)$.
(b) For each (s,t,A), $P(s,\cdot,t,A)$ is $\mathscr{B}(S)$-measurable.
(c) $P[X_t \in A|\mathscr{F}_s^X] = P(s,X_s,t,A)$ (a.s.).
(d) The function $P(s,x,t,A)$ satisfying the properties (a), (b), and (c) is called the *transition probability function* of the Markov process if it further satisfies the Chapman-Kolmogorov equation

$$P(s,x,t,A) = \int_S P(s,x,u,dy)P(u,y,t,A)$$

for all $x \in S$, $A \in \mathscr{B}(S)$, and (s,u,t) such that $s < u < t$.

Markov Property of Solutions

Let us now return to the equation

$$d\xi_t = b(t,\xi_t)\,dt + \sigma(t,\xi_t)\,dW_t, \qquad \xi_0 = \eta. \tag{5.4.2}$$

Denote by (ξ_t) the unique solution of (5.4.2). For $s \geq 0$, let η_s be a random variable such that

$$\eta_s \perp\!\!\!\perp \sigma[W_v - W_u, v \geq u \geq s], \qquad (\eta_0 = \eta)$$

Write $\mathscr{F}_s^t = \mathscr{F}^{\eta_s} \vee \mathscr{F}_z^{W,t}$ $(t > s)$, where $\mathscr{F}_s^{W,t} = \sigma\{W_v - W_u, s \leq u \leq v \leq t\} \vee$ {all P-null sets}. If we take $\eta_s = x$ (a.s.), where $x \in \mathbf{R}^d$, then $\mathscr{F}_s^t = \mathscr{F}_s^{W,t}$. (Despite the inconsistency we shall adhere to the notation $\mathscr{F}_t$ for $\mathscr{F}_0^t$. It is often convenient to write $\xi(t)$ instead of ξ_t.)

Lemma 5.4.1. *Suppose $f: \mathbf{R}^d \times \Omega \to \mathbf{R}$ is a bounded $\mathscr{B}(\mathbf{R}^d) \times \mathscr{F}_s^{W,t}$-measurable function. If Z is a d-dimensional, $\mathscr{F}_s$-measurable random variable, then*

$$E[f(Z,\omega)|\mathscr{F}_s] = g(Z) \qquad \text{(a.s.)}, \tag{5.4.3}$$

where $g(x) = E[f(x,\omega)]$.

Proof. Observe first, that for each fixed x, the random variable $f(x,\cdot) \perp\!\!\!\perp \mathscr{F}_s$. Let

$$\mathscr{S} = \{M: M \in \mathscr{B}(\mathbf{R}^d) \times \mathscr{F}_s^{W,t} \text{ and } f = I_M \text{ satisfies (5.4.3)}\}.$$

If $M = A \times B$, $A \in \mathscr{B}(\mathbf{R}^d)$, $B \in \mathscr{F}_s^{W,t}$,

$$\begin{aligned} E[I_M(Z,\omega)|\mathscr{F}_s] &= I_A(Z)E\{I_B(\omega)|F_s\} = I_A(Z)E(I_B) \\ &= g(Z) \quad \text{with } g(x) = E[I_M(x,\omega)]. \end{aligned}$$

Hence (5.4.3) holds for $f = I_M$, where M is a finite union of disjoint sets

$A_i \times B_i$ $[A_i \in \mathscr{B}(\mathbf{R}^d), B_i \in \mathscr{F}_s^{W,t}]$, that is, $\mathscr{S}$ contains all such sets M which form a field generating the σ-field $\mathscr{B}(\mathbf{R}^d) \times \mathscr{F}_s^{W,t}$. Next let M_n be a monotone sequence with $M_n \in \mathscr{S}$ and limit M. Then

$$E[I_M(Z,\omega)|\mathscr{F}_s] = \lim_{n\to\infty} E[I_{M_n}(Z,\omega)|\mathscr{F}_s] = \lim_{n\to\infty} g_n(Z),$$

where

$$g_n(x) = \int_\Omega I_{M_n}(x,\omega)P(d\omega) \to \int_\Omega I_M(x,\omega)P(d\omega) = g(x).$$

Thus $\lim_{n\to\infty} g_n(Z) = g(Z)$. This shows $\mathscr{S}$ is a monotone class and so $\mathscr{S} = \mathscr{B}(\mathbf{R}^d) \times \mathscr{F}_s^{W,t}$. The equality (5.4.3) holds with f a simple function and hence for any bounded $\mathscr{B}(\mathbf{R}^d) \times \mathscr{F}_s^{W,t}$-measurable f. □

Corollary 5.4.1 (*to Lemma 5.4.1*). *Suppose $f: \mathbf{R}^d \times \Omega \to \mathbf{R}$ is a nonnegative $\mathscr{B}(\mathbf{R}^d) \times \mathscr{F}_s^{W,T}$-measurable function and Z is an $\mathbf{R}^d$-valued, $\mathscr{F}_s$-measurable random vector. Then*

$$E(f(Z(\omega),\omega))) = \int_{\mathbf{R}^d} E(f(x,\omega))PZ^{-1}(dx).$$

PROOF. Let $f_m = f \cdot I_{[f \leq m]}$ and apply Lemma 5.4.1 to each f_m. Setting $g_m(x) = E(f_m(x,\omega))$, we obtain

$$\begin{aligned} Ef_m(Z(\omega),\omega) &= E(E(f_m(Z(\omega),\omega)|\mathscr{F}_s)) \\ &= Eg_m(Z(\omega)) \\ &= \int_{\mathbf{R}^d} g_m(x)PZ^{-1}(dx) \\ &= \int_{\mathbf{R}^d} (Ef_m(x,\omega))PZ^{-1}(dx). \end{aligned}$$

The monotone convergence theorem applied to each side now yields the above assertion. □

Let $\xi_{s,x}(t)$ be the unique solution of the equation

$$\begin{aligned} d\xi_t &= b(t,\xi_t)\,dt + \sigma(t,\xi_t)\,dW_t \qquad (s \leq t \leq T) \\ \xi_s &= x \qquad \text{(a.s.)}. \end{aligned} \tag{5.4.4}$$

It is assumed that the coefficients b and σ satisfy the conditions (5.1.10) of Theorem (5.1.3). The existence of a unique solution of (5.4.4) is established in the same manner as for Equation (5.4.2).

It is necessary in what follows to know that the stochastic differential Equation (5.4.4) has a unique continuous solution $\xi_{s,x}(t,\omega)$ which, for each fixed t $(t \geq s)$, is $\mathscr{B}(\mathbf{R}^d) \times \mathscr{F}_s^{W,t}$-measurable as a function of (x,ω).

Lemma 5.4.2. *There exists a constant C depending only on K and T such that for any x and y in $\mathbf{R}^d$, $s \in [0,T]$,*

$$E\left(\sup_{s \leq t \leq T} |\xi_{s,x}(t) - \xi_{s,y}(t)|^2\right) \leq C|x - y|^2,$$

where $\xi_{s,x}$ and $\xi_{s,y}$ are two continuous solutions of (5.4.4) with initial values x and y, respectively.

PROOF. Fix x, y, and s. For $t \in [s,T]$, define

$$\phi(t) = E \sup_{s \leq u \leq t} |\xi_{s,x}(u) - \xi_{s,y}(u)|^2.$$

Then

$$\begin{aligned} \phi(t) &\leq 3|x - y|^2 + 3E \sup_{s \leq u \leq t} \left| \int_s^u [b(v,\xi_{s,x}(v)) - b(v,\xi_{s,y}(v))]\,dv \right|^2 \\ &\quad + 3E \sup_{s \leq u \leq t} \left| \int_s^u [\sigma(v,\xi_{s,x}(v)) - \sigma(v,\xi_{s,y}(v))]\,dW_v \right|^2 \\ &\leq 3|x - y|^2 + 3(t - s)E \int_s^t |b(v,\xi_{s,x}(v)) - b(v,\xi_{s,y}(v))|^2\,dv \\ &\quad + 12E \int_s^t |\sigma(v,\xi_{s,x}(v)) - \sigma(v,\xi_{s,y}(v))|^2\,dv \\ &\leq 3|x - y|^2 + 3(T + 4)K \int_s^t \phi(v)\,dv. \end{aligned}$$

Notice that ϕ is measurable on $[s,T]$ because the two processes $\xi_{s,x}$ and $\xi_{s,y}$ are continuous and that the use of the martingale inequality made in the above estimate is valid because the initial conditions are square-integrable (cf. proof of Theorem 5.1.1). From Proposition 5.1.1, $\phi(t) \leq 3|x - y|^2 e^{H(t-s)}$, where $H = 3(T + 4)K$. Taking $C = 3e^{HT}$, we have

$$\phi(T) \leq C|x - y|^2. \qquad \square$$

Lemma 5.4.3. *There is a function $\xi_s(x,t,\omega)$ defined on $\mathbf{R}^d \times [s,T] \times \Omega$ and taking values in $\mathbf{R}^d$ with the following properties:*

(i) *For each $x \in \mathbf{R}^d$, the process $\xi_s(x,t,\omega)$ is a continuous solution of (5.4.4).*
(ii) *For each $t \in [s,T]$, the restriction of $\xi_s(x,u,\omega)$ to $\mathbf{R}^d \times [s,t] \times \Omega$ is $\mathscr{B}(\mathbf{R}^d) \times \mathscr{B}[s,t] \times \mathscr{F}_s^{W,t}$-measurable.*

PROOF. For each $a_m^k = (k_1 2^{-m}, \ldots, k_d 2^{-m})$, a point in $\mathbf{R}^d$ with dyadic rational coordinates, choose a process $\xi_{s,a_m^k}(t,\omega)$ which is a solution of (5.4.4) with initial condition a_m^k and, in addition, is continuous for all ω. Hence ξ_{s,a_m^k} is progressively measurable, so that $\xi_{s,a_m^k}(u,\omega)$ restricted to $[s,t] \times \Omega$ is $\mathscr{B}[s,t] \times \mathscr{F}_s^{W,t}$-measurable. For $x \in \mathbf{R}^d$, let $a_m^k(x)$ be such that $k_j 2^{-m} \leq x_j < (k_j + 1)2^{-m}$ $(1 \leq j \leq d)$, where $x = (x_1, \ldots, x_d)$. Define

$$\xi_s^m(x,t,\omega) = \xi_{s,a_m^k(x)}(t,\omega).$$

It is easy, then, to see that for each x and ω, $\xi_s^m(x,t,\omega)$ is continuous in t and that the function $(x,u,\omega) \to \xi_s^m(x,u,\omega)$ from $\mathbf{R}^d \times [s,t] \times \Omega$ into $\mathbf{R}^d$ is $\mathscr{B}(\mathbf{R}^d) \times \mathscr{B}[s,t] \times \mathscr{F}_s^{W,t}$-measurable. Set

$$\xi_s(x,t,\omega) = \limsup_{m \to \infty} \xi_s^m(x,t,\omega). \tag{5.4.5}$$

Let $\xi_{s,x}(t)$ be the unique continuous solution of (5.4.4) corresponding to the initial condition x. Now from Lemma 5.4.2,

$$E\left(\sup_{s \leq t \leq T} |\xi_{s,x}(t) - \xi_{s,a_m^k(x)}(t)|^2 \right) \leq C|x - a_m^k(x)|^2$$
$$\leq Cd2^{-2m}.$$

From the Borel-Cantelli lemma it follows that

$$P\left[\omega\colon \sup_{s \leq t \leq T} |\xi_{s,x}(t,\omega) - \xi_{s,a_m^k(x)}(t,\omega)| > \frac{1}{m} \text{ i.o.} \right] = 0.$$

(i.o. stands for "infinitely often"). Hence for each x,

$$P[\omega\colon \xi_s^m(x,t,\omega) \to \xi_{s,x}(t,\omega) \text{ uniformly on } [s,T]] = 1.$$

By definition (5.4.5) we conclude that for almost all ω, $\xi_s(x,t,\omega) = \xi_{s,x}(t,\omega)$ for every $x \in \mathbf{R}^d$ and $t \in [s,T]$. Thus for every $x \in \mathbf{R}^d$, $\xi_s(x,t,\omega)$ is a solution of (5.4.4) with initial condition x which is a continuous function of t for almost all ω. Moreover, from the definition (5.4.5) it follows that the process $\xi_s(x,t,\omega)$ has the measurability property asserted in (ii). $\square$

The following lemma is a consequence of Lemma 5.4.3.

Lemma 5.4.4. *The stochastic differential equation (5.4.4) has a unique solution* $\xi_{s,x}(t)$ $(t \geq s)$ *which is* $\mathscr{B}(\mathbf{R}^d) \times \mathscr{F}_s^{W,t}$*-measurable as a function of* (x,ω).

Lemma 5.4.5. *Let* $\phi\colon \mathbf{R}^d \to \mathbf{R}$ *be a bounded, Borel function. Then* $f(x,\omega) = \phi[\xi_{s,x}(t,\omega)]$ *is* $\mathscr{B}(\mathbf{R}^d) \times \mathscr{F}_s^{W,t}$*-measurable.*

Proof. Denote by $\mathscr{M}$ the class of all $\mathscr{B}(\mathbf{R}^d)$-sets A for which the assertion of the Lemma is true with $\phi = I_A$. It is enough to show that $\mathscr{M} = \mathscr{B}(\mathbf{R}^d)$, for then the conclusion follows for general ϕ by standard arguments via simple functions and limits of sequences of simple functions. Now if $\phi = I_A$, the truth of the lemma is obvious for $f(x,\omega) = I_A[\xi_{s,x}(t,\omega)]$ because

$$\{(x,\omega)\colon I_A[\xi_{s,x}(t,\omega)] = 1\} = \{(x,\omega)\colon \xi_{s,x}(t,\omega) \in A\} \in \mathscr{B}(\mathbf{R}^d) \times \mathscr{F}_s^{W,t}$$

by Lemma 5.4.4. The proof is complete. □

Introduce the following important notation. For x, s, and t fixed and $t \geq s$ write

$$P(s,x,t,A) = P[\omega\colon \xi_{s,x}(t,\omega) \in A].$$

Since

$$P(s,x,t,A) = \int I_A[\xi_{s,x}(t,\omega)]P(d\omega),$$

from Lemma 5.4.3 it follows that for s, t, and A fixed, $P(s,\cdot,t,A)$ is a Borel (that is, $\mathscr{B}(\mathbf{R}^d)$-measurable) function of x. It is also obvious from its definition that for s, x, and t fixed, $P(s,x,t,\cdot)$ is a probability measure on $\mathscr{B}(\mathbf{R}^d)$.

Let us now consider the equation

$$\begin{aligned} dy_t &= b(t,y_t)\,dt + \sigma(t,y_t)\,dW_t, \qquad t \geq s, \\ y_s &= \eta_s, \end{aligned} \tag{5.4.6}$$

where the random variable η_s is $\mathscr{F}_s$-measurable.

Lemma 5.4.6. *The unique solution of (5.4.6) is given by* $\xi_s(\eta_s(\omega),t,\omega)$, *where* $\xi_s(x,t,\omega)$ *is the solution given in Lemma 5.4.3.*

Proof. For convenience, set $\bar{\sigma}_s(t,\omega) = \sigma(t,\xi_s(\eta_s(\omega),t,\omega))$. First, observe that $\bar{\sigma}(t,\omega) \in \mathscr{L}_2^W[s,T]$. The measurability and $\mathscr{F}_t$-adaptability follow from Lemma 5.4.3. For integrability, let $A = \{(x,\omega)\colon \int_s^T |\xi_s(x,t,\omega)|^2\,dt < \infty\}$ and $\zeta\colon \Omega \to \mathbf{R}^d \times \Omega$ be defined by $\zeta(\omega) = (\eta_s(\omega),\omega)$. Using Condition (5.1.10),

$$\begin{aligned} P\left(\left\{\omega\colon \int_s^T |\bar{\sigma}(t,\omega)|^2\,dt < \infty\right\}\right) &\geq P\left(\left\{\omega\colon \int_s^T |\xi_s(\eta_s(\omega),t,\omega)|^2\,dt < \infty\right\}\right) \\ &= P(\zeta^{-1}(A)) = P\eta_s^{-1} \times P(A) = 1. \end{aligned}$$

The last equality holds since $P(A^x) = 1$ for every x, where A^x is the x section of A. Let

$$J_t(\omega) = \xi_s(\eta_s(\omega),t,\omega) - \eta_s(\omega) - \int_s^t b(u,\xi_s(\eta_s(\omega),u,\omega))\,du.$$

Then (J_t) is a d-dimensional continuous local $\mathscr{F}_t$-martingale and

$$\langle J \rangle_t(\omega) = \int_s^t \bar{\sigma}(u,\omega)\bar{\sigma}^*(u,\omega)\,du. \tag{5.4.7}$$

To show this, let $\tilde{J}_t(x,\omega) = \int_s^t \sigma(u,\xi_s(x,u,\omega))\,dW_u(\omega)$ and for $\theta \in \mathbf{R}^d$, define $\tilde{M}_t(x,\omega) = (\theta,\tilde{J}_t(x,\omega))$, $(\ ,\ )$ denoting the inner product in $\mathbf{R}^d$. Since θ is fixed throughout the argument it will be suppressed in the notation. For each x, $\tilde{M}_t(x,\omega)$ is a continuous local martingale and is, moreover, (t,x,ω)-measurable. Define the stopping times

$$\tau_n(x,\omega) = \begin{cases} \inf\{t \geq s\colon |\tilde{M}_t(x,\omega)| \geq n \text{ or } \langle \tilde{M} \rangle_t(x,\omega) \geq n\} & \\ T & \text{if the set is empty.} \end{cases}$$

Then τ_n is (x,ω)-measurable and $\tilde{M}_t^n(x,\omega) = \tilde{M}_{t \wedge \tau_n(x,\omega)}(x,\omega)$ is a (t,x,ω)-measurable, continuous, bounded martingale. Let $M_t^n(\omega) = \tilde{M}_t^n(\eta_s(\omega),\omega)$. For $T \geq t > t' \geq s$ define $f(x,\omega) = \tilde{M}_t^n(x,\omega) - \tilde{M}_{t'}^n(x,\omega)$. Then f is bounded, $\mathscr{B}(\mathbf{R}^d) \times \mathscr{F}_s^t$-measurable and η_s is $\mathscr{F}_{t'}$-measurable. From Lemma 5.4.1 it follows that $E[M_t^n(\omega) - M_{t'}^n(\omega)|\mathscr{F}_{t'}] = g(\eta_s(\omega))$, where $g(x) = Ef(x,\omega)$. But $Ef(x,\omega) = 0$ for all x. Hence (M_t^n) is a martingale and, being bounded and continuous, is an L^2-martingale. It is similarly shown that $\langle M^n \rangle_t(\omega) = \langle \tilde{M}^n \rangle_t(\eta_s(\omega),\omega)$. Thus, (M_t) is a continuous local martingale; that is, (J_t) is a d-dimensional, continuous martingale with $\langle J \rangle_t(\omega)$ given by (5.4.7). Finally, the continuous local martingale

$$I_t(\omega) = \int_s^t \bar{\sigma}(u,\omega)\,dW_u(\omega)$$

also has $\langle I \rangle_t$ given by the right-hand side of (5.4.7). By a stopping time argument similar to the one just shown above, it is also verified that

$$\langle J,I \rangle_t(\omega) = \int_s^t \bar{\sigma}(u,\omega)\bar{\sigma}^*(u,\omega)\,du.$$

Thus we have $\langle J - I \rangle_t = 0$, that is,

$$J_t(\omega) = I_t(\omega) \quad \text{for all } t \in [s,T] \qquad \text{(a.s.).}$$

□

Let us now consider the function $f(x,\omega)$ of Lemma 5.4.5. For each $x, f(x,\cdot)$, being $\mathscr{F}_s^{W,t}$-measurable, is independent of $\mathscr{F}_s$. Hence if Z is any $\mathscr{F}_s$-measurable random variable $(\mathscr{F}_s = \mathscr{F}^\eta \vee \mathscr{F}_s^W)$, from Lemma 5.4.1 we have

$$E[\phi\{\xi_{s,Z}(t,\omega)\}|\mathscr{F}_s] = g(Z), \tag{5.4.8}$$

where

$$g(x) = \int_\Omega \phi[\xi_{s,x}(t,\omega)]P(d\omega).$$

We now take $Z = \xi(s)$ since $\xi(s)$ is $\mathscr{F}_s$-measurable. From Lemma 5.4.6 it follows that $\xi_{s,\xi(s)}(t) = \xi_t$, the solution of Equation (5.4.2). Equation (5.4.8) then yields

$$E[\phi(\xi_t)|\mathscr{F}_s] = g(\xi_s) \qquad \text{(a.s.).} \tag{5.4.9}$$

Note also that (5.4.9) holds for the process $\xi_{s,x}(t)$:

$$E(\phi[\xi_{s,x}(t)]|\mathscr{F}_{s'}) = g[\xi_{s,x}(s')], \qquad s \leq s' \leq t, \tag{5.4.9a}$$

where $g(x) = \int_\Omega \phi[\xi_{s,x}(t,\omega)]P(d\omega)$. Since $g(x)$ is $B(\mathbf{R}^d)$-measurable, $g(\xi_s)$ is $\mathscr{F}^{\xi_s}$-measurable. From (5.4.9) it follows that

$$\begin{aligned} E[\phi(\xi_t)|\mathscr{F}_s] &= g(\xi_s) = E[E\{\phi(\xi_t)|\mathscr{F}_s\}|\xi_s] \\ &= E[\phi(\xi_t)|\xi_s] \qquad \text{(a.s.).} \end{aligned} \tag{5.4.10}$$

This establishes the Markov property of (ξ_t) with respect to $(\mathscr{F}_t)$. By the definition of $P(s,x,t,A)$ we obtain

$$E\phi[\xi_{s,x}(t)] = \int_{\mathbf{R}^d} P(s,x,t,dy)\phi(y). \tag{5.4.11}$$

Since $g(x) = E\phi[\xi_{s,x}(t)]$, taking $\phi = I_A$ and applying (5.4.10) and (5.4.11), we get

$$P[\xi_t \in A|\mathscr{F}_s] = P[\xi_t \in A|\xi_s] = P(s,\xi_s,t,A) \qquad \text{(a.s.).} \tag{5.4.12}$$

In (5.4.10) choosing $\phi(y) = P(t,y,u,A)$, where $u > t$ and $A \in \mathscr{B}(\mathbf{R}^d)$ are fixed, we get

$$\phi[\xi_{s,x}(t)] = P(t,\xi_{s,x}(t),u,A). \tag{5.4.13}$$

Hence from (5.4.11) we have [using Equation (5.4.9a)]

$$\begin{aligned} \int P(s,x,t,dy)P(t,y,u,A) &= EP(t,\xi_{s,x}(t),u,A) \\ &= EP[\xi_{s,x}(u) \in A|\xi_{s,x}(t)] \\ &= E(E\{I_A[\xi_{s,x}(u)]|\xi_{s,x}(t)\}) = EI_A[\xi_{s,x}(u)] \\ &= P[\xi_{s,x}(u) \in A] = P(s,x,u,A). \end{aligned}$$

Therefore $P(s,x,t,A)$ satisfies the Chapman-Kolmogorov equation

$$\int_{\mathbf{R}^d} P(s,x,t,dy)P(t,y,u,A) = P(s,x,u,A) \tag{5.4.14}$$

for $s < t < u$, $x \in \mathbf{R}^d$, and $A \in \mathscr{B}(\mathbf{R}^d)$.

This shows that $P(s,x,t,A)$ is the transition probability function of the Markov process (ξ_t). We have proved the following result.

Theorem 5.4.1. *The stochastic differential equation*

$$d\xi_t = b(t,\xi_t)\,dt + \sigma(t,\xi_t)\,dW_t, \qquad \xi_0 = \eta,$$

where $\eta \perp\!\!\!\perp F_T^W$, and b and σ satisfy conditions (5.1.10) has a unique solution (ξ_t) which is a continuous Markov process relative to $(\mathscr{F}_t)$ and such that

$$P[\xi_t \in A|\mathscr{F}_s] = P[\xi_t \in A|\xi_s] = P(s,\xi_s,t,A) \qquad \text{(a.s.),} \tag{5.4.15}$$

where $P(s,x,t,A)$ defined by $P[\xi_{s,x}(t) \in A]$ is the transition probability function of the process.

5.5 Extended Generator of $\xi(t)$

Let $\xi(t)$ be the solution of the stochastic differential Equation (5.1.9). We have seen that the transition probability function satisfies

$$P(s,x,t,A) = P[\xi_{s,x}(t) \in A]. \tag{5.5.1}$$

For each s and t $(0 \leq s < t \leq T)$, define $T_s^t f$ for Borel measurable functions $f: \mathbf{R}^d \to \mathbf{R}$ by

$$T_s^s = I \quad \text{(the identity operator)} \tag{5.5.2}$$

$$T_s^t f(x) = \int_{\mathbf{R}^d} f(y) P(s,x,t,dy), \tag{5.5.3}$$

provided the integral on the right exists. Then $T_s^t f(\xi(s)) = E[f(\xi(t))|\mathscr{F}_s]$ (a.s.).

On the class of bounded Borel-measurable functions, T_s^t defines a family of operators satisfying the semigroup property

$$T_s^u f = T_s^t T_t^u f \qquad (0 \leq s \leq t \leq u \leq T). \tag{5.5.4}$$

Equation (5.5.4) is a restatement of the Chapman-Kolmogorov equation. The family $\{T_s^t\}$ is called the *semigroup of the Markov process* $\xi(t)$.

Denote by $C_b^2 = C_b^2(\mathbf{R}^d)$ the class of all real-valued functions on $\mathbf{R}^d$ which are continuous and bounded together with their first- and second-order partial derivatives. Let C_b be the class of bounded, real continuous functions on $\mathbf{R}^d$.

Theorem 5.5.1. *Let $\xi(t)$ be the solution of the stochastic differential Equation (5.1.9), where the coefficients b and σ satisfy condition (5.1.10). Let L_s be the differential operator defined on C_b^2 by*

$$L_s f(x) = \frac{1}{2} \sum_{i,j} a^{ij}(s,x) \frac{\partial^2 f}{\partial x^i \partial x^j}(x) + \sum_i b^i(s,x) \frac{\partial f}{\partial x^i}(x). \tag{5.5.5}$$

Then for each $f \in C_b^2$,

$$L_u f(y) \text{ is jointly measurable in } u \text{ and } y \tag{5.5.6}$$

and

$$\int_{\mathbf{R}^d} |(L_u f)(y)| P(s,x,u,dy) < \infty \tag{5.5.7}$$

for every $x \in \mathbf{R}^d$, $0 \leq s < u \leq T$;

$$T_s^t f(x) = f(x) + \int_s^t T_s^u L_u f(x)\, du \tag{5.5.8}$$

for every $x \in \mathbf{R}^d$, $0 \leq s < t \leq T$.

Furthermore, if b and σ are continuous functions, then

$$\lim_{h \downarrow 0} \frac{T_s^{s+h} f(x) - f(x)}{h} = L_s f(x) \tag{5.5.9}$$

for each $x \in \mathbf{R}^d$ and $0 \leq s < T$.

PROOF. The process $\xi_{s,x}(t)$ satisfies the stochastic equation

$$\xi_{s,x}(t) = x + \int_s^t b[u,\xi_{s,x}(u)]\,du + \int_s^t \sigma[u,\xi_{s,x}(u)]\,dW_u. \qquad (5.5.10)$$

Taking $f(x) = \psi(t,x)$ in formula (4.4.6), we have

$$Ef[\xi_{s,x}(t)] = f(x) + \int_s^t E[L_u f(\xi_{s,x}(u))]\,du.$$

This integral exists by (5.1.10), since $f \in C_b^2$ and $E\sup_{s \le u \le T} |\xi_{s,x}(u)|^2 < \infty$. Hence $T_s^t f(x) = f(x) + \int_s^t T_s^u L_u f(x)\,du$ and we have (5.5.8). When b and σ are continuous, $T_s^u L_u f(x)$ is a continuous function of u in $[s,T]$, this gives (5.5.9). □

Proposition 5.5.1. *Let $f \in C_b^2$. Then for each t*

$$\int_0^T E|L_u f(\xi(u))|\,du < \infty, \qquad (5.5.11)$$

and

$$M_t^f = f[\xi(t)] - f[\xi(0)] - \int_0^t (L_u f)(\xi(u))\,du \qquad (5.5.12)$$

is an $(\mathscr{F}_t)$-martingale.

PROOF. Since $\xi(t)$ is Markov with respect to $(\mathscr{F}_t)$, M_t^f is $\mathscr{F}_t$-measurable. From (5.5.11) we have $E|M_t^f| < \infty$. For $s \le t$,

$$M_t^f - M_s^f = f[\xi(t)] - f[\xi(s)] - \int_s^t L_u f(\xi(u))\,du.$$

Hence

$$\begin{aligned} E[M_t^f - M_s^f|\mathscr{F}_s] &= E[f(\xi(t))|\mathscr{F}_s] - f(\xi(s)) - E\left[\int_s^t L_u f(\xi(u))\,du|\mathscr{F}_s\right] \\ &= T_s^t f[\xi(s)] - f[\xi(s)] - \int_s^t T_s^u L_u f(\xi(s))\,du \qquad \text{(a.s.)} \\ &= 0 \qquad \text{(a.s.)}, \end{aligned}$$

from (5.5.8). □

$\xi(t)$ is a Markov process with *extended generator* $\{L_u\}$ defined on C_b^2, satisfying (5.5.8). The expression on the right side of (5.5.9) defines what is often called the *weak infinitesimal generator*. When b and σ are continuous, these notions coincide on C_b^2.

The Markov process $\xi(t)$ has stationary transition probabilities if

$$P(s,x,t,A) = P(0,x,\, t - s,\, A).$$

We then have $T_s^t = T_0^{t-s}$. Let

$$Q(t - s,\, x,A) = P(0,x,\, t - s,\, A) \quad \text{and} \quad S_{t-s} = T_0^{t-s},$$

the semigroup is now given by $\{S_t\}$ $(t \ge 0)$,

$$S_t f(x) = \int f(y) Q(t,x,dy).$$

The extended generator in this case is defined to be a linear operator A satisfying

$$S_t f(x) = f(x) + \int_0^t S_u Af(x)\,du.$$

Feller Semigroup

Definition 5.5.1. The semigroup $\{T_s^t\}$ has the Feller property, or is called a *Feller semigroup* if, for $f \in C_b$ and $u > 0$, the function

$$(s,x) \to T_s^{s+u}f(x)$$

is continuous.

We note that the above definition reduces to the well-known definition of a Feller semigroup in the case of Markov process with stationary transition probabilities. Our aim now is to establish the Feller property for the process $\xi(t)$ of Theorem 5.5.1.

Let us apply Ito's formula in (5.5.10) taking $f(x) = |x|^2$. Recall that the general version of the formula given in Chapter 4 does not require the boundedness assumption on f. We then obtain

$$|\xi_{s,x}(t) - x|^2 = 2\int_s^t (\xi_{s,x}(u) - x, b[u,\xi_{s,x}(u)])\,du + \int_s^t \sum_{i=1}^d a^{ii}[u,\xi_{s,x}(u)]\,du$$
$$+ 2\sum_{i,k} \int_s^t [\xi_{s,x}^i(u) - x^i]\sigma^{ik}[u,\xi_{s,x}(u)]\,dW_u^k. \tag{5.5.13}$$

Here, (,) denotes the inner product in $\mathbf{R}^d$. Choose a sequence of stopping times $\tau_n \uparrow T$ (a.s.) such that the stochastic integrals

$$\int_s^{t\wedge\tau_n} [\xi_{s,x}^i(u) - x^i]\sigma^{ik}[u,\xi_{s,x}(u)]\,dW_u^k$$

are L^2-martingales. Taking expectations on both sides of (5.5.13) [where t is replaced by $t \wedge \tau_n$ in (5.5.13)], we obtain

$$E|\xi_{s,x}(t\wedge\tau_n) - x|^2 = 2E\int_s^{t\wedge\tau_n} (\xi_{s,x}(u) - x, b[u,\xi_{s,x}(u)])\,du$$
$$+ E\int_s^{t\wedge\tau_n} \sum_{i=1}^d a^{ii}[u,\xi_{s,x}(u)]\,du.$$

It was shown earlier that $E\sup_{0\le t\le T}|\xi_{s,x}(t)|^2 < \infty$. Using this fact and the dominated convergence theorem the left-hand side converges to $E|\xi_{s,x}(t) - x|^2$. Similarly, noting that $|b(u,x)|^2 \le K(1 + |x|^2)$ and $\sum_{i,j=1}^d |a^{ij}(u,x)| \le C_0(1 + |x|^2)$ (where C_0 is a constant), we have the desired convergence on the right-hand side and obtain

$$E|\xi_{s,x}(t) - x|^2 = 2E\int_s^t (\xi_{s,x}(u) - x, b[u,\xi_{s,x}(u)])\,du + E\int_s^t \sum_{i=1}^d a^{ii}[u,\xi_{s,x}(u)]\,du.$$

From now on it is sufficient for our purpose to restrict x to a bounded region, that is, $|x| \leq a$. It is easy to see that for $|x| \leq a$, $E \sup_{s \leq t \leq T} |\xi_{s,x}(t)|^2 \leq C_1 < \infty$, where C_1 is a constant depending only on K, T, and a (but not s). We then have

$$E\left|\int_s^t (\xi_{s,x}(u) - x, b[u,\xi_{s,x}(u)])\, du\right| \leq K^{\frac{1}{2}} E \int_s^t (|\xi_{s,x}(u)| + |x|)(1 + |\xi_{s,x}(u)|)\, du$$
$$\leq C_2(t - s),$$

where $C_2 = K^{\frac{1}{2}}[(a + 1)C_1^{\frac{1}{2}} + C_1 + a]$. A similar bound is easily obtained for the second term of (5.5.13)

$$E \int_s^t \sum_{i=1}^d a^{ii}[u,\xi_{s,x}(u)]\, du \leq C_0(1 + C_1)(t - s).$$

Hence

$$E|\xi_{s,x}(t) - x|^2 \leq C_3[t - s] \qquad (t \geq s) \tag{5.5.14}$$

for all x such that $|x| \leq a$. The constant C_3 depends on K, T, and a. Next, if $s \leq s' \leq t$, from

$$\xi_{s,x}(t) = \xi_{s,x}(s') + \int_{s'}^t b[u,\xi_{s,x}(u)]\, du + \int_{s'}^t \sigma[u,\xi_{s,x}(u)]\, dW_u$$

and

$$\xi_{s',y}(t) = y + \int_{s'}^t b[u,\xi_{s',y}(u)]\, du + \int_{s'}^t \sigma[u,\xi_{s',y}(u)]\, dW_u,$$

we have

$$\xi_{s,x}(t) - \xi_{s',y}(t) = [\xi_{s,x}(s') - y] + \int_{s'}^t \{b[u,\xi_{s,x}(u)] - b[u,\xi_{s',y}(u)]\}\, du$$
$$+ \int_{s'}^t [\sigma[u,\xi_{s,x}(u)] - \sigma[u,\xi_{s',y}(u)]]\, dW_u. \tag{5.5.15}$$

Now it was shown above that

$$E|\xi_{s,x}(s') - y|^2 \leq 2E|\xi_{s,x}(s') - x|^2 + 2|x - y|^2$$
$$\leq 2C_3|s' - s| + 2|x - y|^2 \quad \text{for } |x| \leq a,\ |y| \leq a.$$

Hence taking expectation of the supremum of the squares of both sides of (5.5.15), we have

$$E \sup_{s' \leq u \leq t} |\xi_{s,x}(u) - \xi_{s',y}(u)|^2 \leq A_0[|s' - s| + |x - y|^2]$$
$$+ K_1 E \int_{s'}^t \sup_{s' \leq u \leq t} |\xi_{x,s}(u) - \xi_{s',y}(u)|^2\, du,$$

where A_0 and K_1 are constants. It is easy to see that the above yields a constant A_1 depending on K, T, and a such that

$$E \sup_{s' \leq t \leq T} |\xi_{s,x}(t) - \xi_{s',y}(t)|^2 \leq A_1[|s' - s| + |x - y|^2] \tag{5.5.16}$$

for all values of x and y satisfying $|x| \leq a$, $|y| \leq a$.

Remark 5.5.1. The estimate (5.5.16) can be obtained directly from (5.5.14) and Lemma 5.4.2.

Using the estimate (5.5.16) we have the following result.

Theorem 5.5.2. *The semigroup $\{T_s^t\}$ of the Markov process $\xi(t)$ of Theorem 5.4.1 has the Feller property.*

PROOF. For $u > 0$ and $f \in C_b$,

$$T_s^{s+u}f(x) = Ef[\xi_{s,x}(s+u)]$$

and

$$T_{s'}^{s'+u}f(y) = Ef[\xi_{s',y}(s'+u)].$$

If $s' \geq s$, using (5.5.16), we have the inequality

$$E|\xi_{s',y}(s'+u) - \xi_{s,x}(s+u)| \leq \{A_1(|s'-s| + |x-y|^2)\}^{\frac{1}{2}} + E|\xi_{s,x}(s'+u) - \xi_{s,x}(s+u)|.$$

Similarly, for $s' < s$,

$$E|\xi_{s',y}(s'+u) - \xi_{s,x}(s+u)| \leq \{A_1(|s'-s| + |x-y|^2)\}^{\frac{1}{2}} + E|\xi_{s',y}(s+u) - \xi_{s',y}(s'+u)|.$$

From the stochastic differential equation (5.1.9) and conditions (5.1.10), we have

$$\begin{aligned} E|\xi_{s',y}(s+u) - \xi_{s',y}(s'+u)|^2 &\leq 2K_2 \int_{s'+u}^{s+u} E[1 + |\xi_{s',y}(t)|^2]\, dt \\ &\leq 2K_2\left[1 + \sup_{s' \leq t \leq T} E|\xi_{s',y}(t)|^2\right](s-s'). \end{aligned}$$

From (5.5.14),

$$\begin{aligned} \sup_{s' \leq t \leq T} E|\xi_{s',y}(t)|^2 &\leq 2 \sup_{s' \leq t \leq T} E|\xi_{s',y}(t) - y|^2 + 2|y|^2 \\ &\leq 2C_3(T-s') + 2a^2 \leq 2(C_3T + a^2). \end{aligned}$$

Hence we obtain

$$E|\xi_{s',y}(s+u) - \xi_{s',y}(s'+u)| \leq [2K_2 + 2C_3T + 2a^2)]^{\frac{1}{2}} \cdot |s-s'|^{\frac{1}{2}}.$$

The same bound is obtained also for $E|\xi_{s,x}(s'+u) - \xi_{s,x}(s+u)|$. The assertion of the theorem now follows immediately. □

5.6 Diffusion Processes

A Markov process $\xi(t)$ with transition probability function P is a diffusion process if a d-vector-valued function $b(t,x)$ and a $d \times d$-matrix-valued function $a(t,x)$ exist such that the following conditions hold: For every bounded open neighbourhood U_x of x,

(a) $\displaystyle\lim_{h \downarrow 0} \frac{1}{h} \int_{U_x^c} P(t,x,t+h,dy) = 0$

(b) $\lim_{h\downarrow 0} \frac{1}{h}\int_{U_x} (y^i - x^i)P(t,x,\, t+h,\, dy) = b^i(t,x) \qquad (i = 1,\ldots,d)$

(c) $\lim_{h\downarrow 0} \frac{1}{h}\int_{U_x} (y^i - x^i)(y^j - x^j)P(t,x,\, t+h,\, dy) = a^{ij}(t,x) \qquad (i,j = 1,\ldots,d)$

The vector $b(t,x)$ is called the *drift* and the matrix $a(t,x)$ is called the *diffusion coefficient* of the process.

Theorem 5.6.1. *Suppose b and σ satisfy (5.1.10) and are continuous. The process $\xi(t)$ which is the solution of Equation (5.1.9) is a diffusion process with drift b and diffusion coefficient $a = \sigma\sigma^*$.*

Proof. Let U_x be any bounded open set containing x, and let V_x be an open set containing x and such that its closure $\bar{V}_x \subset U_x$. Let f be a C_b^2 function such that $0 \le f \le 1$, $f = 0$ on $\bar{V}_x$, and $f = 1$ on U_x^c. Then

$$0 \le \frac{1}{h}\int_{U_x^c} P(s,x,s+h,dy) = \frac{1}{h}\left[\int_{U_x^c} f(y)P(s,x,s+h,dy) - f(x)\right]$$

$$\le \frac{1}{h}\left[\int_{\mathbf{R}^d} f(y)P(s,x,\, s+h,\, dy) - f(x)\right].$$

As $h \downarrow 0$, the quantity on the right-hand side tends to $L_s f(x) = 0$ by Theorem 5.5.1, and (a) is proved.

Next, suppose V_x is a bounded open set containing $\bar{U}_x$. Let ϕ be a C_b^2 function with $0 \le \phi \le 1$, $\phi = 0$ on V_x^c, and $\phi = 1$ on $\bar{U}_x$. For $i = 1, \ldots, d$, write $f^i(y) = (y^i - x^i)\phi(y)$. We have

$$\left.\frac{\partial f^i}{\partial y^i}\right|_{y=x} = 1, \qquad \left.\frac{\partial f^i}{\partial y^j}\right|_{y=x} = 0 \qquad (j \ne i),$$

and

$$\left.\frac{\partial^2 f^i}{\partial y^j\,\partial y^k}\right|_{y=x} = 0$$

for all j,k. Hence $L_s f^i(x) = b^i(s,x)$ and we obtain, from Theorem 5.5.1,

$$\lim_{h\downarrow 0}\left[\frac{1}{h}\int f^i(y)P(s,x,\, s+h,\, dy) - f^i(x)\right] = b^i(s,x),$$

or since $f^i(x) = 0$,

$$\lim_{h\downarrow 0}\frac{1}{h}\int (y^i - x^i)\phi(y)P(s,x,\, s+h,\, dy) = b^i(s,x).$$

The open set V_x being bounded, $|y^i - x^i| \le K$, a constant for $y \in V_x$. Hence

$$\left|\frac{1}{h}\int (y^i - x^i)\phi(y)P(s,x,\, s+h,\, dy) - \frac{1}{h}\int_{U_x}(y^i - x^i)P(s,x,\, s+h,\, dy)\right|$$

$$\le \frac{K}{h}\int_{V_x\setminus U_x} P(s,x,\, s+h,\, dy) \le \frac{K}{h}\int_{U_x^c} P(s,x,\, s+h,\, dy) \to 0 \quad \text{as } h \downarrow 0.$$

This proves

$$\lim_{h\downarrow 0}\frac{1}{h}\int_{U_x}(y^i - x^i)P(s,x,s+h,dy) = b^i(s,x).$$

Finally, to verify (c) we proceed similarly by taking

$$f^{ij}(y) = (y^i - x^i)(y^j - x^j)\phi(y). \qquad \square$$

Remark 5.6.1. Suppose that L_s is the differential operator of the last section where a defined on $[0,T] \times \mathbf{R}^d$ is a continuous $d \times d$-matrix-valued functional and b defined on $[0,T] \times \mathbf{R}^d$ is a d-dimensional continuous functional. It will further be assumed that besides being symmetric,

$$a(t,x) \text{ is positive definite for all } (t,x). \tag{5.6.1}$$

The following result may be regarded as a weak converse of Proposition 5.5.1.

Proposition 5.6.1. *Let $\xi(t)$ be a continuous stochastic process defined on some probability space $(\Omega,\mathscr{A},P)$ such that*

$$\int_0^T |b(t,\xi_t)|\,dt + \int_0^T |\sigma(t,\xi_t)|^2\,dt < \infty \qquad (P\text{-a.s.}), \tag{5.6.2}$$

where $\sigma = a^{\frac{1}{2}}$, the positive definite square root of a. Suppose that the following condition is satisfied: for each $f \in C^2$,

$$M_t^f = f[\xi(t)] - f[\xi(0)] - \int_0^t L_u f[\xi(u)]\,du \tag{5.6.3}$$

is a (continuous) local martingale. Then there exists a Wiener martingale $(W_t,\mathscr{F}_t^\xi,P)$ such that

$$\xi(t) = \xi(0) + \int_0^t b[u,\xi(u)]\,du + \int_0^t \sigma[u,\xi(u)]\,dW(u). \tag{5.6.4}$$

Proof. Apply Ito's formula to the continuous local semimartingale [obtained for each component by taking $f(x^1,\ldots,x^d) = x^i$ for each $1 \le i \le d$ in Equation (5.6.3) and denoting M_t^f by M_t^i]

$$\xi(t) = \xi(0) + M_t + \int_0^t b[u,\xi(u)]\,du \tag{5.6.5}$$

and the function $F(x) = e^{(\theta,x)}$. Note that this version of the formula does not assume the boundedness of F or its derivatives. We obtain

$$F[\xi(t)] = F[\xi(0)] + \sum_i \theta^i \int_0^t F[\xi(u)]\,dM_u^i + \int_0^t F[\xi(u)]\cdot(\theta,b(u,\xi(u)))\,du$$

$$+ \tfrac{1}{2}\sum_{i,j}\theta^i\theta^j \int_0^t F[\xi(u)]\,d\langle M^i,M^j\rangle_u. \tag{5.6.6}$$

Now, by hypothesis (5.6.3) holds for $f = F$. Noting that

$$L_u F[\xi(u)] = \{\tfrac{1}{2}(a(u,\xi(u))\theta,\theta) + (\theta,b(u,\xi(u)))\}F[\xi(u)],$$

we find that

$$F[\xi(t)] - F[\xi(0)] - \int_0^t F[\xi(u)]\{\tfrac{1}{2}(a(u,\xi(u))\theta,\theta) + (\theta, b(u,\xi(u)))\}\,du \quad (5.6.7)$$

is a continuous local martingale. It follows that the difference in the two expressions (5.6.6) and (5.6.7) is a continuous local martingale. Hence for all $\theta \in \mathbf{R}^d$,

$$\sum_{i,j} \theta^i\theta^j\left[\int_0^t F[\xi(u)]a^{ij}(u,\xi(u))\,du - \int_0^t F[\xi(u)]\,d\langle M^i,M^j\rangle_u\right]$$

is a continuous local martingale. But the above process is of bounded variation in $[0,T]$ and vanishes for $t = 0$. Since θ is arbitrary, we obtain

$$\langle M^i,M^j\rangle_t = \int_0^t a^{ij}(u,\xi(u))\,du \qquad (i,j = 1,\ldots,d). \qquad (5.6.8)$$

The process $W(t) = \int_0^t a^{-\frac{1}{2}}(u,\xi(u))\,dM_u$ is now seen to be a d-dimensional Wiener process which is $\mathscr{F}_t^\xi$-measurable, and

$$M_t = \int_0^t a^{\frac{1}{2}}(u,\xi(u))\,dW_u. \qquad \square \quad (5.6.9)$$

We have seen in Theorem 5.1.3 that Equation (5.1.9) has a unique strong solution if the coefficients satisfy Lipschitz and growth conditions. The existence of a weak solution can be asserted under much more general conditions.

Theorem 5.6.2. *Let the d-dimensional vector function $b(t,x)$ and the $d \times d$-matrix function $\sigma(t,x)$, defined on $\mathbf{R}_+ \times \mathbf{R}^d$ be bounded and Borel-measurable with respect to (t,x). Then the stochastic differential Equation (5.1.9) with initial condition $\xi_0 = x$ $(x \in \mathbf{R}^d)$ has a weak solution.*

For a proof of this result we refer the reader to Krylov's book.

Remark 5.6.2. If it is assumed that $b(t,x)$ and $\sigma(t,x)$ are bounded and *continuous*, then Stroock and Varadhan have shown the uniqueness of the weak solution. (See Notes for the references.)

5.7 Existence of Moments

Let ξ be a solution of the stochastic differential equation (5.1.1) with initial condition $\xi_0 = \eta$. For any $t \in [0,T]$, set $\|\xi\|_t = \sup_{0\le s\le t}|\xi_s|$. This section is devoted to the proof of a result (Theorem 5.7.1) which ensures the finiteness of $E[e^{c\|\xi\|_T^2}]$ for some $c > 0$. Such a result naturally implies the finiteness of all moments $E|\xi_t|^n$. In addition to assumption (C) on the coefficients of (5.1.1), we impose the following restrictions: Only the case of real-valued processes ($d = 1$) will be considered. There exists a positive constant K such that

$$0 < \sigma^2(t,g,f) \le K \tag{5.7.1}$$

for all $t \in [0,T]$, $g \in D$, and $f \in C_1$. It is convenient here to take K to be the constant which occurs in condition (C) of Section 5.1. For some positive constant c_0,

$$E[e^{c_0\eta^2}] < \infty. \tag{5.7.2}$$

There exists a $\delta > 0$ and $c_1 > 0$ such that

$$E[e^{c_1 Y_\delta}] < \infty, \tag{5.7.3}$$

where

$$Y_\delta = \sup_{\substack{0 \le t' - t \le \delta \\ t,t' \in [0,T] \\ (t<t')}} \int_t^{t'} |L(u,X)|^2\,du.$$

Theorem 5.7.1. *Let the coefficients b and σ of Equation* (5.1.1) *satisfy* (B) *and* (C) *of Section 5.1 and conditions* (5.7.1) *and* (5.7.3). *Further suppose that the initial random variable η satisfies condition* (5.7.2). *If* (ξ_t) *is a solution of* (5.1.1), *then for some positive constant c,*

$$E[e^{c\|\xi\|_T^2}] < \infty. \tag{5.7.4}$$

The following two lemmas are needed in the proof of Theorem 5.7.1.

Lemma 5.7.1. *Let* $f \in \mathscr{M}_2^W[0,T]$ *satisfy the condition*

$$|f_u(\omega)| \le C \qquad \text{(a.s.) for all } u \in [0,T], \tag{5.7.5}$$

where C is a constant. Then, for $0 \le s < t \le T$, *we have*

$$E\left[\int_s^t f_u\,dW_u\right]^{2n} \le \frac{[2(t-s)C^2]^n \Gamma(n+\frac{1}{2})}{\Gamma(\frac{1}{2})}$$
$$\le [2(t-s)C^2]^n n!, \tag{5.7.6}$$

n being any nonnegative integer.

PROOF. We shall prove (5.7.6) for a step function f_u on $[s,t]$. Denoting by $\{t_i\}$ $(i = 0, \ldots, m+1)$ the corresponding partition, we have

$$\int_s^t f_u\,dW_u = \sum_{i=0}^m f_{t_i}[W_{t_{i+1}} - W_{t_i}].$$

For $j = 1, \ldots, m$, the random variables $\sum_{i=0}^{j-1} f_{t_i}[W_{t_{i+1}} - W_{t_i}]$ and f_{t_j} are $\mathscr{F}_{t_j}$-measurable, while $W_t - W_{t_{j+1}}$ and $W_{t_{j+1}} - W_{t_j}$ are independent, Gaussian variables. Then

$$E\left(\left\{\sum_{i=0}^j f_{t_i}[W_{t_{i+1}} - W_{t_i}] + C[W_t - W_{t_{j+1}}]\right\}^{2n} \middle| \mathscr{F}_{t_j}\right)$$
$$= \sum_{k=0}^{2n} \binom{2n}{k} \left\{\sum_{i=0}^{j-1} f_{t_i}[W_{t_{i+1}} - W_{t_i}]\right\}^{2n-k} \mu_k \tag{5.7.7}$$

where μ_k is the kth (conditional) moment of

$$f_{t_j}[W_{t_{j+1}} - W_{t_j}] + C[W_t - W_{t_{j+1}}]. \tag{5.7.8}$$

Since the conditional distribution of (5.7.8) with respect to $\mathscr{F}_{t_j}$ is Gaussian with zero mean and variance equal to $f_{t_j}^2(t_{j+1} - t_j) + C^2(t - t_{j+1})$, $\mu_k = 0$ for k odd and the even moments are monotone increasing in the variance. Hence for k even, $\mu_k \le \mu'_k$, where μ'_k is the kth moment of $C[W_t - W_{t_j}]$. Thus from (5.7.7)

$$\begin{aligned}
&E\left(\left\{\sum_{i=0}^{j} f_{t_i}[W_{t_{i+1}} - W_{t_i}] + C[W_t - W_{t_{j+1}}]\right\}^{2n} \middle| \mathscr{F}_{t_j}\right) \\
&\quad \le \sum_{k=0}^{2n} \binom{2n}{k} \sum_{i=0}^{j-1} \left\{f_{t_i}[W_{t_{i+1}} - W_{t_i}]\right\}^{2n-k} \mu'_k \\
&\quad = E\left(\left\{\sum_{i=0}^{j-1} f_{t_i}[W_{t_{i+1}} - W_{t_i}] + C[W_t - W_{t_j}]\right\}^{2n} \middle| \mathscr{F}_{t_j}\right).
\end{aligned} \tag{5.7.9}$$

Hence, using (5.7.9), we obtain

$$\begin{aligned}
&E\left\{\sum_{i=0}^{m} f_{t_i}[W_{t_{i+1}} - W_{t_i}]\right\}^{2n} \\
&\quad = EE\left(\left\{\sum_{i=0}^{m} f_{t_i}[W_{t_{i+1}} - W_{t_i}]\right\}^{2n} \middle| \mathscr{F}_{t_m}\right) \\
&\quad \le EE\left(\left\{\sum_{i=0}^{m-1} f_{t_i}[W_{t_{i+1}} - W_{t_i}] + C[W_t - W_{t_m}]\right\}^{2n} \middle| \mathscr{F}_{t_m}\right) \\
&\quad = E\left\{\sum_{i=0}^{m-1} f_{t_i}[W_{t_{i+1}} - W_{t_i}] + C[W_t - W_{t_m}]\right\}^{2n} \\
&\quad \le E\left\{\sum_{i=0}^{j} f_{t_i}[W_{t_{i+1}} - W_{t_i}] + C[W_t - W_{t_{j+1}}]\right\}^{2n} \\
&\quad = EE\left(\left\{\sum_{i=0}^{j} f_{t_i}[W_{t_{i+1}} - W_{t_i}] + C[W_t - W_{t_{j+1}}]\right\}^{2n} \middle| \mathscr{F}_{t_j}\right) \\
&\quad \le EE\left(\left\{\sum_{i=0}^{j-1} f_{t_i}[W_{t_{i+1}} - W_{t_i}] + C[W_t - W_{t_j}]\right\}^{2n} \middle| \mathscr{F}_{t_j}\right) \\
&\quad = E\left\{\sum_{i=0}^{j-1} f_{t_i}[W_{t_{i+1}} - W_{t_i}] + C[W_t - W_{t_j}]\right\}^{2n} \\
&\quad \le E(C[W_t - W_s])^{2n} = \frac{[2(t-s)C^2]^n \Gamma(n + \frac{1}{2})}{\Gamma(\frac{1}{2})}.
\end{aligned} \tag{5.7.10}$$

Now let $f_u \in \mathscr{M}_2^W[0,T]$ and $|f_u| \le C$ in $[s,t]$. Then there exists a sequence of step functions $f^N \in \mathscr{M}_2^W[0,T]$ such that $|f_u^N| \le C$ (a.s.) and $\int_s^t |f_u - f_u^N|^2\, du \to 0$ in probability. Then $\int_s^t f_u^N\, dW_u \to \int_s^t f_u\, dW_u$ in probability and a subsequence $\int_s^t f_u^{N'}\, dW_u \to \int_s^t f_u\, dW_u$ (a.s.). From Fatou's lemma and inequality (5.7.10) it follows that

$$E\left[\int_s^t f_u\,dW_u\right]^{2n} \le \liminf_{N'\to\infty} E\left[\int_s^t f_u^{N'}\,dW_u\right]^{2n}$$
$$\le \frac{[2C^2(t-s)]^n\Gamma(n+\frac{1}{2})}{\Gamma(\frac{1}{2})}$$
$$= \frac{[2C^2(t-s)]^n(n-\frac{1}{2})(n-\frac{1}{2}-1)\cdots\frac{1}{2}\Gamma(\frac{1}{2})}{\Gamma(\frac{1}{2})}$$
$$\le [2C^2(t-s)]^n n!,$$

and (5.7.6) is proved.

Lemma 5.7.2. *Let condition (5.7.1) be satisfied. Fix $t,t' \in [0,T]$ such that $t < t'$. Then for any positive λ for which $8KT\lambda < 1$, $E[\exp(\lambda \sup_{t\le u\le t'} |\int_t^u \sigma_v\,dW_v|^2)]$ is finite and has an upper bound B which does not depend on t and t'.*

PROOF. Letting M_u be the continuous version of the stochastic integral $\int_0^u \sigma_v\,dW_v$, applying the martingale inequality (2.2.7) to $(M_u, \mathscr{F}_u)$ $(0 \le u \le T)$ and using the elementary inequality $|a+b|^p \le 2^{p-1}(|a|^p + |b|^p)$ $(p \ge 1)$, we have

$$E\left[\sup_{t\le u\le t'}\left|\int_t^u \sigma_v\,dW_v\right|^p\right] \le 2^{p-1}\left\{E\left[\sup_{t\le u\le t'}|M_u|^p\right] + E|M_t|^p\right\}$$
$$\le 2^{p-1}\left(\frac{p}{p-1}\right)^p E|M_{t'}|^p + 2^{p-1}E|M_t|^p. \tag{5.7.11}$$

From Lemma 5.7.1, taking $p = 2n$ and $C = K^{\frac{1}{2}}$, we obtain

$$E|M_{t'}|^{2n} \le [2t'K]^n n!$$

and

$$E|M_t|^{2n} \le [2tK]^n n!.$$

Substituting in inequality (5.7.11), we have

$$E\left[\sup_{t\le u\le t'}\left|\int_t^u \sigma_v\,dW_v\right|^{2n}\right] \le 2^{2n-1}n!\left[\left(\frac{2n}{2n-1}\right)^{2n}(2t'K)^n + (2tK)^n\right]$$
$$\le (1+e)(8KT)^n n!.$$

Hence

$$E\left[\exp\left(\lambda \sup_{t\le u\le t'}\left|\int_t^u \sigma_v\,dW_v\right|^2\right)\right] = \sum_{n=0}^{\infty}\frac{\lambda^n}{n!}E\left[\sup_{t\le u\le t'}\left|\int_t^u \sigma_v\,dW_v\right|^{2n}\right]$$
$$\le (1+e)\sum_{n=0}^{\infty}(8KT\lambda)^n = B < \infty$$

if $8KT\lambda < 1$. □

PROOF OF THEOREM 5.7.1. Take $t \in [0,T]$ and δ to be the positive number mentioned in condition (5.7.3) and small enough so that $4\delta^2 K < 1$. Let N be the integer defined by the inequality $(N-1)\delta < T \leq N\delta$. If $u \in [t, t+\delta]$, we have the following inequality from Eq. (5.1.1):

$$|\xi_u - \xi_t|^2 \leq 2\left[\int_t^u |b_v|\, dv\right]^2 + 2\left|\int_t^u \sigma_v\, dW_v\right|^2.$$

Now, from condition (C),

$$\left[\int_t^u |b_v|\, dv\right]^2 \leq \delta \int_t^u K[1 + \|\xi\|_v^2 + |L_v|^2]\, dv.$$

Here we have used the abbreviations b_v, σ_v, and L_v for $b(v,X,\xi)$, $\sigma(v,X,\xi)$, and $L(v,X)$. Using the fact that for $v \geq t$,

$$\|\xi\|_v = \max\left[\|\xi\|_t, \sup_{t \leq s \leq v} |\xi_s|\right],$$

we have

$$\begin{aligned} |\xi_u|^2 &\leq 2|\xi_t|^2 + 2|\xi_u - \xi_t|^2 \\ &\leq 2|\xi_t|^2 + 4\delta K\left[\delta\left\{1 + \|\xi\|_t^2 + \sup_{t \leq u \leq t+\delta} |\xi_u|^2\right\} + \int_t^{t+\delta} |L_v|^2\, dv\right] \\ &\quad + 4 \sup_{t \leq u \leq t+\delta} \left|\int_t^u \sigma_v\, dW_v\right|^2. \end{aligned} \tag{5.7.12}$$

Taking the supremum over $[t, t+\delta]$ of the quantity on the left-hand side of (5.7.12), we get

$$\begin{aligned} (1 - 4\delta^2 K) \sup_{t \leq u \leq t+\delta} |\xi_u|^2 &\leq 4\delta^2 K + (2 + 4\delta^2 K)\|\xi\|_t^2 + 4\delta K \int_t^{t+\delta} |L_v|^2\, dv \\ &\quad + 4 \sup_{t \leq u \leq t+\delta} \left|\int_t^u \sigma_v\, dW_v\right|^2. \end{aligned} \tag{5.7.13}$$

Let $\alpha_1 = 4\delta^2 K/(1 - 4\delta^2 K)$, $\alpha_2 = (2 + 4\delta^2 K)/(1 - 4\delta^2 K)$, $\alpha_3 = 4\delta K/(1 - 4\delta^2 K)$, $\alpha_4 = 4/(1 - 4\delta^2 K)$, and let r be an arbitrary positive number. Then

$$\begin{aligned} &E\left[\exp\left(r \sup_{t \leq u \leq t+\delta} |\xi_u|^2\right)\right] \\ &\quad \leq e^{\alpha_1 r} E\left[\exp\left(\alpha_2 r \|\xi\|_t^2 + \alpha_3 r \int_t^{t+\delta} |L_v|^2\, dv + \alpha_4 r \sup_{t \leq u \leq t+\delta} \left|\int_t^u \sigma_v\, dW_v\right|^2\right)\right]. \end{aligned}$$

Applying the Schwarz and Hölder inequalities to the expected value on the right-hand side, we obtain

$$\begin{aligned} E\left[\exp\left(r \sup_{t \leq u \leq t+\delta} |\xi_u|^2\right)\right] &\leq e^{\alpha_1 r} \left\{ E[e^{3\alpha_2 r \|\xi\|_t^2}] E\left[\exp\left(3\alpha_3 r \int_t^{t+\delta} |L_v|^2\, dv\right)\right] \right. \\ &\quad \left. \times E\left[\exp\left(3\alpha_4 r \sup_{t \leq u \leq t+\delta} \left|\int_t^u \sigma_v\, dW_v\right|^2\right)\right] \right\}^{1/3}. \end{aligned} \tag{5.7.14}$$

Let $a = \min(c_0/3\alpha_2, c_1/3\alpha_3, \lambda/3\alpha_4)$. Observe that by condition (5.7.3) $E[\exp(c_1 \int_t^{t+\delta} |L_v|^2 \, dv)] \le E[e^{c_1 Y_\delta}] < \infty$ and that by Lemma 5.7.2,

$$E\left[\exp\left(\lambda \sup_{t \le u \le t+\delta} \left|\int_t^u \sigma_v \, dW_v\right|^2\right)\right] \le B \quad \text{for } t, t+\delta \text{ in } [0,T].$$

Hence taking $r = a$ and $t = 0$ in (5.7.14), we get

$$\begin{aligned} E[e^{a\|\xi\|_\delta^2}] &\le e^{\alpha_1 a}\left\{E[e^{c_0|\eta|^2}]E\left[\exp\left(c_1 \int_0^\delta |L_v|^2 \, dv\right)\right]B\right\}^{1/3} \\ &= e^{\alpha_1 a}(K_0 AB)^{1/3} < \infty \end{aligned} \tag{5.7.15}$$

using (5.7.3), where $K_0 = E[e^{c_0|\eta|^2}]$ and $A = E[e^{c_1 Y_\delta}]$. Taking $t = \delta$ in (5.7.14), we have

$$\begin{aligned} E\left[\exp\left(r \sup_{\delta \le u \le 2\delta} |\xi_u|^2\right)\right] \le e^{\alpha_1 r}\Bigg\{E[e^{3\alpha_2 r\|\xi\|_\delta^2}]E\left[\exp\left(3\alpha_3 r \int_\delta^{2\delta} |L_v|^2 \, dv\right)\right] \\ \times E\left[\exp\left(3\alpha_4 r \sup_{\delta \le u \le 2\delta} \left|\int_\delta^u \sigma_v \, dW_v\right|^2\right)\right]\Bigg\}^{\frac{1}{3}}. \end{aligned}$$

If we choose $r = a/3\alpha_2$ ($<a$) in the above inequality,

$$E\left[\exp\left(\frac{a}{3\alpha_2} \sup_{\delta \le u \le 2\delta} |\xi_u|^2\right)\right] \le K_0^{\frac{1}{9}}(e^{3\alpha_1 a}AB)^{\frac{4}{9}} < \infty. \tag{5.7.16}$$

Since $\|\xi\|_{2\delta}^2 = \max(\|\xi\|_\delta^2, \sup_{\delta \le u \le 2\delta}|\xi_u|^2)$ it follows from (5.7.15) and (5.7.16) that

$$E[e^{a_1\|\xi\|_{2\delta}^2}] < \infty, \tag{5.7.17}$$

where $a_1 = a/3\alpha_2$. Proceeding similarly, let $a_1, \ldots, a_{i-1}$ be positive numbers such that

$$E[e^{a_{j-1}\|\xi\|_{j\delta}^2}] < \infty \quad \text{for } j = 2, \ldots, i. \tag{5.7.18}$$

Now taking $t = i\delta$ and $r = a_i = \min((a_{i-1})/3\alpha_2, a)$ in (5.7.14), we obtain

$$E\left[\exp\left(a_i \sup_{i\delta \le u \le (i+1)\delta} |\xi_u|^2\right)\right] \le e^{\alpha_1 a}\{E[e^{a_{i-1}\|\xi\|_{i\delta}^2}]AB\}^{\frac{1}{3}} \tag{5.7.19}$$

which is finite because of (5.7.18). Noting that $a_i < a_{i-1}$ and making use of (5.7.18) and (5.7.19), we obtain

$$E[e^{a_i\|\xi\|_{(i+1)\delta}^2}] < \infty. \tag{5.7.20}$$

Hence (5.7.18) holds for $j = 2, \ldots, N-1$. For the last subinterval $[(N-1)\delta, T]$ only a slight change in the above argument is needed. Observe that (5.7.14) holds with $t = (N-1)\delta$ and $t + \delta$ replaced by T. Define $c = \min(a_{N-2}/3\alpha_2, a)$ and choose $r = c$ in (5.7.14). Exactly as above we then get

$$\begin{aligned} E\left[\exp\left(c \sup_{(N-1)\delta \le u \le T} |\xi_u|^2\right)\right] &\le e^{\alpha_1 a}(AB)^{\frac{1}{3}}\{E[e^{a_{N-2}\|\xi\|_{(N-1)\delta}^2}]\}^{\frac{1}{3}} \\ &< \infty. \end{aligned}$$

Combining this with (5.7.18) with $j = N - 1$ we finally obtain $E[e^{c\|\xi\|_T^2}] < \infty$ and the proof is complete. $\square$

For a diffusion process which is a solution of the stochastic differential Equation (5.1.9) the statement of the theorem assumes a simpler form since condition (5.7.3) no longer plays a part.

Theorem 5.7.2. *Let conditions (5.1.10), (5.7.1), and (5.7.2) be satisfied. Then the diffusion process* (ξ_t) *which is the unique solution of (5.1.9) has the property*

$$E[e^{c\|\xi\|_T^2}] < \infty \tag{5.7.21}$$

for some positive constant c.

Remark 5.7.1. The boundedness condition on σ cannot be removed as the following example shows.

EXAMPLE 5.7.1. Consider the stochastic equation

$$d\xi_t = \xi_t \, dW_t$$

with $\xi_0 = 1$ (a.s.). Here $b(t,x) \equiv 0$ and $\sigma(t,x) = x$, so that (5.7.1) is violated. Condition (5.7.2), however, is trivially satisfied for every $c_0 > 0$. The solution is $\xi_t = e^{W_t - \frac{1}{2}t}$ and hence

$$\begin{aligned} E[e^{c|\xi_t|^2}] &= E[e^{ce^{2W_t - t}}] \\ &= \infty \quad \text{for every } c > 0. \end{aligned}$$

6 Functionals of a Wiener Process

6.1 Introduction

The purpose of this chapter is to derive representations of square-integrable functionals on Wiener space. This is a topic of importance in the theory of nonlinear prediction and filtering. The three main results in the literature derive for a square-integrable functional of a Wiener process (see definition below)

(a) An L^2-convergent expansion in terms of Hermite functionals—Cameron-Martin
(b) An L^2-convergent expansion in terms of multiple Wiener integrals—Ito
(c) An Ito stochastic integral representation.

These results were originally obtained in different contexts, each representing a different phase in the development of the subject. A unified theory is presented here. The first step in our development will be Ito's original definition of the multiple integral. Next, we shall give the Cameron-Martin theorem and a second (and alternative) definition of multiple Wiener integrals via the theory of symmetric tensor products of Hilbert spaces. The latter theory will be developed for general Gaussian processes using the idea of reproducing kernel Hilbert spaces. In the concluding section we obtain the Ito integral representation. Two proofs of this are given, one based on (b) and the other an independent proof.

6.2 The Multiple Wiener Integral

First we recall the familiar definition of the ordinary Wiener integral. Let (W_t), $t \in [0,1]$ be a standard Wiener process (hence sample continuous by the definition given in Chapter 2) defined on a complete probability space

$(\Omega,\mathscr{A},P)$. Let $\mathscr{F}^W = \mathscr{F}_1^W = \sigma\{W_s\colon 0 \le s \le 1\} \vee \{P\text{-null sets}\}$. Then the triplet $(\Omega,\mathscr{F}^W,P)$ will be called a *Wiener space* of the (Wiener) process (W_t). When the probability parameter is to be put in evidence, we shall write $W_t(\omega)$ instead of W_t. Let $L_1(W)$ be the linear (Hilbert) space of the process (W_t), that is, the closed linear subspace of $L^2 = L^2(\Omega,\mathscr{F}^W,P)$ generated by $\{W_s, 0 \le s \le 1\}$. Consider first the familiar definition of the ordinary Wiener integral. For $\phi \in L^2[0,1]$, the integral $\int_0^1 \phi(t)\,dW_t$ is defined as follows: let $\Pi\colon 0 = t_0 < t_1 < \cdots < t_n = 1$ be a partition of $[0,1]$, and let

$$\phi(t) = \sum c_i 1_{(t_{i-1},t_i]}(t)$$

where $(t_{i-1},t_i]$ denotes the interval $t_{i-1} < x \le t_i$. Let $\phi(0)$ be arbitrary. We define

$$I_1(\phi) = \int_0^1 \phi(t)\,dW_t = \sum_{i=1}^{n} c_i[W_{t_i} - W_{t_{i-1}}].$$

I_1 has the following properties. If ϕ and ψ are step functions as above and c is a real number, then

(i) $I_1(c\phi) = cI_1(\phi)$.
(ii) $I_1(\phi + \psi) = I_1(\phi) + I_1(\psi)$.
(iii) $E(I_1(\phi)) = 0$.
(iv) $E(|I_1(\phi)|^2) = \sum_{i=1}^n |c_i|^2(t_i - t_{i-1}) = \int_0^1 |\phi(t)|^2\,dt$.
(v) $E(I_1(\phi)I_1(\psi)) = \int_0^1 \phi(t)\psi(t)\,dt$.

All these properties are easy to prove. We note that in the proof of (iv) and (v) the fact that $[W_t{:}0 \le t \le 1]$ is a process with orthogonal increments is used.

Properties (i) to (v) of I_1 show that I_1 is an isometry from the linear space of step functions in $L^2[0,1]$ into $L_1(W)$, and hence it can be extended to an isometry from $L^2[0,1]$ onto $L_1(W)$. Thus $I_1(\phi)$ (called the *Wiener integral*) is defined for every ϕ in $L^2[0,1]$, and if we write $I_1(\phi) = \int_0^1 \phi(t)\,dW(t)$, then clearly (i) to (v) continue to hold for ϕ in $L^2[0,1]$. Let us write, with a slight abuse of notation, $W(B) = I_1(1_B)$, where $B \in \mathscr{B} = \mathscr{B}[0,1]$. Then $W(\cdot)$ is a Gaussian random measure on $\mathscr{B}$, that is,

(i) For each $B \in \mathscr{B}$, $W(B)$ is a Gaussian random variable with $E(W(B)) = 0$.
(ii) $E(|W(B)|^2) = m(B)$, where m denotes Lebesgue measure.
(iii) If $B_1,B_2,B_3,\ldots,$ are disjoint Borel sets, then

$$W\left(\bigcup_{i=1}^{\infty} B_i\right) = \sum_{i=1}^{\infty} W(B_i) \qquad (L^2\text{-convergence}).$$

One can define the ordinary Wiener integral in terms of Gaussian random measure. Let $X(\cdot)$ be a Gaussian random measure such that $E(X(B)X(C)) = m(B \cap C)$.

For a simple function $\phi = \sum_{i=1}^{n} c_i 1_{B_i}$, we write

$$\int_0^1 \phi(t)X(dt) = \sum_{i=0}^{n} c_i X(B_i).$$

Then

$$\phi \to \int_0^1 \phi(t)X(dt)$$

is a linear isometry between the linear space of simple functions in $L^2[0,1]$ and the Hilbert space $L_1(X)$ spanned by $[X(B): B \in \mathscr{B}]$. This extends to an isometry between $L^2[0,1]$ and $L_1(X)$. It is clear that if W_t is the process $W_t = X([0,t])$, then

$$\int_0^1 \phi(t)\, dW_t = \int_0^1 \phi(t)X(dt).$$

We now give a third method of defining the standard Wiener process and the associated Wiener integral. Let $(\xi_n)_{n=1}^{\infty}$ be a sequence of independent random variables each normally distributed with mean zero and variance one. Let $(\phi_n)_{n=1}^{\infty}$ be a CONS in $L^2[0,1]$, and let

$$W_t = \sum_{n=1}^{\infty} (1_{[0,t]}, \phi_n)\xi_n,$$

where $(\ , \ )$ denotes the inner product in $L^2[0,1]$. Recall from Remark 2.1.2 that $[W_t, 0 \le t \le 1]$ is a standard Wiener process. It is easy to verify that if $\phi \in L^2[0,1]$, then

$$\int_0^1 \phi(t)\, dW_t = \sum_{n=1}^{\infty} (\phi, \phi_n)\xi_n.$$

We turn now to the discussion of the multiple Wiener integral. Let $\mathbf{T}$ denote a closed bounded interval on the real line, and let $\{W_t, t \in \mathbf{T}\}$ be a Wiener process such that $E(W_t - W_s)^2 = t - s$ and for all t, $E(W_t) = 0$. Let $L_1(W)$ be the Hilbert space spanned by $[W_t: t \in \mathbf{T}]$. Let m denote the Lebesgue measure on T, and let m^p denote the product measure on $\mathbf{T} \times \mathbf{T} \times \cdots \times \mathbf{T} = \mathbf{T}^p$. Let $\pi: A_1, A_2, \ldots, A_n$ be a partition of $\mathbf{T}$ into Borel measurable subsets. By a special elementary function on $\mathbf{T}^p$ we mean a function f of the form

$$f = \sum_{i_1 \cdots i_p = 1}^{n} a_{i_1 i_2 \cdots i_p} 1_{A_{i_1} \times A_{i_2} \times \cdots \times A_{i_p}},$$

where $a_{i_1 \cdots i_p}$ are 0 unless $i_1, i_2, \ldots, i_p$ are all distinct. We note that if π_1: $B_1, B_2, \ldots, B_m$ is a measurable partition of T which is a refinement of $\pi: A_1, \ldots, A_n$, then f can also be expressed as $\sum_{i_1 \cdots i_p = 1}^{m} b_{i_1 \cdots i_p} 1_{B_{i_1} \times \cdots \times B_{i_p}}$. If f and g are two special elementary functions, then $f + g$ is also a special elementary function and so is a constant times f. Let S_p denote the linear space of all special elementary functions. Before we define multiple Wiener integral, we need the following.

Lemma 6.2.1. S_p *is dense in* $L^2(\mathbf{T}^p, m^p)$.

PROOF. Let F denote the set of all those $t = (t_1, \ldots, t_p)$ in $\mathbf{T}^p$ such that at least two coordinates are equal. Then F is a closed set. Further by Fubini's theorem it is easy to see that $m^p(F) = 0$. Hence any function $f \in L^2(\mathbf{T}^p, m^p)$ can be approximated arbitrarily closely by functions of the form

$$\sum_{k=1}^{n} a_k 1_{A_k},$$

where A_k's are Borel sets in $\mathbf{T}^p - F$. It is therefore enough to show that each Borel set A in $\mathbf{T}^p - F$ is approximable arbitrarily closely in $L^2(\mathbf{T}^p, m^p)$ by a special elementary function. Let $\varepsilon > 0$ and let $C \subset A$ be a closed set such that $m^p(A - C) < \varepsilon$. Since C is a closed set of $\mathbf{T}^p - F$, C is at a positive distance from F, and it is clear how to approximate 1_C by special elementary functions made out of cubes in p dimensions. Hence S_p is dense in $L^2(\mathbf{T}^p, m^p)$. □

Definition 6.2.1. Let $f = \sum_{i_1 \cdots i_p = 1}^{n} a_{i_1 \cdots i_p} 1_{A_{i_1} \times A_{i_2} \times \cdots \times A_{i_p}}$ be a special elementary function. By the *multiple Wiener integral* of f we mean

$$I_p(f) = \sum_{i_1 \cdots i_p = 1}^{n} a_{i_1 \cdots i_p} W(A_{i_1}) \cdots W(A_{i_p}),$$

where $W(A)$ denotes the random measure of A given by $[W(t), t \in T]$ on Borel subsets of $\mathbf{T}$. The reader may show that I_p is well-defined and linear on S_p. Before we extend I_p from S_p to all of $L^2(\mathbf{T}^p, m^p)$, we set down the properties of I_p on S_p which continue to hold when I_p is extended to $L^2(\mathbf{T}^p, m^p)$.

Properties of I_p

1. For any $f \in S_p$ let $\tilde{f}$ denote its symmetrization, that is,

$$\tilde{f}(\underset{\sim}{t}) = \frac{1}{p!} \sum_{\pi} f_\pi(\underset{\sim}{t}), \qquad \underset{\sim}{t} = (t_1, \ldots, t_p),$$

where $f_\pi(t_1, \ldots, t_p) = f(t_{\pi(1)}, \ldots, t_{\pi(p)})$ and π is a permutation of $(1,2,3, \ldots, p)$. Then

$$I_p(f) = I_p(\tilde{f}).$$

To see this, note that for each permutation π,

$$W(A_{i_1}) W(A_{i_2}) \cdots W(A_{i_p}) = W(A_{i_{\pi^{-1}(1)}}) \cdots W(A_{i_{\pi^{-1}(p)}})$$

and

$$I_p(\tilde{f}_\pi) = \sum_{i_1 \cdots i_p = 1}^{n} a_{i_1 \cdots i_p} W(A_{i_{\pi^{-1}(1)}}) \cdots W(A_{i_{\pi^{-1}(p)}}).$$

Hence $I_p(f) = I_p(\tilde{f})$.

2. $E(I_p(f)) = 0$. For

$$E(I_p(f)) = \sum_{i_1 \cdots i_p} a_{i_1 \cdots i_p} E(W(A_{i_1}) \cdots W(A_{i_p})).$$

When $i_1, \ldots, i_p$ are all distinct, $A_{i_1}, \ldots, A_{i_p}$ are all disjoint. Since $a_{i_1 \cdots i_p} = 0$ unless all $i_1, \ldots, i_p$ are distinct,

$$E(I_p(f)) = \sum_{i_1 \cdots i_p} a_{i_1 \cdots i_p} E(W(A_{i_1})) \cdots E(W(A_{i_p})) = 0.$$

3. $E(I_p(f)I_p(g)) = E(I_p(\tilde{f})I_p(\tilde{g})) = p!(\tilde{f},\tilde{g})$, where $(\ ,\ )$ denotes the inner product in $L^2(\mathbf{T}^p, m^p)$. Since $I_p(f) = I_p(\tilde{f})$, it is clear that

$$E(I_p(f)I_p(g)) = E(I_p(\tilde{f})I_p(\tilde{g})).$$

Now

$$\begin{aligned}
E(I_p(f)I_p(g)) &= E(\textstyle\sum a_{i_1 \cdots i_p} W(A_{i_1}) \cdots W(A_{i_p}) \sum b_{i_1 \cdots i_p} W(A_{i_1}) \cdots W(A_{i_p})) \\
&= E\left[\left(\sum_{i_1 < \cdots < i_p} \left[\sum_{(j) \sim (i)} a_{j_1, \ldots, j_p}\right] W(A_{i_1}) \cdots W(A_{i_p})\right)\right. \\
&\quad \times \left.\left(\sum_{i_1 < \cdots < i_p} \left[\sum_{(j) \sim (i)} b_{j_1, \ldots, j_p}\right] W(A_{i_1}) \cdots W(A_{i_p})\right)\right]. \\
&= \sum_{i_1 < \cdots < i_p} \left(\sum_{(j) \sim (i)} a_{j_1, \ldots, j_p}\right)\left(\sum_{(j) \sim (i)} b_{j_1, \ldots, j_p}\right) m(A_{i_1}) \cdots m(A_{i_p}) \\
&= \frac{1}{p!} \sum_{(i_1, \ldots, i_p)} \left(\sum_{(j) \sim (i)} a_{j_1, \ldots, j_p}\right)\left(\sum_{(j) \sim (i)} b_{j_1, \ldots, j_p}\right) \\
&\quad \times m(A_{i_1}) \cdots m(A_{i_p}) \\
&= p! \sum_{(i_1, \ldots, i_p)} \left[\frac{1}{p!} \sum_{(j) \sim (i)} a_{j_1, \ldots, j_p}\right]\left[\frac{1}{p!} \sum_{(j) \sim (i)} b_{j_1, \ldots, j_p}\right] \\
&\quad \times m(A_{i_1}) \cdots m(A_{i_p}) \\
&= p!(\tilde{f},\tilde{g}).
\end{aligned}$$

$(j) \sim (i)$ means that $(j) = (j_1, \ldots, j_p)$ is a permutation of $(i) = (i_1, \ldots, i_p)$.

4. $\|I_p(\tilde{f})\|^2 = p!\|\tilde{f}\|^2$. This follows immediately from property 3 by taking $f = g$.

Extension of $I_p(f)$. Since $I_p(f) = I_p(\tilde{f})$, we have $\|I_p(f)\| = \|I_p(\tilde{f})\| = \sqrt{p!}\|\tilde{f}\| \leq \sqrt{p!}\|f\|$. Thus I_p is a uniformly continuous function on S_p with values in the complete metric space $L^2(\Omega)$. Hence I_p has a unique extension to $L^2(\mathbf{T}^p, m^p)$.

Definition 6.2.2. I_p thus defined on $L^2(\mathbf{T}^p, m^p)$ is called the *multiple Wiener integral.* I_p again has the properties 1 to 4.

6.3 Hilbert Spaces Associated with a Gaussian Process

Let (X_t), $t \in \mathbf{T}$, be a real Gaussian process defined on a probability space $(\Omega,\mathscr{A},P)$ and define the σ-field $\mathscr{F}^X = \sigma\{X_t, t \in \mathbf{T}\} \vee \{P\text{-null sets}\}$. The mean function $E(X_t)$ will be assumed to be 0 and the covariance function $E(X_tX_s)$ will be denoted by $R(t,s)$. The index set $\mathbf{T}$ will be assumed to be a complete, separable metric space, and R is assumed continuous. The following facts are well known: R determines a Hilbert space $H(R)$ called the *reproducing kernel Hilbert space* (RKHS) of R or of the process (X_t). For each $t \in \mathbf{T}$, $R(\cdot,t) \in H(R)$ and $(f,R(\cdot,t)) = f(t)$ for each $f \in H(R)$ where $(\ ,\ )$ is the inner product in $H(R)$. The separability of $H(R)$ follows easily from the assumptions on $\mathbf{T}$ and R. As in the case of the Wiener process, we define $L_1(W)$ to be the closed linear subspace of $L^2(\Omega,\mathscr{F}^X,P)$ spanned by all finite, real linear combinations $\sum_{i=1}^n c_i X_{t_i}$. The linear map

$$\sum_{i=1}^{n} c_i X_{t_i} \xrightarrow{\psi_1} \sum_{i=1}^{n} c_i R(\cdot,t_i)$$

extends to a congruence (that is, isometric isomorphism) between $L_1(X)$ and $H(R)$. This congruence has played an important part in the investigation of linear estimation and prediction problems. The spaces $L_1(X)$ and $H(R)$ and their respective tensor products form the building blocks for a study of the nonlinear theory. The next few sections of this chapter will be devoted to preparatory results on tensor and symmetric tensor products of Hilbert spaces and to the homogeneous chaos of a Gaussian process, that is, a certain direct sum decomposition of L^2 from which the Cameron-Martin result and Ito's result follow. The treatment given here closely follows [28].

6.4 Tensor Products and Symmetric Tensor Products of Hilbert Spaces

Let H_1 and H_2 be real separable Hilbert spaces with inner products $(\ ,\ )_1$ and $(\ ,\ )_2$. We shall assume that the reader is familiar with the definition of the tensor- (or direct-) product Hilbert space of H_1 and H_2, written $H_1 \otimes H_2$.

For any two elements h_1 and h_2 ($h_i \in H_i$), let $h_1 \otimes h_2$ denote their tensor product, which is an element of $H_1 \otimes H_2$. Let $(\ ,\ )$ denote the inner product for $H_1 \otimes H_2$. Some useful facts about a tensor product Hilbert space are given without proof in the following result.

Lemma 6.4.1. (i) *If h_i and g_i are elements of H_i, then*

$$(h_1 \otimes h_2, g_1 \otimes g_2) = (h_1,g_1)_1(h_2,g_2)_2.$$

(ii) *The set of all finite linear combinations of the form*

$$\sum_{k=1}^{n} c_k(h_1^k \otimes h_2^k)$$

where the c_k are real, $h_1^k \in H_1$, and $h_2^k \in H_2$, is dense in $H_1 \otimes H_2$.

(iii) *Let $\{f_\alpha\}$ be a set that spans H_1, that is, such that H_1 is the closed linear manifold spanned by the f_α. Similarly, let $\{g_\beta\}$ be a set that spans H_2. Then the set of elements $\{f_\alpha \otimes g_\beta\}$ spans $H_1 \otimes H_2$.*

(iv) *Let $\{e_i\}_1^\infty$ be a complete orthonormal system* (CONS) *in H_1 and $\{f_j\}_1^\infty$ be a* CONS *in H_2. Then $\{e_i \otimes f_j\colon i, j = 1,2, \ldots\}$ is a* CONS *in $H_1 \otimes H_2$.*

Both the definition of the tensor product and Lemma 6.4.1 generalize in an obvious way to the case of any finite number of Hilbert spaces H_i $(i = 1, \ldots, p)$. Then tensor-product Hilbert space in that case is written $H_1 \otimes \cdots \otimes H_p$. When all the spaces are the same, say, H, we introduce the shorter notation $\bigotimes^p H$.

Let $h_1 \otimes \cdots \otimes h_p$ be an element of $\bigotimes^p H$. Define

$$\sigma(h_1 \otimes \cdots \otimes h_p) = \frac{1}{p!} \sum_\pi h_{\pi_1} \otimes \cdots \otimes h_{\pi_p}, \tag{6.4.1}$$

where $\pi = (\pi_1, \ldots, \pi_p)$ is a permutation of the integers $(1, \ldots, p)$.

The symmetric tensor-product Hilbert space $\sigma[\bigotimes^p H]$ is the closed linear subspace of $\bigotimes^p H$ generated by elements of the form

$$\sum_{k=1}^{n} c_k \sigma(h_1^k \otimes \cdots \otimes h_p^k).$$

In fact, it can be shown that σ defined by Definition 6.2.1 can be extended to define a projection operator on $\bigotimes^p H$ whose range is $\sigma[\bigotimes^p H]$. We shall follow these ideas more closely for $H = H(R)$, the RKHS of the Gaussian process (X_t), $t \in \mathbf{T}$.

First we introduce the following notation. Write $\mathbf{T}^p = \mathbf{T} \times \cdots \times \mathbf{T}$ and let $\bigotimes^p R$ denote the following covariance function on $\mathbf{T}^p \times \mathbf{T}^p$:

$$\bigotimes^p R(s_1, \ldots, s_p; t_1, \ldots, t_p) = R(s_1, t_1) \cdots R(s_p, t_p) \tag{6.4.2}$$

for $(s_1, \ldots, s_p)$, $(t_1, \ldots, t_p)$ in $\mathbf{T}^p$. Letting $H(\bigotimes^p R)$ be the RKHS generated by $\bigotimes^p R$, we have the following lemma. The symbol $\cong$ stands for an isometric isomorphism.

Lemma 6.4.2. $\bigotimes^p H(R) \cong H(\bigotimes^p R)$.

PROOF. By Lemma 6.4.1 (ii), the set of elements

$$\{R(\cdot, t_1) \otimes \cdots \otimes R(\cdot, t_p), (t_1, \ldots, t_p) \in \mathbf{T}^p\}$$

spans $\bigotimes^p H(R)$. Also, by the definition of a RKHS, the set

$$\{(\bigotimes^p R)(\cdot\,; t_1, \ldots, t_p), (t_1, \ldots, t_p) \in \mathbf{T}^p\}$$

spans $H(\bigotimes^p R)$. From (i) of Lemma 6.4.1 we have

$$\begin{aligned}
&(R(\cdot,s_1) \otimes \cdots \otimes R(\cdot,s_p), R(\cdot,t_1) \otimes \cdots \otimes R(\cdot,t_p)) \\
&\quad = \prod_{i=1}^{p} (R(\cdot,s_i), R(\cdot,t_i)) \\
&\quad = \prod_{i=1}^{p} R(s_i,t_i) \\
&\quad = ((\textstyle\bigotimes^p R)(\cdot;s_1,\ldots,s_p), \textstyle\bigotimes^p R(\cdot;t_1,\ldots,t_p)).
\end{aligned} \tag{6.4.3}$$

The inner products appearing at the extreme ends of the relation (6.4.3) refer, as is obvious from the context, to the Hilbert spaces $\bigotimes^p H(R)$ and $H(\bigotimes^p R)$, respectively.

Unless the context makes it necessary, we shall use the same symbol for the inner product of different Hilbert spaces. The same applies also to the norm. From (6.4.3) the correspondence

$$R(\cdot,t_1) \otimes \cdots \otimes R(\cdot,t_p) \leftrightarrow (\textstyle\bigotimes^p R)(\cdot;t_1,\ldots,t_p) \tag{6.4.4}$$

can be extended to a congruence (that is, an isometric isomorphism) between $\bigotimes^p H(R)$ and $H(\bigotimes^p R)$. □

In what follows we write K in place of $\bigotimes^p R$.

Lemma 6.4.3. *Let $\mathscr{P}$ be defined on $H(K)$ by*

$$(\mathscr{P}F)(t_1,\ldots,t_p) = \frac{1}{p!} \sum_{\pi} F_\pi(t_1,\ldots,t_p), \tag{6.4.5}$$

where π is a permutation of $(1,\ldots,p)$ and

$$F_\pi(t_1,\ldots,t_p) = F(t_{\pi_1},\ldots,t_{\pi_p}), \qquad F \in H(K). \tag{6.4.6}$$

Then $\mathscr{P}$ is an orthogonal projection operator on $H(K)$ whose range S is the closed linear subspace of $H(K)$ consisting of all symmetric functions in $H(K)$. $\mathscr{P}F$ (denoted also by $\tilde{F}$) is called the symmetrization of F.

Proof. We first show that $F_\pi \in H(K)$. If $F \in H(K)$, there exists a sequence $\xi^{(m)} = \sum_{k=1}^{n_m} C_{m,k} K(\cdot;u_1^{m,k},\ldots,u_p^{m,k})$ which converges to F in $H(K)$-norm. Taking inner products with $K(\cdot;t_{\pi_1},\ldots,t_{\pi_p})$ gives

$$\begin{aligned}
F(t_{\pi_1},\ldots,t_{\pi_p}) &= \lim_{m\to\infty} \sum_{k=1}^{n_m} C_{m,k} K(u_1^{m,k},\ldots,u_p^{m,k}; t_{\pi_1},\ldots,t_{\pi_p}) \\
&= \lim_{m\to\infty} \sum_{k=1}^{n_m} C_{m,k} \prod_{i=1}^{p} R(u_i^{m,k},t_{\pi_i}) \\
&= \lim_{m\to\infty} \sum_{k=1}^{n_m} C_{m,k} K(u_{\pi_1^{-1}}^{m,k},\ldots,u_{\pi_p^{-1}}^{m,k}; t_1,\ldots,t_p) \\
&= \lim_{m\to\infty} (\xi_\pi^{(m)}, K(\cdot;t_1,\ldots,t_p)),
\end{aligned}$$

where each $\xi_\pi^{(m)} = \sum_{k=1}^{n_m} C_{m,k} K(\cdot\,; u_{\pi_1}^{m,k}, \ldots, u_{\pi_p}^{m,k})$ clearly belongs to $H(K)$. Next, note that for any m and m',

$$
\begin{aligned}
(\xi_\pi^{(m)}, \xi_\pi^{(m')}) &= \sum_{k=1}^{n_m} \sum_{k'=1}^{n_{m'}} C_{m,k} C_{m',k'} \prod_{i=1}^{p} R(u_{\pi_i}^{m,k}, u_{\pi_i}^{m',k'}) \\
&= \sum_{k=1}^{n_m} \sum_{k'=1}^{n_{m'}} C_{m,k} C_{m',k'} \prod_{i=1}^{p} R(u_i^{m,k}, u_i^{m',k'}) \\
&= (\xi^{(m)}, \xi^{(m')}). \qquad (6.4.7)
\end{aligned}
$$

The sequence $(\xi_\pi^{(m)})$ will then be, like $(\xi^{(m)})$, a Cauchy sequence in $H(K)$ converging to some element F'. We then have

$$
\begin{aligned}
F_\pi(t_1, \ldots, t_p) &= \lim_{m \to \infty} (\xi_\pi^{(m)}, K(\cdot\,; t_1, \ldots, t_p)) \\
&= (F', K(\cdot\,; t_1, \ldots, t_p)) \\
&= F'(t_1, \ldots, t_p).
\end{aligned}
$$

Hence $F_\pi = F' \in H(K)$, and from (6.4.7), taking the limit as $m,m' \to \infty$,

$$\|F_\pi\| = \|F\|.$$

For each permutation π, the operator $\mathscr{Q}_\pi(F) = F_\pi$ is linear. Also $\mathscr{Q}_\pi$ is easily seen to have an inverse, namely $\mathscr{Q}_{\pi^{-1}}$. Each $\mathscr{Q}_\pi$ is thus an isomeric isomorphism of $H(K)$ onto itself, with adjoint equal to its inverse. By definition,

$$\mathscr{P} = \frac{1}{p!} \sum_\pi \mathscr{Q}_\pi.$$

Thus

$$\mathscr{P}^* = \frac{1}{p!} \sum_\pi \mathscr{Q}_\pi^* = \frac{1}{p!} \sum_\pi \mathscr{Q}_{\pi^{-1}} = \mathscr{P}.$$

Also from the definition, we obtain for all π,

$$(\mathscr{P}F)_\pi = \mathscr{P}F,$$

so that

$$\mathscr{P}^2 F = \frac{1}{p!} \sum_\pi (\mathscr{P}F)_\pi = \mathscr{P}F.$$

$\mathscr{P}$ is therefore a projection as claimed. The identification of the subspace S as its range is immediate, and the lemma is proved. □

Let

$$K_{\mathscr{P}}(\cdot\,; t_1, \ldots, t_p) = \mathscr{P}[K(\cdot\,; t_1, \ldots, t_p)]. \qquad (6.4.8)$$

Lemma 6.4.4.

$$\sigma\left(\bigotimes^p H(R)\right) \cong S = H(K_{\mathscr{P}}).$$

Proof. It follows from Lemma 6.4.3 that for every $(t_1, \ldots, t_p) \in \mathbf{T}^p$, $\mathscr{P}[K(\cdot\,; t_1, \ldots, t_p)]$ belongs to S, and it is easy to see that the set of all such functions spans S. Furthermore, $K_{\mathscr{P}}$ is a covariance function (symmetric and positive definite) on $\mathbf{T}^p \times \mathbf{T}^p$. Since the functions $K_{\mathscr{P}}(\cdot\,;t_1,\ldots,t_p)$ span $H(K_{\mathscr{P}})$, we need only show that the two inner products of $H(K)$ and $H(K_{\mathscr{P}})$ coincide on this generating class.

$$(K_{\mathscr{P}}(\cdot\,;t_1,\ldots,t_p), K_{\mathscr{P}}(\cdot\,;s_1,\ldots,s_p))_{H(K_{\mathscr{P}})} = K_{\mathscr{P}}(s_1,\ldots,s_p;\, t_1,\ldots,t_p)$$

and

$$\begin{aligned}(\mathscr{P}K(\cdot\,;t_1,\ldots,t_p), \mathscr{P}K(\cdot\,;s_1,\ldots,s_p))_{H(K)} &= (\mathscr{P}^2K(\cdot\,;t_1,\ldots,t_p), K(\cdot\,;s_1,\ldots,s_p))_{H(K)} \\ &= (\mathscr{P}K(\cdot\,;t_1,\ldots,t_p), K(\cdot\,;s_1,\ldots,s_p))_{H(K)} \\ &= [\mathscr{P}K(\cdot\,;t_1,\ldots,t_p)](s_1,\ldots,s_p).\end{aligned}$$

These are equal by the defining equation (6.4.8). Hence

$$S = H(K_{\mathscr{P}}).$$

To show that $\sigma[\bigotimes^p H(R)] \cong S$, observe that the set of elements

$$\{\sigma[R(\cdot,t_1) \otimes \cdots \otimes R(\cdot,t_p)]\},$$

where σ is defined in (6.4.1), spans $\sigma[\bigotimes^p H(R)]$. Letting ϕ denote the congruence relation of Lemma 6.4.2, it can be seen from (6.4.4) that

$$\begin{aligned}\phi(\sigma[R(\cdot,t_1) \otimes \cdots \otimes R(\cdot,t_p)]) &= \phi\left(\frac{1}{p!}\sum_{\pi} R(\cdot,t_{\pi_1}) \otimes \cdots \otimes R(\cdot,t_{\pi_p})\right) \\ &= \frac{1}{p!}\sum_{\pi} \bigotimes^p R(\cdot\,;t_{\pi_1},\ldots,t_{\pi_p}) \\ &= \mathscr{P}[K(\cdot\,;t_1,\ldots,t_p)].\end{aligned}$$

Formula (6.4.8) and the fact that $\{K_{\mathscr{P}}(\cdot\,;t_1,\ldots,t_p)\}$ spans S complete the proof of the theorem. □

Remark 6.4.1. It is clear from the above proof that the projection operators σ of (6.4.1) and $\mathscr{P}$ of (6.4.5) are connected by the relation

$$\sigma = \phi^{-1}\mathscr{P}\phi.$$

From the congruence of $L_1(X)$ and $H(R)$, we deduce the following corollary to Lemmas 6.4.2 and 6.4.4.

Corollary 6.4.1.

$$\bigotimes^p L_1(X) \cong H(\bigotimes^p R)$$

and

$$\sigma[\bigotimes^p L_1(X)] \cong \sigma[\bigotimes^p H(R)] \cong H(K_{\mathscr{P}}). \tag{6.4.9}$$

We shall denote the congruence relations between $\sigma[\bigotimes^p L_1(X)]$ and either of the other two Hilbert spaces by the same symbol ψ_p.

6.5 CONS in $\sigma[\otimes^p H(R)]$ and $\sigma[\otimes^p L_1(X)]$

We shall assume throughout that $\{e_i\}_1^\infty$ is a fixed (but arbitrary) CONS in $H(R)$ and that $\{\xi_i\}_1^\infty$ are the random variables over $(\Omega,\mathscr{F}^X,P)$ such that ξ_i is the element of $L_1(X)$ which corresponds to e_i in $H(R)$ under the isomorphism ψ_1 between the two Hilbert spaces. Let $p \geq 1$ be any integer. From Lemma 6.4.1 (iv) (extended to p-fold tensor products), it is clear that the elements $e_{i_1} \otimes \cdots \otimes e_{i_p}$, as $i_1, \ldots, i_p$ range independently from 1 to ∞, form a CONS for $\bigotimes^p H(R)$. It is more convenient to arrange the elements of this CONS in the following manner. Let $\lambda_1, \ldots, \lambda_r$ be the distinct integers in the sequence $(i_1, \ldots, i_p)$ with λ_1 occurring n_1 times, λ_2 occurring n_2 times, etc., and $\lambda_1 < \lambda_2 < \cdots < \lambda_r$. Then $n_i > 0$ and $n_1 + \cdots + n_r = p$. The following lemma gives a description of a CONS in $\sigma(\bigotimes^p H(R))$ and $\sigma(\bigotimes^p L_1(X))$ in terms of $\{e_i\}$ and $\{\xi_i\}$. Define

$$e_{\lambda_1, \ldots, \lambda_r}^{n_1, \ldots, n_r} = \frac{\sqrt{p!}}{(n_1! \cdots n_r!)^{\frac{1}{2}}} \sigma(e_{i_1} \otimes \cdots \otimes e_{i_p}). \tag{6.5.1}$$

As defined earlier,

$$\sigma(e_{i_1} \otimes \cdots \otimes e_{i_p}) = \frac{1}{p!} \sum_{(j) \sim (i)} e_{j_1} \otimes \cdots \otimes e_{j_p}, \tag{6.5.2}$$

where (j) is a permutation $(j_1, \ldots, j_p)$ of $(i) = (i_1, \ldots, i_p)$. We define the elements $\xi_{\lambda_1, \ldots, \lambda_r}^{n_1, \ldots, n_r}$ in a similar manner.

Lemma 6.5.1. (i) *The elements* $\{e_{\lambda_1, \ldots, \lambda_r}^{n_1, \ldots, n_r}, \sum_1^r n_i = p, 1 \leq \lambda_1 < \cdots, \text{and } n_i > 0\}$ *form a* CONS *in* $\sigma(\bigotimes^p H(R))$;

(ii) *The elements* $\{\xi_{\lambda_1, \ldots, \lambda_r}^{n_1, \ldots, n_r}, \sum_1^r n_i = p, 1 \leq \lambda_1 < \cdots, \text{and } n_i > 0\}$ *form a* CONS *in* $\sigma(\bigotimes^p L_1(X))$.

PROOF. (i) There are $N = p!/(n_1! \cdots n_r!)$ distinct permutations of $(i) = (i_1, \ldots, i_p)$. Denoting by $C_1, \ldots, C_N$ the classes representing the distinct permutations, we have

$$\begin{aligned} \left\| \sum_{(j) \sim (i)} e_{j_1} \otimes \cdots \otimes e_{j_p} \right\|^2 &= \left\| \sum_{i=1}^N \sum_{(j) \in C_i} e_{j_1} \otimes \cdots \otimes e_{j_p} \right\|^2 \\ &= \sum_{i=1}^N (n_1! \cdots n_r!)^2 = p!(n_1! \cdots n_r!). \end{aligned} \tag{6.5.3}$$

Formulas (6.5.1) to (6.5.3) give

$$\|e_{\lambda_1, \ldots, \lambda_r}^{n_1, \ldots, n_r}\| = 1. \tag{6.5.4}$$

Also, if $(i) = (i_1, \ldots, i_p) \neq (k_1, \ldots, k_p)$, then

$$\left(\sum_{(j) \sim (i)} e_{j_1} \otimes \cdots \otimes e_{j_p}, \sum_{(m) \sim (k)} e_{m_1} \otimes \cdots \otimes e_{m_p} \right)$$

is clearly 0, so that it follows from (6.5.1) that the set $\{e^{n_1,\ldots,n_r}_{\lambda_1,\ldots,\lambda_r}\}$ is orthonormal. Let $h \in \sigma[\bigotimes^p H(R)]$ be such that for all $(n_1, \ldots, n_r)$, $(\lambda_1, \ldots, \lambda_r)$,

$$(h, e^{n_1,\ldots,n_r}_{\lambda_1,\ldots,\lambda_r}) = 0. \tag{6.5.5}$$

Again from (6.5.1) and the definition of $\sigma[\bigotimes^p H(R)]$, the left-hand side of (6.5.5) equals

$$(\sigma h, e_{i_1} \otimes \cdots \otimes e_{i_p}) = (h, e_{i_1} \otimes \cdots \otimes e_{i_p}).$$

Hence

$$(h, e_{i_1} \otimes \cdots \otimes e_{i_p}) = 0 \tag{6.5.6}$$

for all sequences $(i_1, \ldots, i_p)$ of positive integers. Equation (6.5.6) implies that $h = 0$ and (i) is proved. The proof of (ii) follows from (i) upon using the congruence ψ_p established in Corollary 6.4.1. □

6.6 Homogeneous Chaos

First we shall define the following linear subspaces of $L^2(\Omega, \mathscr{F}^X, P)$.

Let $G_0 = \{1\}$ be the closed linear subspace spanned by the constant random variables, and for $p \geq 1$ let $\hat{G}_p$ be the linear subspace of all polynomials in $\{\xi_i\}_1^\infty$ of degree not exceeding p. Let us write $G_p = \hat{G}_p \ominus \hat{G}_{p-1}$, the set of all polynomials in $\hat{G}_p$ orthogonal to every polynomial in $\hat{G}_{p-1}$. Finally, let $\bar{G}_p$ be the closed linear subspace spanned by G_p. In all of this we set $\bar{G}_0 = \hat{G}_0 = G_0$.

Definition 6.6.1. The subspace $\bar{G}_p$ of L^2 is called the pth *homogeneous chaos*. G_p is called the pth *polynomial chaos*.

To simplify the notation in the next lemma we write a polynomial θ_p of degree at most p in the variables $u_1, \ldots, u_r$ as

$$\theta_p[u_1, \ldots, u_r] = \sum a_{m_1,\ldots,m_r} u_1^{m_1} \cdots u_r^{m_r},$$

where the m_i are nonnegative integers and

$$a_{m_1,\ldots,m_r} = 0 \quad \text{if } m_1 + \cdots + m_r > p.$$

Let $h_n(x)$ denote the nth normalized Hermite polynomial in x, that is,

$$h_n(x) = (-1)^n \frac{1}{\sqrt{n!}} e^{x^2/2} \left(\frac{d}{dx}\right)^n e^{-x^2/2}. \tag{6.6.1}$$

Lemma 6.6.1. *A random variable γ belongs to $G_p (p \geq 1)$ if and only if it is of the form*

$$\gamma(\omega) = \sum a_{m_1,\ldots,m_r} h_{m_1}[\xi_{\lambda_1}(\omega)] \cdots h_{m_r}[\xi_{\lambda_r}(\omega)] \tag{6.6.2}$$

for some choice of distinct integers $\lambda_1, \ldots, \lambda_r$. In the expression on the right-hand side of (6.6.2), the summation is over $m_i \geq 0$, r and $\lambda_1, \ldots, \lambda_r$ being fixed,

$h_n(x)$ is the n*th normalized Hermite polynomial given in (6.6.1), and the coefficients $a_{m_1,\ldots,m_r}$ satisfy*

$$a_{m_1,\ldots,m_r} = 0 \qquad \text{if } m_1 + \cdots + m_r \neq p. \tag{6.6.3}$$

PROOF. Suppose $\gamma \in G_p$. Then since $\gamma \in \hat{G}_p$, γ is a polynomial θ_p of degree less than or equal to p in some r of the random variables $\{\xi_i\}_1^\infty$, say, $\xi_{\lambda_1},\ldots,\xi_{\lambda_r}$, that is,

$$\theta_p(\xi_{\lambda_1},\ldots,\xi_{\lambda_r}) = \sum c_{m_1,\ldots,m_r}\xi_{\lambda_1}^{m_1}\cdots\xi_{\lambda_r}^{m_r}, \tag{6.6.4}$$

where $m_i \geq 0$ and $m_1 + \cdots + m_r \leq p$. Clearly, the polynomial $\theta_p(x_1,\ldots,x_r)$ in the r real variables $x_1,\ldots,x_r$ belongs to $L^2[(-\infty,\infty)^r, v_r]$, where

$$v_r(dx_1,\ldots,dx_r) = (2\pi)^{-r/2}e^{-\frac{1}{2}(x_1^2+\cdots+x_r^2)}\,dx_1\cdots dx_r.$$

Since the space $L^2[(-\infty,\infty)^r, v_r]$ is spanned by the complete orthonormal family $\{h_{n_1}(x_1)\cdots h_{n_r}(x_r), n_i \geq 0\}$, it follows that

$$\theta_p(x_1,\ldots,x_r) = \sum_{\substack{n_i=0\\(i=1,\ldots,r)}}^{\infty} a_{n_1,\ldots,n_r}h_{n_1}(x_1)\cdots h_{n_r}(x_r). \tag{6.6.5}$$

From (6.6.4) and (6.6.5) it follows that

$$a_{n_1,\ldots,n_r} = \sum_{\substack{m_i\geq 0\\ m_1+\cdots+m_r\leq p}} c_{m_1,\ldots,m_r}\prod_{i=1}^{r}\left(\frac{1}{\sqrt{2\pi}}\int_{-\infty}^{\infty} e^{-\frac{1}{2}x_i^2}x_i^{m_i}h_{n_i}(x_i)\,dx_i\right). \tag{6.6.6}$$

If $n_1 + \cdots + n_r > p$, there is at least one m_i $(i = 1,\ldots,r)$ for which $m_i < n_i$, so that

$$\frac{1}{\sqrt{2\pi}}\int_{-\infty}^{\infty} e^{-\frac{1}{2}x_i^2}x_i^{m_i}h_{n_i}(x_i)\,dx_i = 0.$$

Hence every term in the sum on the right-hand side of (6.6.6) vanishes and we obtain

$$a_{n_1,\ldots,n_r} = 0 \quad \text{whenever } n_1 + \cdots + n_r > p. \tag{6.6.7}$$

On the other hand, if $n_1 + \cdots + n_r < p$,

$$\begin{aligned} a_{n_1,\ldots,n_r} &= E[\theta_p(\xi_{\lambda_1},\ldots,\xi_{\lambda_r})h_{n_1}(\xi_{\lambda_1})\cdots h_{n_r}(\xi_{\lambda_r})] \\ &= 0, \end{aligned} \tag{6.6.8}$$

since θ_p is orthogonal to all polynomials in $\hat{G}_{p-1}$ and $h_{n_1}(\xi_{\lambda_1})\cdots h_{n_r}(\xi_{\lambda_r})$ is a polynomial of degree $n_1 + \cdots + n_r \leq p - 1$. Formulas (6.6.7) and (6.6.8) prove (6.6.2) and (6.6.3). Conversely, if γ is a random variable of the form (6.6.2) where (6.6.3) holds, then clearly $\gamma \in \hat{G}_p$ and γ is orthogonal to $\hat{G}_{p-1}$ and hence belongs to G_p. This completes the proof of the lemma. □

The remaining theorems in this section comprise our general discussion of homogeneous chaos for Gaussian processes. The central theorem (Theorem 6.6.1) gives the homogeneous chaos decomposition of the L^2-space of a Gaussian process. The other theorems establish congruences between several

Hilbert spaces of interest. We have tried to present them in a form which brings out clearly the part played by the RKHS of the process and exploits the "duality" between the tensor products of the space of random variables and the corresponding tensor products of the RKHS.

The classical result of Cameron and Martin for the Wiener space, as well as Ito's version of the same in terms of multiple Wiener integrals, follows as an immediate consequence of Theorem 6.6.1.

Let H_p ($p = 0,1,2, \ldots$) be Hilbert spaces with inner products $(\ , \ )_p$. We shall denote by $\sum_{p \geq 0} \oplus H_p$ their orthogonal (external) direct-sum Hilbert space.

Lemma 6.6.2.

$$\sum_{p \geq 0} \oplus \sigma\left[\bigotimes^p L_1(X)\right] \overset{\psi}{\cong} \sum_{p \geq 0} \oplus \sigma\left[\bigotimes^p H(R)\right]. \tag{6.6.9}$$

Furthermore, let u be an element of the Hilbert space on the left-hand side of (6.6.9), and let

$$\psi(u) = F.$$

Then

$$u = \sum_{p \geq 0} \sum_{\substack{n_1 + \cdots + n_r = p \\ n_i \geq 1}} \sum_{\lambda_1 < \lambda_2 < \cdots < \lambda_r} (F_p, e_{\lambda_1, \ldots, \lambda_r}^{n_1, \ldots, n_r})_p \, \xi_{\lambda_1, \ldots, \lambda_r}^{n_1, \ldots, n_r},$$

where

$$F_p \in \sigma\left[\bigotimes^p H(R)\right] \qquad (p = 0,1, \ldots) \tag{6.6.10}$$

are uniquely determined and

$$\sum_{p \geq 0} \sum_{\substack{n_1 + \cdots + n_r = p \\ n_i \geq 1}} \sum_{\lambda_1 < \lambda_2 < \cdots < \lambda_r} (F_p, e_{\lambda_1, \ldots, \lambda_r}^{n_1, \ldots, n_r})_p^2 < \infty.$$

PROOF. Formula (6.6.9) follows from Corollary 6.4.1. Let

$$u \in \sum_{p \geq 0} \oplus \sigma[\otimes^p L_1(X)].$$

Then

$$u = \sum_{p \geq 0} u_p,$$

where

$$u_p \in \sigma\left[\bigotimes^p L_1(X)\right]$$

are uniquely determined by u,

$$u_p \perp u_q \ (p \neq q), \tag{6.6.11}$$

and

$$\sum_{p \geq 0} \|u_p\|_p^2 < \infty.$$

{Note that (6.6.11) means that u_p and u_q are orthogonal as elements of the Hilbert space $\sum_{p \geq 0} \oplus \sigma[\otimes^p L_1(X)]$.}

From Lemma 6.5.1 (ii) it follows that for $p \geq 1$,

$$u_p = \sum_{\substack{n_1 + \cdots + n_r = p \\ n_i \geq 1}} \sum_{\lambda_1 < \cdots < \lambda_r} (u_p, \xi_{\lambda_1, \ldots, \lambda_r}^{n_1, \ldots, n_r})_p \xi_{\lambda_1, \ldots, \lambda_r}^{n_1, \ldots, n_r}.$$

Let

$$F_p = \psi_p(u_p),$$

where ψ_p is the congruence introduced in Corollary 6.4.1. Clearly F_p satisfies (6.6.10). Also from (6.4.9) and Lemma 6.5.1(i) we have

$$(u_p, \xi_{\lambda_1, \ldots, \lambda_r}^{n_1, \ldots, n_r})_p = (F_p, e_{\lambda_1, \ldots, \lambda_r}^{n_1, \ldots, n_r})_p.$$

□

Lemma 6.6.3. *For each* $p \geq 1$,

$$\sigma[\bigotimes^p L_1(X)] \overset{J_p}{\cong} \bar{G}_p. \tag{6.6.12}$$

PROOF. Recalling the definition given in Section 6.5 of $\xi_{\lambda_1, \ldots, \lambda_r}^{n_1, \ldots, n_r}$ in terms of the ξ_i, define J_p first for $\xi_{\lambda_1, \ldots, \lambda_r}^{n_1, \ldots, n_r}$ by

$$J_p(\xi_{\lambda_1, \ldots, \lambda_r}^{n_1, \ldots, n_r}) = h_{n_1}(\xi_{\lambda_1}) \cdots h_{n_r}(\xi_{\lambda_r}). \tag{6.6.13}$$

For finite linear combinations of $\xi_{\lambda_1, \ldots, \lambda_r}^{n_1, \ldots, n_r}$ define J_p by extending (6.6.13) in the usual way. It then follows from Lemma 6.6.1 that the J_p thus obtained is an isometric isomorphism between the linear manifold spanned by the $\xi_{\lambda_1, \ldots, \lambda_r}^{n_1, \ldots, n_r}$ and G_p. J_p then extends to become a congruence between the closed linear manifolds respectively spanned by $\{\xi_{\lambda_1, \ldots, \lambda_r}^{n_1, \ldots, n_r}\}$ and $\{h_{n_1}(\xi_{\lambda_1}) \cdots h_{n_r}(\xi_{\lambda_r})\}$. This completes the proof of (6.6.12). □

The following result can easily be proved by applying Lemma 6.6.3 and using the definition of direct sum. The proof is omitted.

Lemma 6.6.4.

$$\sum_{p \geq 0} \oplus \sigma[\bigotimes^p L_1(X)] \overset{J}{\cong} \sum_{p \geq 0} \oplus \bar{G}_p.$$

Let us recall that since $\{\xi_i\}_1^\infty$ is a CONS in $L_1(X)$ and $\{e_i\}$ with $e_i = \psi_1(\xi_i)$ is a CONS in $H(R)$, we have

$$X(t,\omega) = \sum_{i=1}^{\infty} \xi_i(\omega) e_i(t) \tag{6.6.14}$$

with the series coverging (a.s.) for each fixed t. From Equation (6.6.14) and since each ξ_i is $\mathscr{F}^X$-measurable, it follows that

$$\mathscr{F}^\xi = \mathscr{F}^X, \tag{6.6.15}$$

where $\mathscr{F}^\xi$ denotes the σ-field generated by the random variables $\{\xi_i(\omega)\}_1^\infty$ augmented by the P-null sets of Ω.

Theorem 6.6.1.

$$L^2(\Omega,\mathscr{F}^X,P) = \sum_{p\geq 0} \oplus \bar{G}_p \cong \sum_{p\geq 0} \oplus \sigma[\bigotimes^p L_1(X)]. \tag{6.6.16}$$

For any $u \in L^2$, the (L^2-convergent) expansion

$$u = \sum_{p\geq 0} \sum_{n_1+\cdots+n_r=p} \sum_{\lambda_1<\cdots<\lambda_r} (F_p, e^{n_1,\ldots,n_r}_{\lambda_1,\ldots,\lambda_r})_p h_{n_1}(\xi_{\lambda_1})\cdots h_{n_r}(\xi_{\lambda_r}) \tag{6.6.17}$$

holds, where

$$F_p \in \sigma[\bigotimes^p H(R)].$$

PROOF. The second assertion in (6.6.16) has already been shown in Lemma 6.6.4. To prove the first part, let $u \in L^2$ be such that

$$u \perp \sum_{p\geq 0} \oplus \bar{G}_p.$$

Then for every integer N,

$$E\{uh_{n_1}(\xi_1)\cdots h_{n_N}(\xi_N)\} = 0$$

for all $n_1, \ldots, n_N \geq 0$. Hence, taking conditional expectations, we have

$$E\{E(u|\xi_1,\ldots,\xi_N)h_{n_1}(\xi_1)\cdots h_{n_N}(\xi_N)\} = 0 \tag{6.6.18}$$

for all nonnegative integers $n_1, \ldots, n_N$. Since

$$E(u|\xi_1,\ldots,\xi_N) \in L^2(\Omega,\sigma(\xi_1,\ldots,\xi_N),P),$$

it follows from (6.6.18) that

$$E(u|\xi_1,\ldots,\xi_N) = 0 \qquad \text{(a.s.)}, \tag{6.6.19}$$

N being arbitrary. Making $N \to \infty$ in (6.6.19), a standard martingale theorem gives

$$E(u|\mathscr{F}^\xi) = 0 \qquad \text{(a.s.)},$$

which implies

$$u = 0 \qquad \text{(a.s.)}$$

Since (6.6.15) holds, (6.6.16) is proved. The relation (6.6.17) can be easily shown to follow from Lemmas 6.6.2 and 6.6.4. □

Let $\mathscr{A}$ be the class of all infinite sequences $a^{n_1,\ldots,n_r}_{\lambda_1,\ldots,\lambda_r}$ such that

$$\sum_{p\geq 0} \sum_{\substack{n_1+\cdots+n_r=p \\ n_i\geq 1}} \sum_{\lambda_1<\cdots<\lambda_r} (a^{n_1,\ldots,n_r}_{\lambda_1,\ldots,\lambda_r})^2 < \infty. \tag{6.6.20}$$

The enumeration system used for $p \geq 1$ is invalid for $p = 0$. We shall ignore this in the notation, however, tacitly assuming the summation ranges over exactly one index, say a_0, when $p = 0$. Define the real-valued function ϕ on $H(R)$ by

$$\phi(m) = \sum_{p\geq 0} \sum_{n_1+\cdots+n_r=p} \sum_{\lambda_1<\cdots<\lambda_r} a^{n_1,\ldots,n_r}_{\lambda_1,\ldots,\lambda_r} \prod_{i=1}^r \frac{(m,e_{\lambda_i})^{n_i}}{\sqrt{n_i!}} \tag{6.6.21}$$

$[m \in H(R)]$, where $\{a_{\lambda_1,\ldots,\lambda_r}^{n_1,\ldots,n_r}\}$ belongs to $\underline{\mathscr{A}}$. It is not hard to verify that the correspondence

$$\phi \sim \{a_{\lambda_1,\ldots,\lambda_r}^{n_1,\ldots,n_r}\}$$

is one-to-one by showing that $\phi(m) = 0$ for all m in $H(R)$ implies that all the coefficients $a_{\lambda_1,\ldots,\lambda_r}^{n_1,\ldots,n_r}$ vanish. Let $\underline{\mathscr{F}}$ be the class of all functions ϕ on $H(R)$ given by (6.6.21). If $\psi \in \underline{\mathscr{F}}$ and

$$\psi \sim \{b_{\lambda_1,\ldots,\lambda_r}^{n_1,\ldots,n_r}\}$$

and if we define $[\quad,\quad]$ by

$$[\phi,\psi] = \sum_{p \geq 0} \sum_{n_1+\cdots+n_r=p} \sum_{\lambda_1<\cdots<\lambda_r} a_{\lambda_1,\ldots,\lambda_r}^{n_1,\ldots,n_r} \cdot b_{\lambda_1,\ldots,\lambda_r}^{n_1,\ldots,n_r} \tag{6.6.22}$$

[the series converges by Schwarz's inequality and (6.6.20)], then it is easy to see that $[\quad,\quad]$ defines an inner product in $\underline{\mathscr{F}}$.

Theorem 6.6.2. *$\underline{\mathscr{F}}$ with the inner product $[\quad,\quad]$ is a* RKHS *with a kernel Γ_R on $H(R) \times H(R)$ given by*

$$\Gamma_R(m,m_0) = \exp\{(m,m_0)\} \qquad (m,m_0 \in H(R)).$$

We give here only an indication of the proof. First, completeness is proved from (6.6.22) in the usual way. After some computation, it is verified that

$$\Gamma_R(m,m_0) = \sum_{p \geq 0} \sum_{n_1+\cdots+n_r=p} \sum_{\lambda_1<\cdots<\lambda_r} \prod_{i=1}^{r} \frac{(m,e_{\lambda_i})^{n_i}(m_0,e_{\lambda_i})^{n_i}}{n_i!}. \tag{6.6.23}$$

Setting

$$\gamma_{\lambda_1,\ldots,\lambda_r}^{n_1,\ldots,n_r} = \prod_{i=1}^{r} \frac{(m_0,e_{\lambda_i})^{n_i}}{\sqrt{n_i!}}, \; \gamma_0 = 1,$$

it follows immediately from (6.6.23) that the sequence $\{\gamma_{\lambda_1,\ldots,\lambda_r}^{n_1,\ldots,n_r}\} \in \underline{\mathscr{A}}$ and that $\Gamma_R(\cdot,m_0) \in \underline{\mathscr{F}}$ for every $m_0 \in H(R)$ since

$$\Gamma_R(m,m_0) = \sum_{p \geq 0} \sum_{n_1+\cdots+n_r=p} \sum_{\lambda_1<\cdots<\lambda_r} \gamma_{\lambda_1,\ldots,\lambda_r}^{n_1,\ldots,n_r} \prod_{i=1}^{r} \frac{(m,e_{\lambda_i})^{n_i}}{\sqrt{n_i!}}.$$

Finally, if $\phi \in \underline{\mathscr{F}}$, from (6.6.22) we have for every $m_0 \in H(R)$ the reproducing property

$$\begin{aligned}[\phi,\Gamma_R(\quad,m_0)] &= \sum_{p \geq 0} \sum_{n_1+\cdots+n_r=p} \sum_{\lambda_1<\cdots<\lambda_r} a_{\lambda_1,\ldots,\lambda_r}^{n_1,\ldots,n_r} \prod_{i=1}^{r} \frac{(m_0,e_{\lambda_i})^{n_i}}{\sqrt{n_i!}} \\ &= \phi(m_0).\end{aligned}$$

□

We shall denote the RKHS of the above theorem by $H(\Gamma_R)$. For fixed $m \in H(R)$, introduce the abbreviation $m^{\otimes p}$ for $m \otimes \cdot\overset{p}{\cdot}\cdot \otimes m$. Then the series $\sum_{p \geq 0} (m^{\otimes p}/\sqrt{p!})$ defines an element of $\sum_{p \geq 0} \oplus [\bigotimes^p H(R)]$ which we denote by $\exp[\bigotimes m]$.

Lemma 6.6.5. *For every* $m \in H(R)$,

$$\exp[\bigotimes m] \in \sum_{p \geq 0} \oplus \sigma[\bigotimes^p H(R)]. \tag{6.6.24}$$

Furthermore, if $m_0 \in H(R)$,

$$(\exp[\bigotimes m], \exp[\bigotimes m_0]) = \exp\{(m, m_0)\}. \tag{6.6.25}$$

PROOF. Clearly, since $m^{\otimes p} \in \sigma[\bigotimes^p H(R)]$, (6.6.24) is proved since

$$\sum_{p \geq 0} \frac{1}{p!} \|m^{\otimes p}\|_p^2 = \sum_{p \geq 0} \frac{\|m\|^{2p}}{p!} = e^{\|m\|^2} < \infty.$$

Similarly, Equation (6.6.25) follows from

$$(\exp[\bigotimes m], \exp[\bigotimes m_0]) = \sum_{p \geq 0} \left(\frac{m^{\otimes p}}{\sqrt{p!}}, \frac{m_0^{\otimes p}}{\sqrt{p!}}\right)_p = \sum_{p \geq 0} \frac{(m, m_0)^p}{p!}. \qquad \square$$

We may easily compute the coordinates of $\exp[\bigotimes m]$ relative to the basis $\{e_{\lambda_1, \ldots, \lambda_r}^{n_1, \ldots, n_r}\}$ from (6.5.1):

$$\begin{aligned}
\left(\frac{m^{\otimes p}}{\sqrt{p!}}, e_{\lambda_1, \ldots, \lambda_r}^{n_1, \ldots, n_r}\right)_p &= \left(\frac{m^{\otimes p}}{\sqrt{p!}}, \frac{\sqrt{p!}}{\sqrt{n_1! \cdots n_r!}} \sigma(e_{\lambda_1}^{\otimes n_1} \otimes \cdots \otimes e_{\lambda_r}^{\otimes n_r})\right)_p \\
&= \left(\prod_{i=1}^r n_i!\right)^{-\frac{1}{2}} (\sigma(m^{\otimes p}), e_{\lambda_1}^{\otimes n_1} \otimes \cdots \otimes e_{\lambda_r}^{\otimes n_r})_p \\
&= \left(\prod_{i=1}^r n_i!\right)^{-\frac{1}{2}} (m^{\otimes p}, e_{\lambda_1}^{\otimes n_1} \otimes \cdots \otimes e_{\lambda_r}^{\otimes n_r})_p \\
&= \prod_{i=1}^r \frac{(m, e_{\lambda_i})^{n_i}}{\sqrt{n_i!}}.
\end{aligned}$$

Theorem 6.6.3.

$$H(\Gamma_R) \cong \sum_{p \geq 0} \oplus \sigma[\bigotimes^p H(R)]. \tag{6.6.26}$$

PROOF. We first show that the family

$$\{\exp[\bigotimes m], m \in H(R)\} \quad \text{spans} \quad \sum_{p \geq 0} \oplus \sigma[\bigotimes H(R)].$$

Let

$$F = \sum_{p \geq 0} \sum_{n_1 + \cdots + n_r = p} \sum_{\lambda_1 < \cdots < \lambda_r} a_{\lambda_1, \ldots, \lambda_r}^{n_1, \ldots, n_r} e_{\lambda_1, \ldots, \lambda_r}^{n_1, \ldots, n_r}$$

be such that

$$(F, \exp[\bigotimes m]) = 0 \quad \text{for all } m \in H(R). \tag{6.6.27}$$

The left-hand side of (6.6.27) equals

$$\sum_{p \geq 0} \sum_{n_1 + \cdots + n_r = p} \sum_{\lambda_1 < \cdots < \lambda_r} a_{\lambda_1, \ldots, \lambda_r}^{n_1, \ldots, n_r} \prod_{i=1}^r \frac{(m, e_{\lambda_i})^{n_i}}{\sqrt{n_i!}}.$$

Hence from (6.6.27) and the remark concerning the one-to-one-ness of the correspondence, (6.6.26) yields $F = 0$. Finally from (6.6.25) we have, for all $m, m_0 \in H(R)$,

$$(\exp[\otimes m], \exp[\otimes m_0]) = [\Gamma_R(\cdot, m), \Gamma_R(\cdot, m_0)].$$

The conclusion of the theorem now follows by standard arguments upon noting the fact that $\{\exp[\otimes m], m \in H(R)\}$ spans the Hilbert space on the right-hand side of (6.6.26) and that $\{\Gamma_R(\cdot, m), m \in H(R)\}$ spans $H(\Gamma_R)$. □

Theorems 6.6.1 and 6.6.3 are the main results of this section which investigate the structure of the L^2 space of a Gaussian process. We shall now make a few comments on the theorem of Cameron and Martin together with a version of it which uses the multiple Wiener integral due to Ito. The notation has been changed to conform to ours.

Recall the function space $(C[0,1], \mathscr{A}, \mu_W)$ of Chapter 2, with the coordinate process $W_t(x) = x(t)$, μ_W being Wiener measure. We consider the space of square-integrable Wiener functionals, $L^2(C[0,1], \mathscr{A}, \mu_W)$. This is a special case of the space $L^2(\Omega, \mathscr{F}^X, P)$ considered above. Write $\mathbf{T} = [0,1]$, $m^p =$ Lebesgue measure on $\mathbf{T}^p$, and $L^2(\mathbf{T}^p) = L^2(\mathbf{T}^p, m^p)$ (real-valued functions). Let $\hat{L}^2(\mathbf{T}^p)$ be the closed linear subspace of all symmetric functions in $L^2(\mathbf{T}^p)$. Write C for $C[0,1]$.

Theorem 6.6.4. *Let $\{\phi_i(t)\}_1^\infty$ be a CONS in $L^2(\mathbf{T})$. Then any $L^2(C, \mathscr{A}, \mu_W)$-functional u has the development*

$$u = \sum_{p \geq 0} \sum_{n_1 + \cdots + n_r = p} \sum_{\lambda_1 < \lambda_2 < \cdots < \lambda_r} a_{\lambda_1, \ldots, \lambda_r}^{n_1, \ldots, n_r} \prod_{i=1}^r h_{n_i}\left[\int_0^1 \phi_{\lambda_i}(t)\, dW_t\right]. \tag{6.6.28}$$

Also

$$u = \sum_{p \geq 0} I_p(f_p) = \sum_{p \geq 0} I_p(\tilde{f}_p),$$

where $f_p \in L^2(T^p)$ is given by

$$\begin{aligned} f_p(t_1, \ldots, t_p) = & \sum_{n_1 + \cdots + n_r = p} \sum_{\lambda_1 < \cdots < \lambda_r} a_{\lambda_1, \ldots, \lambda_r}^{n_1, \ldots, n_r} \sqrt{\frac{p!}{n_1! \cdots n_r!}} \\ & \phi_{\lambda_1}(t_1) \cdots \phi_{\lambda_1}(t_{n_1}) \phi_{\lambda_2}(t_{n_1+1}) \cdots \phi_{\lambda_2}(t_{n_1+n_2}) \cdots \\ & \phi_{\lambda_r}(t_{n_1+\cdots+n_{r-1}+1}) \cdots \phi_{\lambda_r}(t_p), \end{aligned} \tag{6.6.29}$$

$I_p(f_p)$ is the multiple Wiener integral of degree p and $\tilde{f}_p$ is the symmetrization of f_p. [See p. 137.]

Proof. It is obvious that Theorem 6.6.1 is applicable to $L^2(C, \mathscr{A}, \mu_w)$. From (6.6.17) we may write $u = \sum_{p \geq 0} u_p$, where

$$u_p = \sum_{n_1 + \cdots + n_r = p} \sum_{\lambda_1 < \cdots < \lambda_r} (F_p, e_{\lambda_1, \ldots, \lambda_r}^{n_1, \ldots, n_r})_p \prod_{i=1}^r h_{n_i}(\xi_{\lambda_i}), \tag{6.6.30}$$

where $F_p \in \sigma[\bigotimes^p H(R)]$ is uniquely determined by u_p. Since for the standard Wiener process, $R(t,s) = \min(t,s)$ $(t,s \in \mathbf{T})$, it is well known that $H(R)$ consists of absolutely continuous functions F, $F(0) = 0$, with square-integrable derivatives, and if $F(t) = \int_0^t f(u)\,du$, $G(t) = \int_0^t g(u)\,du$, then $(F,G)_{H(R)} = (f,g)_{L^2(\mathbf{T})}$. Hence, $\{\phi_i\}_1^\infty$, where $e_i(t) = \int_0^t \phi_i(u)\,du$, is a CONS in $L^2(\mathbf{T})$. It is easy to see that the correspondence θ_1 given by $\theta_1 \phi_i = e_i$ $(i = 1,2,\ldots)$ (and defined by linearity for finite linear combinations) extends to a congruence between $L^2(\mathbf{T})$ and $H(R)$. It follows that for any integer p,

$$L^2(\mathbf{T}^p) \cong \bigotimes^p L^2(\mathbf{T}) \cong \bigotimes^p H(R)$$

and

$$\hat{L}^2(\mathbf{T}^p) \cong \sigma[\bigotimes^p L^2(\mathbf{T})] \cong \sigma[\bigotimes^p H(R)]. \tag{6.6.31}$$

Under the congruence defined by (6.6.31) the image in $\hat{L}^2(\mathbf{T}^p)$ of

$$e_{\lambda_1,\ldots,\lambda_r}^{n_1,\ldots,n_r} \text{ is } \phi_{\lambda_1,\ldots,\lambda_r}^{n_1,\ldots,n_r}, \text{ where } \{\phi_{\lambda_1,\ldots,\lambda_r}^{n_1,\ldots,n_r}\} \text{ is the CONS in } \hat{L}^2(\mathbf{T}^p)$$

obtained by symmetrizing and normalizing the sequence

$$\{\phi_{\lambda_1}(t_1)\cdots\phi_{\lambda_1}(t_{n_1})\phi_{\lambda_2}(t_{n_1+1})\cdots\phi_{\lambda_2}(t_{n_1+n_2})\cdots\phi_{\lambda_r}(t_{n_1+\cdots+n_{r-1}+1})\cdots\phi_{\lambda_r}(t_p)\}$$

(here $(\lambda_1 < \cdots < \lambda_r)$ runs over all finite sequences of distinct integers and $n_i \geq 1$ such that $n_1 + \cdots + n_r = p$). Also let the image in $\hat{L}^2(\mathbf{T}^p)$ of F_p be $\tilde{f}_p$. It then follows that

$$\begin{aligned}(F_p, e_{\lambda_1,\ldots,\lambda_r}^{n_1,\ldots,n_r})_p &= (\tilde{f}_p, \phi_{\lambda_1,\ldots,\lambda_r}^{n_1,\ldots,n_r})_{L^2(\mathbf{T}^p)} \\ &= a_{\lambda_1,\ldots,\lambda_r}^{n_1,\ldots,n_r}, \quad \text{say.}\end{aligned} \tag{6.6.32}$$

The random variables ξ_i in $L_1(W)$ are given by the Wiener integrals

$$\xi_i = \int_0^1 \phi_i(t)\,dW_t. \tag{6.6.33}$$

Substituting from (6.6.32) and (6.6.33) in (6.6.30) and using (6.6.17), we obtain the Cameron-Martin expansion (6.6.28) of Theorem 6.6.4. From the way it is defined, it also follows that $\tilde{f}_p$ is the symmetrization of the function f_p defined by (6.6.29).

From the congruence ψ_p between $\sigma[\bigotimes^p L_1(W)]$ and $\sigma[\bigotimes^p H(R)]$ established in Corollary 6.4.1 and from (6.6.31) we have

$$\hat{L}^2(\mathbf{T}^p) \cong \sigma[\bigotimes^p L_1(W)]. \tag{6.6.34}$$

Assertion (6.6.34) suggests that it is appropriate to define a multiple Wiener integral of degree $p\,(p \geq 1)$ to be any element of $\sigma[\bigotimes^p L_1(W)]$.

For every function $\tilde{f}_p$ in $\hat{L}^2(\mathbf{T}^p)$ $(p \geq 1)$ by (6.6.34) there is a unique element η_p in $\sigma[\bigotimes^p L_1(W)]$, and hence by the congruence (6.6.12) established in Lemma 6.6.3, there exists a unique random variable $u_p(x)$ in $\bar{G}_p$ which gives a "concrete" realization of the multiple Wiener integral of degree p, and we write

$$I_p(f_p) = I_p(\tilde{f}_p) = u_p. \tag{6.6.35}$$

I_0 (constant function) is defined to be the same constant function. From (6.6.28), $u_p(x)$ has the expansion

$$u_p = \sum_{n_1+\cdots+n_r=p} \sum_{\lambda_1<\cdots<\lambda_r} a_{\lambda_1,\ldots,\lambda_r}^{n_1,\ldots,n_r} \prod_{i=1}^{r} h_{n_i}\left[\int_0^1 \phi_{\lambda_i}(t)\,dW_t\right]. \quad (6.6.36) \quad \square$$

Equations (6.6.35) and (6.6.36) define an isometric isomorphism:

$$I = (I_p): \sum_{p\geq 0} \oplus \hat{L}^2[0,1]^p \to \sum_{p\geq 0} \oplus \bar{G}_p = L^2(\Omega,\mathscr{F}^W,P),$$

which we have called the *multiple Wiener integral* in Theorem 6.6.4. Each isometry I_p is in fact the normalization (by the factor $1/\sqrt{p!}$) of the (nonisometric) multiple Wiener integral first given by Ito and presented above in Section 6.2.

Remark 6.6.1. Theorem 6.6.4 (and the subsequent definition of the multiple Wiener integral) is, of course, valid for any choice of a probability model that serves as "Wiener space." Traditionally, the latter term is taken to mean the probability space $(C,\mathscr{A},\mu_W)$, but the Cameron-Martin expansion (6.6.28) and the definition of the multiple Wiener integral given in (6.6.35) and (6.6.36) continue to hold (with appropriate notational changes) under the assumptions made in Section 6.2. The theorems of the next section will be formulated in terms of the Wiener space $(\Omega,\mathscr{F}^W,P)$ and the (Wiener) process $W_t(\omega)$ defined on it.

Remark 6.6.2. Expansion (6.6.28) of Theorem 6.6.4 differs from the form in which it is given by Cameron and Martin or by Ito. In the work of the latter author it is customarily given as

$$\sum A_{\lambda_1,\ldots,\lambda_r}^{n_1,\ldots,n_r} \prod_{i=1}^{r} H_{n_i}\left[\frac{1}{\sqrt{2}}\int_0^1 \phi_{\lambda_i}(t)\,dW_t\right],$$

where $A_{\lambda_1,\ldots,\lambda_r}^{n_1,\ldots,n_r}$ are suitable coefficients and $H_n(x)$ is the Hermite polynomial of degree n. Now the latter is usually defined by

$$H_n(x) = (-1)^n e^{x^2}\left(\frac{d}{dx}\right)^n e^{-x^2}.$$

It is easy to verify that

$$h_n(x) = \left(\frac{2^{-n/2}}{\sqrt{n!}}\right) H_n\left(\frac{x}{\sqrt{2}}\right).$$

Hence comparing coefficients in the two expansions for $u(x)$, we see that

$$A_{\lambda_1,\ldots,\lambda_r}^{n_1,\ldots,n_r} = \frac{2^{-p/2} a_{\lambda_1,\ldots,\lambda_r}^{n_1,\ldots,n_r}}{(n_1!\cdots n_r!)^{\frac{1}{2}}} \qquad (n_1+\cdots+n_r=p).$$

Remark 6.6.3. We have not proved the assertion that the isometry I_p introduced in this section is the normalization of the integral of Section 6.2.

This has been shown by Ito ([21], Theorem 3.3.1, p. 162). We rely upon this fact in the proof of Theorem 6.7.1 of the next section, where I_p will always denote the integral of Section 6.2.

6.7 Stochastic (Ito) Integral Representation

To deduce our main result, viz., that every functional $F \in L^2(\Omega,\mathscr{F}^W,P)$ with $E(F) = 0$ is an Ito stochastic integral, we first need two auxiliary propositions.

Lemma 6.7.1. *Let $I_p(f)$ for $f \in L^2([0,1]^p)$ be the multiple Wiener integral of order $p (p \geq 1)$. For $0 \leq t \leq 1$, let $\mathscr{F}_t^W$ be the augmentation by P-null sets of the σ-field generated by $W_s, 0 \leq s \leq t$. Then*

$$E[I_p(f)|\mathscr{F}_t^W] = \int_0^t \overset{p}{\cdots} \int_0^t f(u_1, \ldots ,u_p)\, dW_{u_1} \cdots dW_{u_p} \qquad \text{(a.s.)}, \quad (6.7.1)$$

where the right-hand side denotes the multiple Wiener integral of the function f restricted to $[0,t]^p$.

Proof. Suppose first that f is of the form $f(u_1, \ldots ,u_p) = 1_{T_1}(u_1) \cdots 1_{T_p}(u_p)$, where $T_1, \ldots ,T_p$ are disjoint measurable subsets of $[0,1]$.

$$\begin{aligned} E[I_p(f)|\mathscr{F}_t^W] &= E[W(T_1) \cdots W(T_p)|\mathscr{F}_t^W] \\ &= W(T_1 \cap [0,t]) \cdots W(T_p \cap [0,t]) \\ &= \int_0^t \cdots \int_0^t f(u_1, \ldots ,u_p)\, dW_{u_1} \cdots dW_{u_p}. \end{aligned} \qquad (6.7.2)$$

The maps

$$f \to f \cdot 1_{[0,t]^p} \to I_p(f \cdot 1_{[0,t]^p})$$

and

$$f \to I_p(f) \to E(I_p(f)|\mathscr{F}_t^W)$$

are continuous on $L^2([0,1]^p)$ and coincide on special elementary functions as shown above. Hence they are identical, proving the lemma. □

Lemma 6.7.2. *The multiple integral $I_p(f)$ can be expressed as the Ito stochastic integral of a process $f^*(u,\omega) \in \mathscr{M}_2^W[0,1]$ (see Chapter 3). That is,*

$$I_p(f) = \int_0^1 f^*(u,\omega)\, dW_u, \qquad (6.7.3)$$

where f^* is jointly measurable, $\mathscr{F}_u^W$-adapted, and

$$\int_0^1 E[f^*(u)]^2\, du < \infty. \qquad (6.7.4)$$

Proof. $I_p(f)$ can be written as an iterated stochastic integral.

$$I_p(f) = p! \int_0^1 \left(\int_0^{u_p} \cdots \int_0^{u_3} \int_0^{u_2} \tilde{f}(u_1, \ldots ,u_p)\, dW_{u_1}\, dW_{u_2} \cdots dW_{u_{p-1}} \right) dW_{u_p}. \qquad (6.7.5)$$

This formula is readily verified if f is a special elementary function and then extended to general f by the usual technique of approximation and the properties of multiple Wiener integral and the Ito stochastic integral. Writing u for u_p and $f^*(u,\omega)$ for the expression in parentheses in Equation (6.7.5), it is seen that the right side of (6.7.5) is an Ito integral since the random integrand obviously is $\mathscr{F}_u^W$-measurable (for each u) and satisfies (6.7.4). □

Theorem 6.7.1. *Let F be an L^2-functional on $(\Omega,\mathscr{F}^W,P)$ with $E(F) = 0$. Then*

$$F(\omega) = \int_0^1 \Phi(u,\omega)\, dW_u \qquad \text{(a.s.)}, \tag{6.7.6}$$

where the right side is an Ito stochastic integral and

$$\int_0^1 E[\Phi(u,\omega)]^2\, du < \infty. \tag{6.7.7}$$

Also from

$$F = \sum_{p=1}^{\infty} I_p(f_p), \tag{6.7.8}$$

we obtain

$$E[F|\mathscr{F}_t^W] = \sum_{p=1}^{\infty} \int_0^t \cdots \int_0^t f_p(u_1, \ldots, u_p)\, dW_{u_1} \cdots dW_{u_p}. \tag{6.7.9}$$

PROOF. From (6.7.8)

$$E\left[F - \sum_{p=1}^{N} I_p(f_p)\right]^2 \to 0 \qquad (N \to \infty).$$

Write $I_p(f_p;t)$ for the multiple Wiener integral on the right-hand side of (6.7.9). Since $E[I_p(f_p)|\mathscr{F}_t^W] = I_p(f_p;t)$, by Lemma 6.7.1 we have upon setting

$$\sigma_{N,t} = E(F|\mathscr{F}_t^W) - \sum_{p=1}^{N} I_p(f_p;t),$$

$$\begin{aligned} E(\sigma_{N,t}^2) &= E\left[E\left\{\left(F - \sum_1^N I_p(f_p)\right)\middle|\mathscr{F}_t^W\right\}\right]^2 \\ &\leq E\left[F - \sum_1^N I_p(f_p)\right]^2 \to 0, \end{aligned}$$

proving (6.7.9). Next, from Lemma 6.7.2, writing $I_p(f_p)(\omega) = \int_0^1 f_p^*(u,\omega)\, dW_u$, we obtain from (6.7.8)

$$E\left[F(\omega) - \int_0^1 \Phi_N(u,\omega)\, dW_u\right]^2 \to 0,$$

where

$$\Phi_N(u,\omega) = \sum_{p=1}^{N} f_p^*(u,\omega).$$

Hence $E\int_0^1 [\Phi_N(u,\omega) - \Phi_M(u,\omega)]^2\,du \to 0$ $(M,N \to \infty)$, and so there exists an $\mathscr{F}_u^W$-adapted process $\Phi(u,\omega)$ (from Proposition 1.1.4) such that

$$E\int_0^1 [\Phi_N(u,\omega) - \Phi(u,\omega)]^2\,du \to 0,$$

and the required result (6.7.6) [and (6.7.7.)] follows. □

Let us now consider the time interval to be $[0,T]$ instead of $[0,1]$. Note that now $\mathscr{F}^W = \mathscr{F}_T^W$.

Theorem 6.7.2. *Let $(M_t,\mathscr{F}_t^W)$ with $M_0 = 0$ be an L^2-martingale on the Wiener space $(\Omega,\mathscr{F}^W,P)$. Then (M_t) has a continuous modification, say, $(\tilde{M}_t)$, which is given by an Ito stochastic integral*

$$\tilde{M}_t(\omega) = \int_0^t \Phi(u,\omega)\,dW_u(\omega) \qquad \text{(a.s.)} \tag{6.7.10}$$

for every t, where $\Phi(u,\omega)$ is jointly measurable, $\Phi(u,\cdot)$ is $\mathscr{F}_u^W$-measurable, and

$$\int_0^T E[\Phi(u,\omega)]^2\,du < \infty.$$

Proof. From Theorem 6.7.1, since M_T is an L^2-functional on Wiener space, we have

$$M_T(\omega) = \int_0^T \Phi(u,\omega)\,dW_u,$$

where Φ has the required properties. Using a property of Ito integrals (Theorem 3.3.1), we have

$$M_t(\omega) = E[M_T|\mathscr{F}_t^W](\omega) = \int_0^t \Phi(u,\omega)\,dW_u(\omega) \qquad \text{(a.s.)}$$

for every t. The continuous version of the stochastic integral process on the right-hand side (whose existence was established in Section 3.3) is the required continuous modification $(\tilde{M}_t)$ of (M_t). □

We now give an alternative proof of this important result which is not based on the theorems of Ito and Cameron and Martin but does use the Ito formula. The vector-valued case is considered.

Let $W_t = (W_t^1,\ldots,W_t^N)$ be a standard N-dimensional Wiener process on $(\Omega,\mathscr{A},P)$, and let $\mathscr{F}_t^W$ be the σ-field defined as before except that we consider the family $\{W_s^j, 0 \le s \le t, j = 1,\ldots,N\}$. The probability space $(\Omega,\mathscr{F}^W,P)$ is then the Wiener space of the N vector Wiener process (W_t).

Theorem 6.7.3. *A square-integrable martingale $(M_t,\mathscr{F}_t^W)$ $(0 \le t \le T)$ with $M_0 = 0$ has a continuous modification $(\tilde{M}_t)$ given by*

$$\tilde{M}_t = \sum_{j=1}^N \int_0^t \Phi_j(s)\,dW_s^j, \tag{6.7.11}$$

where the process $\Phi(s) = (\Phi_1(s), \ldots, \Phi_N(s))$ has the usual adaptability and measurability properties and

$$\int_0^T E|\Phi(s)|^2\, ds = \sum_{j=1}^N \int_0^T E(\Phi_j(s))^2\, ds < \infty. \tag{6.7.12}$$

PROOF. Let $m_j \in L^2[0,T]$ $(1 \le j \le N)$ and write

$$\beta_m(t) = \exp\left[\sum_{j=1}^N \int_0^t m_j(u)\, dW_u^j - \frac{1}{2}\sum_{j=1}^N \int_0^t m_j^2(u)\, du\right]:$$

By Ito's formula (the simple version of the formula needed here can be independently obtained easily since the m_j's are nonrandom) we have

$$\beta_m(T) = 1 + \sum_{j=1}^N \int_0^T \beta_m(u) m_j(u)\, dW_u^j. \tag{6.7.13}$$

From

$$\beta_m^2(u) = \beta_{2m}(u) \exp\left[\sum_1^N \int_0^u m_j^2(s)\, ds\right]$$

and the fact that $E[\beta_{2m}(u)] \le 1$ for all $u \in [0,T]$, it follows that

$$E\beta_m^2(u) \le \exp\left[\sum_1^N \int_0^T m_j^2(s)\, ds\right] < \infty.$$

Let $\mathscr{M}$ denote the class of all random variables M_T in $L^2 = L^2(\Omega, \mathscr{F}^W, P)$ given by the right-hand side of (6.7.11) (*with* $t = T$) and (6.7.12). If $Y \in L^2$, $E(Y) = 0$, and $\perp \mathscr{M}$, we must have

$$E[Y\beta_m(T)] = 0 \quad \text{for all } m = (m_1, \ldots, m_N)$$

with $m_j \in L^2[0,T]$. By taking the m_j's to be suitable step functions, we immediately see that $E(Y|W_{t_1}, \ldots, W_{t_n}) = 0$ for arbitrary (t_j) such that $0 \le t_1 < \cdots < t_n < T$. But this implies $Y = 0$.

The proof is completed as in Theorem 6.7.2. □

Remark 6.7.1. We may now generalize Theorem 6.7.1 to the case where W is N-dimensional. For $F \in L^2(\Omega, \mathscr{F}^W, P)$, take a continuous version given by Theorem 6.7.3 of the L^2-martingale

$$M_t = E[F - E(F)|\mathscr{F}_t^W].$$

Then (6.7.11) for $t = T$ gives a representation of $F - E(F)$ in terms of stochastic integrals. For $F \in L^2(\Omega, \mathscr{F}^W, P)$, we then have

$$F(\omega) = E(F) + \sum_{j=1}^N \int_0^T \Phi_j(u,\omega)\, dW_u^j \qquad (P\text{-a.s.}), \tag{6.7.14}$$

where $\Phi = (\Phi_j)$ satisfies (6.7.12).

Remark 6.7.2. For the reader who is familiar with the notion of separability of stochastic processes (see [8] or [41]) Theorems 6.7.2 and 6.7.3 can be reworded by including separability as a part of the hypothesis. Here is a restatement of Theorem 6.7.3.

Theorem 6.7.3.′ *Every separable, square-integrable martingale* $(M_t,\mathscr{F}_t^W)$ $(0 \leq t \leq T)$ *with* $M_0 = 0$ *has an Ito stochastic integral representation*

$$M_t = \sum_{j=1}^{N} \int_0^t \Phi_j(s)\,dW_s^j,$$

where $\Phi_j(s)$ $(j = 1, \ldots, N)$ *satisfy the conditions of Theorem 6.7.3.*

6.8 A Generalization of Theorem 6.7.3

The martingale representation given in Theorem 6.7.3 has a useful and interesting generalization. The result proved below is a special case of the representation theorem of L^2-martingales due to Kunita and Watanabe [36]. The proof is based on ideas of [36] and particularly on the ideas due to Cornea and Licea (see [43]). Let $(\mathscr{F}_t)$, $(0 \leq t \leq T)$ be a right-continuous family of sub σ-fields of $\mathscr{A}$, each $\mathscr{F}_t$ containing all P-null sets of $\mathscr{A}$. With each $M_T \in L^2(\Omega,\mathscr{F}_T,P)$ associate the right-continuous L^2-martingale (M_t), where (M_t) is a right-continuous version of $(E(M_T|\mathscr{F}_t))$. Conversely, every right-continuous L^2-martingale $(M_t,\mathscr{F}_t)$ is of the form $M_t = E(M_T|\mathscr{F}_t)$, where $M_T \in L^2(\Omega,\mathscr{F}_T,P)$. Define $L_0^2 = \{M_T \in L^2(\Omega,\mathscr{F}_T,P)\colon M_0 = 0\}$.

Theorem 6.8.1. *Let* $W_t = (W_t^1, \ldots, W_t^N)$ *be an N-dimensional Wiener martingale with respect to* $(\mathscr{F}_t)$. *Then every right-continuous* L^2*-martingale* $(M_t,\mathscr{F}_t)$ *with* $M_0 = 0$ *has the following representation:*

$$M_t = Y_t + Z_t, \tag{6.8.1}$$

where

$$Y_t = \sum_{j=1}^{N} \int_0^t \Phi_j(s)\,dW_s^j \tag{6.8.2}$$

(the continuous version of the stochastic integrals being taken) and $\Phi(s,\omega) = (\Phi_1(s,\omega), \ldots, \Phi_N(s,\omega))$ *is jointly measurable,* $(\mathscr{F}_s)$*-adapted, with*

$$\sum_{j=1}^{N} \int_0^T E(\Phi_j(s))^2\,ds < \infty, \tag{6.8.3}$$

and where

$$(Z_t)\ (Z_0 = 0) \text{ is a (right-continuous) } L^2\text{-martingale with respect to } (\mathscr{F}_t) \tag{6.8.4}$$

such that

$$\langle Y,Z\rangle_t = 0 \qquad \text{(a.s.) for all } t \in [0,T]. \tag{6.8.5}$$

PROOF. Let $H = \{\xi_T \in L_0^2\colon \xi_T = \sum_{j=1}^N \int_0^T g_j(s)\,dW_s^j$, $g_j(s)$ satisfying the measurability and integrability conditions mentioned in (6.8.2) and (6.8.3).$\}$ It is easy to see that H is a closed linear subspace of L_0^2. For $\xi_T \in H$ and any stopping time τ for $(\mathscr{F}_t)$, let $\xi_t^\tau = \xi_{t \wedge \tau}$. Then H has the following property: For every $(\mathscr{F}_t)$ stopping time τ, $\xi_T \in H$ implies $\xi_T^\tau \in H$. This property may be considered as *stability* of H with respect to stopping times and H is consequently called a *stable subspace* of L_0^2 [43].

Denoting by $H^\perp$ the orthogonal complement of H in L_0^2 we first show that $H^\perp$ is also a stable subspace. Assume $\xi_T \in H$ and $\eta_T \in H^\perp$. If τ is a stopping time, since $\xi_T^\tau \in H$,

$$E(\xi_T^\tau \eta_T^\tau) = E[E(\xi_T^\tau \eta_T | \mathscr{F}_\tau)] = E(\xi_T^\tau \eta_T) = 0.$$

Hence

$$E(\xi_T \eta_T^\tau) = E[E(\xi_T \eta_T^\tau | \mathscr{F}_\tau)] = E(\xi_T^\tau \eta_T^\tau) = 0,$$

that is, $\eta_T^\tau \perp \xi_T$. Therefore $\eta_T^\tau \in H^\perp$ and so $H^\perp$ is stable.

Suppose now that $(M_t, \mathscr{F}_t)$ is a right-continuous L^2-martingale with $M_0 = 0$. Then $M_T \in L_0^2$. Let Y_T be the orthogonal projection of M_T onto H. Setting $Z_T = M_T - Y_T$, we immediately obtain expressions (6.8.1) through (6.8.4). Since both H and $H^\perp$ are stable, for any $L_T \in H$, we have $E(L_T^\tau Z_T^\tau) = 0$ for every stopping time τ. From property 11 of Section 2.2, we conclude that $(L_t Z_t)$ is a martingale with respect to $(\mathscr{F}_t)$. From the relation

$$2L_t Z_t = [(L_t + Z_t)^2 - \{\langle L \rangle_t + \langle Z \rangle_t\}] - [L_t^2 - \langle L \rangle_t] - [Z_t^2 - \langle Z \rangle_t]$$

and the definition of $\langle L \rangle_t$ and $\langle Z \rangle_t$ it follows that the first quantity in square brackets is a martingale. But $\langle L + Z \rangle_t$ is the unique, natural, increasing process such that $(L_t + Z_t)^2 - \langle L + Z \rangle_t$ is a martingale. Hence

$$\langle L + Z \rangle_t = \langle L \rangle_t + \langle Z \rangle_t.$$

A similar argument shows $\langle L - Z \rangle_t = \langle L \rangle_t + \langle Z \rangle_t$. By the definition of $\langle L,Z \rangle_t$ we then have

$$4\langle L,Z \rangle_t = \langle L + Z \rangle_t - \langle L - Z \rangle_t = 0.$$

Thus we have shown that if $L_T \in H$ and $L = (L_t)$ is the corresponding martingale, then

$$\langle L,Z \rangle_t = 0 \qquad \text{(a.s.) for every } t. \tag{6.8.6}$$

We obtain (6.8.5) from (6.8.6) upon taking $L_T = Y_T$. □

Corollary 6.8.1. *If $\mathscr{F}_t = \mathscr{F}_t^W$ it follows from Theorem 6.7.3 that $Z = 0$ in Equation* (6.8.1).

Corollary 6.8.2. *For $j = 1, \ldots, N$,*

$$\langle M,W^j \rangle_t = \int_0^t \Phi_j(s)\, ds. \tag{6.8.7}$$

PROOF. The representation (6.8.1) yields

$$\langle M,W^j \rangle_t = \langle Y,W^j \rangle_t + \langle Z,W^j \rangle_t.$$

From (6.8.2) and Chapter 3 we have

$$\langle Y,W^j\rangle_t = \int_0^t \Phi_j(s)\,ds.$$

Next, since $W_T^j \in H$, $\langle Z,W^j\rangle_t = 0$ follows from (6.8.6), and the proof is complete. □

Corollary 6.8.3. *Suppose that $(W_t,\mathcal{G}_t)$ is an N-dimensional Wiener martingale, where $(\mathcal{G}_t)$ is an increasing family which is not assumed to be right-continuous. If $(M_t,\mathcal{G}_t)$ is an L^2-martingale with $M_0 = 0$ and (M_t) is right-continuous, then M_t has the representation (6.8.1) and the statements (6.8.2) to (6.8.5) are true with the change that the integrand process Φ_s in (6.8 2) is measurable and $(\mathcal{G}_s)$-adapted and that (Z_t) in (6.8.4) is an L^2-martingale with respect to $(\mathcal{G}_t)$.*

Proof. The Wiener processes W_t^j and the right-continuous process M_t are L^2-martingales with respect to the right-continuous family $(\mathcal{G}_{t+})$. The proof is almost identical with that of Theorem 6.8.1. We have only to take $\mathcal{F}_t = \mathcal{G}_{t+}$ in the proof except H is defined with the $(\mathcal{F}_t)$-adaptability replaced by the $(\mathcal{G}_t)$-adaptability of the integrand process $g(t) = (g_1(t), \ldots, g_N(t))$. □

Remark 6.8.1. Formula (6.8.7) holds for the martingale $(M_t,\mathcal{G}_t)$ of Corollary 6.8.3.

Corollary 6.8.3 and Remark 6.8.1 will be used in Chapter 8.

7 Absolute Continuity of Measures and Radon-Nikodym Derivatives

As before, let $(\Omega,\mathscr{A},P)$ be a complete probability space. Throughout this chapter it is assumed that $(\mathscr{F}_t)$ $t \in \mathbf{R}_+$ or $[0,T]$ is an increasing right-continuous family of σ-fields such that $\mathscr{F}_0$ contains all P-null sets.

7.1 Exponential Supermartingales, Martingales, and Girsanov's Theorem

We begin by establishing several auxiliary results.

Let Y_t be of the form

$$Y_t = \exp\{M_t - \tfrac{1}{2}\langle M\rangle_t\}, \tag{7.1.1}$$

where $(M_t,\mathscr{F}_t)$ is a continuous local martingale with $M_0 = 0$.

Lemma 7.1.1. *$(Y_t,\mathscr{F}_t)$ is a positive, continuous local martingale and a supermartingale. It is a martingale if and only if $E(Y_t) = 1$ for each t [or, equivalently, if and only if $E(Y_T) = 1$]. Assume $t \in [0,T]$.*

PROOF. By Ito's formula, $Y_t = 1 + \int_0^t Y_s\,dM_s$. Therefore, $(Y_t,\mathscr{F}_t)$ is a continuous local martingale.

There exists a sequence of stopping times $\tau_n \uparrow T$ (a.s.) such that $Y_t^n = Y_{t\wedge\tau_n}$ is an $(\mathscr{F}_t,P)$ martingale, that is, for $s < t$, $Y_s^n = E[Y_t^n|\mathscr{F}_s]$. Since $Y_s^n \geq 0$ and $Y_s^n \to Y_s$ (a.s.) as $n \to \infty$ for every s, Fatou's lemma gives $Y_s \geq E[Y_t|\mathscr{F}_s]$ (a.s.), and hence, $(Y_t,\mathscr{F}_t)$ is a supermartingale. The last assertion of the lemma follows since $Y_0 = 1$. □

Let $(W_t, \mathscr{F}_t)$ be a Wiener martingale and let (f_t) be a measurable, $(\mathscr{F}_t)$ adapted process such that

$$\int_0^T f_t^2\,dt < \infty \qquad \text{(a.s.)}.$$

A familiar example of an exponential supermartingale, obtained by taking $M_t = \int_0^t f_s\,dW_s$ in Lemma 7.1.1, is given by

$$Y_t = \exp\left\{\int_0^t f_s\,dW_s - \frac{1}{2}\int_0^t f_s^2\,ds\right\}.$$

For our purpose this is a very important type of supermartingale. To bring the role of $f = (f_s)$ more in evidence we henceforth write

$$\xi_s^t(f) = \exp\left\{\int_s^t f_u\,dW_u - \frac{1}{2}\int_s^t f_u^2\,du\right\},$$

so that $Y_t = \xi_0^t(f)$. By Lemma 7.1.1, $(\xi_0^t(f), \mathscr{F}_t)$ is a martingale if and only if $E\xi_0^T(f) = 1$. It is of interest to obtain conditions on f which ensure the latter equality because such exponential martingales turn out to be Radon-Nikodym derivatives in the theory of absolute continuity of measures induced by a wide class of stochastic processes. The following easy result will be very helpful in the proofs of the theorems of this chapter. A more general criterion will be given later.

Lemma 7.1.2. *If there exists a constant C such that*

$$\int_0^T f_t^2 \le C \qquad \text{(a.s.)},$$

then

$$E(\xi_0^T(f)) = 1.$$

PROOF. Let

$$\tau_n(\omega) = \begin{cases} \inf\left\{t \le T : \xi_0^t(f) \ge n \text{ or } \int_0^t f_s^2\,ds \ge n\right\} \\ T \qquad \text{if the above set is empty.} \end{cases}$$

Setting $f_s^{(n)}(\omega) = f_s(\omega) I_{[0,\tau_n(\omega)]}(s)$, we have $\xi_0^t(f^{(n)}) = 1 + \int_0^t \xi_0^s(f^{(n)}) f_s^{(n)}\,dW_s$, and $E\xi_0^t(f^{(n)}) = 1$ for all t and each n. Let $\varepsilon > 0$ (in fact, we can take $\varepsilon = 1$). It is easy to see that

$$\begin{aligned}
[\xi_0^t(f^{(n)})]^{1+\varepsilon} &= \exp\left[(1+\varepsilon)\int_0^t f_s^{(n)}\,dW_s - \frac{1}{2}(1+\varepsilon)^2\int_0^t (f_s^{(n)})^2\,ds\right] \\
&\quad \cdot \exp\left[\frac{1}{2}(\varepsilon+\varepsilon^2)\int_0^t (f_s^{(n)})^2\,ds\right] \\
&\le \xi_0^t(\psi^{(n)}) e^{\frac{1}{2}C(\varepsilon+\varepsilon^2)}, \qquad (7.1.2)
\end{aligned}$$

where $\psi_s^{(n)} = (1+\varepsilon)f_s^{(n)}$. Since $(\xi_0^t(\psi^{(n)}))(0 \le t \le T)$ is a supermartingale with $\xi_0^0(\psi^{(n)}) = 1$, we have $E\xi_0^t(\psi^{(n)}) \le 1$. From (7.1.2)

$$E[\xi_0^t(f^{(n)})]^{1+\varepsilon} \le e^{\frac{1}{2}C(\varepsilon+\varepsilon^2)} \tag{7.1.3}$$

for all n. Hence (7.1.3) implies the uniform integrability of the sequence $(\xi_0^t(f^{(n)}))$, and since $\xi_0^t(f^{(n)}) \to \xi_0^t(f)$ (a.s.) as $n \to \infty$, we obtain $E\xi_0^t(f) = \lim_{n\to\infty} E\xi_0^t(f^{(n)}) = 1$. □

In the next lemma the parameter set is taken to be $\mathbf{R}_+$.

Lemma 7.1.3. *Suppose that P and Q are probability measures on $(\Omega,\mathscr{A})$. Let $Q \ll P$ (that is Q is absolutely continuous with respect to P) relative to $\mathscr{F}_t$ for each $t \in \mathbf{R}_+$. Denote the corresponding Radon-Nikodym derivative by L_t: $dQ = L_t\, dP[\mathscr{F}_t]$. Further assume that $(L_t)(t \in \mathbf{R}_+)$ is uniformly P-integrable.*

Then an $(\mathscr{F}_t)$-adapted continuous process (M_t) is a Q-local martingale if (M_tL_t) is a P-local martingale.

PROOF. Since $(L_t,\mathscr{F}_t)$ is a P-martingale, we may assume we are working with a right-continuous version. Because of the P-uniform integrability of (L_t), $\lim_{t\to\infty} L_t = L_\infty$ exists (P-a.s.) and $L_t \to L_\infty$ also in $L^1(P)$. Then $dQ = L_\infty\, dP$ relative to $\mathscr{F}_\infty = \bigvee_t \mathscr{F}_t$. By the optional sampling theorem, if τ is any stopping time, setting $L_\tau = L_\infty$ on the set $[\tau = \infty]$, we have $L_\tau = E(L_\infty | \mathscr{F}_\tau)$, that is, if $A \in \mathscr{F}_\tau$, $\int_A L_\infty\, dP = \int_A L_\tau\, dP$. Hence $dQ = L_\tau\, dP[\mathscr{F}_\tau]$.

Suppose now that (M_tL_t) is a P-local martingale. Then there exists a sequence (T_n) of stopping times such that $T_n \uparrow \infty$ (P-a.s.) and $(M_{t\wedge T_n}L_{t\wedge T_n},\mathscr{F}_t)$ is a P-martingale. For $s < t$ and $A \in \mathscr{F}_{s\wedge T_n}$, we have

$$\int_A M_{t\wedge T_n}L_{t\wedge T_n}\, dP = \int_A M_{s\wedge T_n}L_{s\wedge T_n}\, dP,$$

that is,

$$\int_A M_{t\wedge T_n}\, dQ = \int_A M_{s\wedge T_n}\, dQ.$$

Hence $(M_{t\wedge T_n},\mathscr{F}_{t\wedge T_n})$ is a Q-martingale and furthermore, $T_n \uparrow \infty$ (Q-a.s.). From Lemma 2.6.1 it follows that $(M_{t\wedge T_n},\mathscr{F}_t)$ is a Q-martingale, that is $(M_t,\mathscr{F}_t)$ is a Q-local martingale. □

Lemma 7.1.4. *Let $(L_t,\mathscr{F}_t,P)$, $t \in [0,T]$, be a positive, continuous local martingale such that $L_0 = 1$ (P-a.s.). Then*

$$L_t = \exp[N_t - \tfrac{1}{2}\langle N\rangle_t],$$

where $(N_t,\mathscr{F}_t,P)$ is a continuous local martingale with $N_0 = 0$. Furthermore, the above representation for L_t uniquely determines N_t.

PROOF. The condition that (L_t) is positive means that

$$P(L_t > 0,\ t \in [0,T]) = 1. \tag{7.1.4}$$

We may then apply Ito's formula to $\log L_t$, getting

$$\log L_t = \int_0^t L_s^{-1}\, dL_s - \frac{1}{2}\int_0^t L_s^{-2}\, d\langle L\rangle_s.$$

If we define $N_t = \int_0^t L_s^{-1}\, dL_s$, clearly (N_t) is a continuous local $(\mathscr{F}_t,P)$ martingale and the second term in the right-hand side of the above equality is $-\frac{1}{2}\langle N\rangle_t$, thus giving the desired expression for L_t. The uniqueness of N_t follows from Theorem 2.6.2. □

The application of Ito's formula made above is a slight variant of Theorem 4.2.3. Here we apply the formula to the function $F(x) = \log x$ $(x > 0)$, whereas in Theorem 4.2.3 F is a C^2 function defined over $\mathbf{R}^1$. The necessary modification of the proof does not present any difficulty and is left to the reader.

Remark 7.1.1. Alternatively, we may apply Ito's formula to

$$L_t \exp[-N_t + \tfrac{1}{2}\langle N\rangle_t],$$

(N being defined as above) and show that it equals 1 (P-a.s.) for all t.

The following useful remark is essentially contained in a part of the argument used in Lemma 7.1.3.

Lemma 7.1.5. *Let $\{\tau_n\}$ be a sequence of stopping times of the family $(\mathscr{F}_t)$ $t \in [0,T]$. For each n, let (M_t^n) be a sample-continuous, $(\mathscr{F}_{t\wedge\tau_n})$-adapted process such that $(M_t^n L_{t\wedge\tau_n})$ is a continuous $(\mathscr{F}_t,P)$-martingale. Then (M_t^n) is a continuous $(\mathscr{F}_t,P)$-martingale.*

Proof. By assumption, if $s \leq t$ and $A \in \mathscr{F}_{\tau_n \wedge s}$, we have

$$\int_A M_t^n L_{t\wedge\tau_n}\, dP = \int_A M_s^n L_{s\wedge\tau_n}\, dP. \tag{7.1.5}$$

Now if σ is any stopping time bounded by T, then [since (L_t), $t \in [0,T]$ is uniformly P-integrable] as in the proof of Lemma 7.1.3, we have $Q \ll P$ relative to $\mathscr{F}_\sigma$ with corresponding $dQ/dP = L_\sigma$. Hence (7.1.5) becomes

$$\int_A M_t^n\, dQ = \int_A M_s^n\, dQ$$

which proves that $(M_t^n, \mathscr{F}_{t\wedge\tau_n}, Q)$ is a martingale. □

Remark 7.1.2. Using appropriate stopping times, one can modify Lemma 7.1.4 when condition (7.1.4) fails to hold, in other words, when $P\{\inf_{0\leq t\leq T} L_t \geq 0\} = 1$. Let

$$\tau_n = \begin{cases} \inf\left\{t \leq T : L_t \leq \dfrac{1}{n} \text{ or } L_t \geq n\right\} & \\ \infty & \text{if the above set is empty} \end{cases}$$

and

$$\tau = \begin{cases} \inf\{t \le T : L_t = 0\} & \\ \infty & \text{if the above set is empty.} \end{cases} \tag{7.1.6}$$

Then,

$$\tau_n \le \tau_{n+1}, \tau_n \uparrow \tau \qquad (P\text{-a.s.}) \tag{7.1.7}$$

and

$$t < \tau_n \quad \text{implies} \quad \frac{1}{n} < L_t < n. \tag{7.1.8}$$

From Lemma 7.1.4 applied to $L_t^n = L_{t \wedge \tau_n}$, we obtain $L_t^n = \exp[N_t^n - \frac{1}{2}A_t^n]$, where

$$N_t^n = \int_0^t (L_s^n)^{-1}\, dL_s^n \quad \text{and} \quad A_t^n = \langle N^n \rangle_t. \tag{7.1.9}$$

Since $N_t^n(\omega) = N_t^k(\omega)$ if $t < \tau_n(\omega)$ for all $k \ge n$, it follows that there exists a process $N_t(\omega)$, defined for $t < \tau(\omega)$ and such that [from (7.1.9)]

$$N_{t \wedge \tau_n} = N_{t \wedge \tau_n}^n = \int_0^{t \wedge \tau_n} (L_s)^{-1}\, dL_s. \tag{7.1.10}$$

Similarly, there exists $A_t(\omega)$ defined for $t < \tau(\omega)$ such that

$$A_{t \wedge \tau_n} = A_{t \wedge \tau_n}^n.$$

Next observe that since $L_\tau = 0$ and (L_t) is a nonnegative supermartingale, $L_t = 0$ for $t \ge \tau$. Since $\tau_n \uparrow \tau$, we have $L_t^n \to L_t$ if $t < \tau(\omega)$, and we obtain the following modification of Lemma 7.1.4.

$$L_t = \begin{cases} \exp[N_t - \frac{1}{2}A_t] & \text{if } t < \tau \\ 0 & \text{if } t \ge \tau. \end{cases} \tag{7.1.11}$$

We shall be dealing with situations of this sort later in this chapter when proving extended versions of Girsanov's theorem.

Theorem 7.1.1. *Let us make the following assumptions:*

(a) $(M_t,\mathscr{F}_t,P)$ *is a sample-continuous,* $\mathbf{R}^N$*-valued stochastic process with* $M_0 = 0$.
(b) $\{A_t, t \ge 0\}$ *is a sample-continuous,* $N \times N$*-matrix-valued,* $(\mathscr{F}_t)$*-adapted, increasing process.*

Then the following statements are equivalent:

(I) $(M_t,\mathscr{F}_t)$ *is a continuous local martingale with* $\langle M \rangle = A$.
(II) *For all* $\theta \in \mathbf{R}^N$, $(M_t^\theta,\mathscr{F}_t)$ *is a continuous, local martingale, where* $M_t^\theta = \exp[(\theta,M_t) - \frac{1}{2}(A_t\theta,\theta)]$.

Proof. An application of Ito's formula shows that (I) implies (II). We first prove that (II) implies (I) for $N = 1$ using the following lemma. The general case is easily reduced to the case $N = 1$ as we shall see below.

Lemma 7.1.6. *Let $Y_t^\theta = \exp[\theta M_t - \frac{1}{2}\theta^2 A_t]$ $(\theta \in \mathbf{R})$ be a continuous martingale and suppose that for some open neighborhood I of $\theta = 0$ and for all t, we have,* (P-a.s.),

(i) $|Y_t^\theta| \le a$.
(ii) $|(d/d\theta)Y_t^\theta| \le b$.
(iii) $|(d^2/d\theta^2)Y_t^\theta| \le c$, *where a,b,c are nonrandom constants depending on I but not on t.*

Then $(M_t,\mathscr{F}_t)$ and $(M_t^2 - A_t, \mathscr{F}_t)$ are (continuous) martingales.

PROOF. The first assertion is a consequence of (i) and (ii). If $s \le t$ and $A \in \mathscr{F}_s$,

$$\int_A E\left[\left(\frac{d}{d\theta} Y_t^\theta\right)_{\theta=0}\middle|\mathscr{F}_s\right]dP = \int_A \left(\frac{d}{d\theta} Y_t^\theta\right)_{\theta=0} dP$$

$$= \left(\frac{d}{d\theta}\int_A Y_t^\theta\, dP\right)_{\theta=0}$$

$$= \left(\frac{d}{d\theta}\int_A Y_s^\theta\, dP\right)_{\theta=0} = \int_A \left(\frac{d}{d\theta} Y_s^\theta\right)_{\theta=0} dP.$$

Hence $E[((d/d\theta)Y_t^\theta)_{\theta=0}|\mathscr{F}_s] = ((d/d\theta)Y_s^\theta)_{\theta=0}$ (P-a.s.), that is, $E(M_t|\mathscr{F}_s) = M_s$. The second assertion follows similarly if we use (iii) and note that $((d^2/d\theta^2)Y_t^\theta)_{\theta=0} = M_t^2 - A_t$. □

PROOF OF THEOREM 7.1.1. (*continued*). Define the stopping times $T_n = \inf\{t : t \le T, |M_t| \ge n \text{ or } A_t \ge n\}$ and $= T$ if the set just described is empty. Then $M_{T_n\wedge t}^\theta = \exp[\theta M_{T_n\wedge t} - \frac{1}{2}\theta^2 A_{T_n\wedge t}]$, and for $\theta \in I$ we have

$$|M_{T_n\wedge t}^\theta| \le \exp[n|\theta| + \tfrac{1}{2}\theta^2 n] \le a_n(I)$$

$$\left|\frac{d}{d\theta} M_{T_n\wedge t}^\theta\right| \le |M_{T_n\wedge t}^\theta||M_{T_n\wedge t} - \theta A_{T_n\wedge t}|$$

$$\le a_n(I)n(1 + |\theta|) \le b_n(I),$$

$$\left|\frac{d^2}{d\theta^2} M_{T_n\wedge t}^\theta\right| = |M_{T_n\wedge t}^\theta|[(M_{T_n\wedge t} - \theta A_{T_n\wedge t})^2 + A_{T_n\wedge t}]$$

$$\le a_n(I)\{n^2(1 + |\theta|)^2 + n\} \le c_n(I).$$

By Lemma 7.1.6 it follows that $(M_{T_n\wedge t},\mathscr{F}_t)$ and $(M_{T_n\wedge t}^2 - A_{T_n\wedge t},\mathscr{F}_t)$ are (continuous) martingales. If $N > 1$, set $X_t = \sum_{i=1}^N \theta^i M_t^i$ and $B_t = (A_t\theta,\theta) = \sum_{i,j} A_t^{ij}\theta^i\theta^j$, where $\theta = (\theta^1, \ldots, \theta^N)$ and A_t^{ij} are the entries of the matrix A_t. The one-dimensional version of Theorem 7.1.1 then applies to X_t and B_t. Statement (II) implies that for every real α and $\theta \in \mathbf{R}^N$, $\exp[(\alpha\theta,M_t) - \frac{1}{2}(\alpha A_t\theta,\alpha\theta)]$, that is, $\exp[\alpha X_t - \frac{1}{2}\alpha^2 B_t]$ is a continuous local martingale. Hence $(X_t,\mathscr{F}_t)$ is a continuous local martingale with $\langle X,X\rangle = B$; that is,

$\sum_{i=1}^{N} \theta^i M_t^i$ is a continuous local martingale with $\langle \sum_i \theta^i M_t^i, \sum_i \theta^i M_t^i \rangle = \sum_{i,j} A_t^{ij} \theta^i \theta^j$ for all $\theta \in \mathbf{R}^N$. This completes the proof of Theorem 7.1.1. □

Theorem 7.1.2. *Let $M_t = (M_t^1, \ldots, M_t^N)$ be an N-dimensional, continuous local $(\mathscr{F}_t, P)$ martingale. Let $\tilde{P}$ be a probability measure such that $\tilde{P} \equiv P$, (that is, mutually absolutely continuous) relative to $\mathscr{F}_T$ and let L_t be the Radon-Nikodym derivative $d\tilde{P}/dP$ relative to $\mathscr{F}_t$, assumed to be sample continuous. Let $V_t = (V_t^1, \ldots, V_t^N)$, where $V_t^i = \langle M^i, N \rangle_t$, N being the continuous local martingale of Lemma 7.1.4. (We need $L_0 = 1$ here.) Define*

$$\tilde{M}_t = M_t - V_t. \tag{7.1.12}$$

Then $(\tilde{M}_t, \mathscr{F}_t, \tilde{P})$ is an N-dimensional, continuous local martingale with

$$\langle \tilde{M} \rangle = \langle M \rangle. \tag{7.1.13}$$

Proof. From Lemma 7.1.4 we have

$$L_t = \exp[N_t - \tfrac{1}{2}\langle N \rangle_t].$$

Fix an arbitrary $\theta \in \mathbf{R}^N$, and write $A_t = \langle M \rangle_t$. In view of Theorem 7.1.1 it suffices to show that $\exp\{(\theta, \tilde{M})_t - \frac{1}{2}(A_t\theta, \theta)\}$ is an $(\mathscr{F}_t, \tilde{P})$ continuous local martingale. By Lemma 7.1.3 this follows if we can show that

$$\exp\{(\theta, \tilde{M})_t - \tfrac{1}{2}(A_t\theta, \theta)\} \exp\{N_t - \tfrac{1}{2}\langle N \rangle_t\} \tag{7.1.14}$$

is a continuous local $(\mathscr{F}_t, P)$-martingale. Let us simplify the quantities in the exponents. For the real-valued local martingale $(\theta, M)_t$ we have

$$\langle (\theta, M) \rangle_t = \sum_{i,j} \theta^i \theta^j \langle M^i, M^j \rangle_t = (A_t\theta, \theta),$$

and

$$\langle (\theta, M), N \rangle_t = \sum_i \theta^i \langle M^i, N \rangle_t = \sum_i \theta^i V_t^i = (\theta, V)_t.$$

Hence

$$\begin{aligned}
(\theta, \tilde{M})_t &- \tfrac{1}{2}(A_t\theta, \theta) + N_t - \tfrac{1}{2}\langle N \rangle_t \\
&= (\theta, M)_t + N_t - \tfrac{1}{2}\langle (\theta, M) \rangle_t - \tfrac{1}{2}\langle N \rangle_t - (\theta, V)_t \\
&= (\theta, M)_t + N_t - \tfrac{1}{2}[\langle (\theta, M) \rangle_t + 2\langle (\theta, M), N \rangle_t + \langle N \rangle_t] \\
&= (\theta, M)_t + N_t - \tfrac{1}{2}\langle (\theta, M) + N \rangle_t.
\end{aligned}$$

Thus (7.1.14) becomes

$$\exp\{[(\theta, M) + N]_t - \tfrac{1}{2}\langle (\theta, M) + N \rangle_t\}$$

which is a continuous local $(\mathscr{F}_t, P)$-martingale from Lemma 7.1.1. This concludes the proof. We have assumed $M_0 = 0$ for convenience. □

The theorem just proved generalizes a famous result due to Girsanov which we now derive. In the following results we use Prop. 1.14 when necessary.

Theorem 7.1.3 (Girsanov). *Let $(W_t,\mathscr{F}_t,P)$ be an N-dimensional Wiener martingale and (f_t) a measurable, $(\mathscr{F}_t)$-adapted process such that*

$$\int_0^T |f_t|^2\,dt < \infty \qquad (P\text{-a.s.}).$$

(Here $|x|$ denotes the norm of $x \in \mathbf{R}^N$). Let

$$\xi_0^t(f) = \exp\left[\int_0^t (f_s,dW_s) - \frac{1}{2}\int_0^t |f_s|^2\,ds\right],$$

where $\int_0^t (f_s,dW_s) = \sum_{i=1}^N \int_0^t f_s^i\,dW_s^i$. Assume that

$$E\xi_0^T(f) = 1. \tag{7.1.15}$$

Let $\tilde{P}$ be the probability measure given by

$$d\tilde{P} = \xi_0^T(f)\,dP. \tag{7.1.16}$$

If $\tilde{W}_t = W_t - \int_0^t f_s\,ds$, then $(\tilde{W}_t,\mathscr{F}_t,\tilde{P})$ is an N-dimensional Wiener martingale.

Proof. The conditions of Theorem 7.1.2 are met. Writing $N_t = \int_0^t (f_s,dW_s)$ and $M_t^i = W_t^i$, we have $V_t^i = \langle M^i,N\rangle_t = \langle W^i,\sum_{j=1}^N \int_0 f_s^j\,dW_s^j\rangle_t = \int_0^t f_s^i\,ds$, so that $\tilde{W}_t = \tilde{M}_t$. It follows from Theorem 7.1.2 that $(\tilde{W}_t)$ is a continuous local $(\mathscr{F}_t,\tilde{P})$-martingale with $\langle \tilde{W}\rangle_t = tI$, I being the $N \times N$ identity matrix. The result now follows from the characterization of the Wiener process proved in Section 4.4. □

The following result is an extension of Girsanov's theorem. We consider real-valued processes ($N = 1$).

Theorem 7.1.4. *Let $(W_t,\mathscr{F}_t,P)$, $t \in [0,T]$, be a Wiener martingale and let $\tilde{P}$ be a probability measure on $\mathscr{F}_T$ such that*

$$d\tilde{P} = L_T\,dP, \tag{7.1.17}$$

where

$$L_t = 1 + \int_0^t b_s\,dW_s, \tag{7.1.18}$$

(b_t) being a measurable, $(\mathscr{F}_t)$-adapted process such that

$$\int_0^T b_t^2\,dt < \infty \qquad (P\text{-a.s.}). \tag{7.1.19}$$

On the probability space $(\Omega,\mathscr{F}_t,\tilde{P})$ define

$$\tilde{W}_t = W_t - \int_0^t \hat{L}_s b_s\,ds$$

for all t in $[0,T]$, where

$$\hat{L}_s(\omega) = \begin{cases} \dfrac{1}{L_s(\omega)} & \text{if } L_s(\omega) > 0, \\ 0 & \text{if } L_s(\omega) = 0. \end{cases}$$

Then $(\tilde{W}_t,\mathscr{F}_t,\tilde{P})$ is a Wiener martingale.

PROOF. Let us first show the validity of the definition of $(\tilde{W}_t)$. As in Lemma 7.1.5 let

$$\tau = \begin{cases} \inf\{t \le T: L_t = 0\} \\ \infty & \text{if } \inf_{0 \le t \le T} L_t > 0. \end{cases}$$

Let $A = \{\omega: \inf_{0 \le t \le T} L_t(\omega) > 0\}$ and define $L_\tau = L_T$ on the set $[\tau = \infty]$. Then

$$\tilde{P}(A^c) = \int_{[\tau \le T]} L_T \, dP = \int_{[\tau \le T]} L_\tau \, dP = 0,$$

so that

$$\tilde{P}(A) = 1.$$

We also have $\tilde{P}(\int_0^T b_t^2 \, dt < \infty) = 1$. Hence on the set $A \cap [\int_0^T b_t^2 \, dt < \infty]$ [and hence ($\tilde{P}$-a.s.)],

$$\int_0^T (\hat{L}_t b_t)^2 \, dt \le \frac{1}{\left(\inf_{0 \le t \le T} L_t\right)^2} \int_0^T b_t^2 \, dt < \infty.$$

Define the stopping times

$$\tau_n = \begin{cases} \inf\left\{t \le T: \int_0^t b_s^2 \, ds \ge n \text{ or } L_t \le \frac{1}{n}\right\} \\ \infty & \text{if this set is empty.} \end{cases}$$

Clearly $\tau_n \le \tau_{n+1}$ and $\tau_n \uparrow \tau$ (P-a.s.). Define

$$Y_t^n = \exp\left[\theta\left(W_{t \wedge \tau_n} - \int_0^{t \wedge \tau_n} f_s \, ds\right) - \tfrac{1}{2}\theta^2 (t \wedge \tau_n)\right],$$

where θ is a fixed real number and $f_s = \hat{L}_s b_s$. Since $L_{t \wedge \tau_n} \ge 1/n$, from (7.1.18) and Ito's formula

$$\log L_{t \wedge \tau_n} = \int_0^t \left(\frac{b_s}{L_{s \wedge \tau_n}}\right) I_{[0,\tau_n]}(s) \, dW_s - \frac{1}{2} \int_0^t \left(\frac{b_s}{L_{s \wedge \tau_n}} I_{[0,\tau_n]}(s)\right)^2 ds,$$

where $I_{[0,\tau_n]}$ denotes the indicator function of $[0,\tau_n]$. Since $(b_s/L_{s \wedge \tau_n}) I_{[0,\tau_n]}(s) = \hat{L}_s b_s$ if $s < \tau_n$, and 0 otherwise, we have

$$L_{t \wedge \tau_n} = \exp\left[\int_0^{t \wedge \tau_n} f_s \, dW_s - \frac{1}{2} \int_0^{t \wedge \tau_n} f_s^2 \, ds\right].$$

Note that the stochastic integral $\int_0^{t \wedge \tau_n} f_s \, dW_s$ exists since

$$\int_0^t f_s^2 I_{[0,\tau_n]}(s) \, ds < \infty \qquad (P\text{-a.s.}).$$

Hence

$$Y_t^n L_{t \wedge \tau_n} = \exp\left[\int_0^{t \wedge \tau_n} (f_s + \theta) \, dW_s - \frac{1}{2} \int_0^{t \wedge \tau_n} (f_s + \theta)^2 \, ds\right].$$

Furthermore, from the definition of τ_n,

$$\int_0^{t\wedge\tau_n}(f_s+\theta)^2\,ds \le 2\int_0^{t\wedge\tau_n}(f_s^2+\theta^2)\,ds$$
$$\le 2\theta^2 T + 2n^3.$$

From Lemmas 7.1.1 and 7.1.2 it follows that $Y_t^n L_{t\wedge\tau_n}$ is a continuous $(\mathscr{F}_t,P)$-martingale and from Lemma 7.1.5, $(Y_t^n,\mathscr{F}_t,\tilde{P})$ is a continuous martingale. Since $\tau_n \uparrow \tau$ (P-a.s.) and $\tilde{P}[\tau=\infty]=1$, we have $\tau_n\uparrow\infty$ ($\tilde{P}$-a.s.). Making $n\to\infty$ in the expression for Y_t^n, we find that

$$Y_t = \exp\left[\theta\left(W_t - \int_0^t f_s\,ds\right) - \tfrac{1}{2}\theta^2 t\right]$$

is a continuous local $(\mathscr{F}_t,\tilde{P})$-martingale. The conclusion that

$$\left(W_t - \int_0^t f_s\,ds, \mathscr{F}_t, \tilde{P}\right)$$

is a Wiener martingale now follows from Theorem 7.1.1. □

Remark 7.1.3. Since (L_t) is a nonnegative $(\mathscr{F}_t,P)$-supermartingale, we have $L_t=0$ for $t\ge\tau$ (since $L_\tau=0$). Let us set $M_t^n=\int_0^{t\wedge\tau_n} f_s\,dW_s$. Then as we have seen above,

$$L_{t\wedge\tau_n} = \exp[M_t^n - \tfrac{1}{2}\langle M^n\rangle_t];$$

thus $L_{t\wedge\tau_n}=\exp[M_t^n-\tfrac{1}{2}A_t^n]$, where $A_t^n=\langle M^n\rangle_t$. It is easy to see that $M_t^n(\omega)=M_t^k(\omega)$ if $t<\tau_n(\omega)$ for all $k\ge n$, so that there exists a process $M_t(\omega)$ defined for $t<\tau(\omega)$ such that

$$M_t(\omega)=M_t^n \quad \text{for } t<\tau_n(\omega).$$

Similarly, there exists $A_t(\omega)(t<\tau(\omega))$ such that

$$A_t(\omega)=A_t^n \quad \text{for } t<\tau_n(\omega).$$

We thus arrive at the following expression for the Radon-Nikodym derivative L_t.

$$L_t = \begin{cases} \exp[M_t - \tfrac{1}{2}A_t] & \text{if } t<\tau, \\ 0 & \text{if } t\ge\tau, \end{cases}$$

where M_t and A_t are as defined above.

Finally, if $P[\tau\le T]=0$, then $P\ll\tilde{P}$ on $\mathscr{F}_t$. Since then $P[\tau\le t]=0$ for all t, the formula for L_t takes the form

$$L_t = \exp\left[\int_0^t f_s\,dW_s - \frac{1}{2}\int_0^t f_s^2\,ds\right] \qquad (0\le t\le T).$$

Theorem 7.1.2 can be recast to yield a more general version of the original Girsanov theorem (Theorem 7.1.3).

Theorem 7.1.5. *Let M_t be an N-dimensional, continuous local $(\mathscr{F}_t,P)$-martingale. Let $(N_t,\mathscr{F}_t,P)$ be a continuous local martingale such that $N_0=0$ and*

$E(Y_T) = 1$, where

$$Y_t = \exp[N_t - \tfrac{1}{2}\langle N\rangle_t].$$

Then $(\tilde{M}_t)$, where $\tilde{M}_t = M_t - V_t$ and $V_t^i = \langle M^i,N\rangle_t$, is a continuous local $(\mathscr{F}_t,\tilde{P})$-martingale, $\tilde{P}$ being the probability measure defined by $d\tilde{P} = Y_T\,dP$, with $\langle\tilde{M}\rangle = \langle M\rangle$.

7.2 Sufficient Conditions for the Validity of Girsanov's Theorem

Let (N_t) and (Y_t) be as in Theorem 7.1.5.

Theorem 7.2.1. *Suppose that for some positive number δ,*

$$E\exp[(1+\delta)\langle N\rangle_T] < \infty. \tag{7.2.1}$$

Then

$$E(Y_T) = 1.$$

PROOF. Fix $t_1,t_2 \in [0,T]$, $t_1 < t_2$. We consider temporarily the interval $[t_1,t_2]$. For $t_1 \le t \le t_2$ define $M_t = N_t - N_{t_1}$ and $A_t = \langle N\rangle_t - \langle N\rangle_{t_1}$.

Then $(M_t,\mathscr{F}_t, t_1 \le t \le t_2, P)$ is a continuous local martingale with $\langle M\rangle = A$, and $M_{t_1} = A_{t_1} = 0$. Write $\eta_t = \exp\{M_t - \frac{1}{2}A_t\}$ and define the stopping times $\tau_n = \inf\{t\colon t_1 \le t \le t_2\colon \eta_t \ge n \text{ or } A_t \ge n\}$. Then $\eta_t^n \equiv \eta_{t\wedge\tau_n}$ is a martingale and $E(\eta_t^n) = 1$. For $\varepsilon > 0$ (and setting $M_t^n = M_{t\wedge\tau_n}$, $A_t^n = A_{t\wedge\tau_n}$), $(\eta_t^n)^{1+\varepsilon} = [e^{(1+\varepsilon)M_t^n - \frac{1}{2}(1+\varepsilon)_3A_t^n}][e^{\frac{1}{2}\varepsilon(1+\varepsilon)(2+\varepsilon)A_t^n}]$. Applying Hölder's inequality, we obtain

$$E(\eta_t^n)^{1+\varepsilon} \le (E[e^{(1+\varepsilon)^2M_t^n - \frac{1}{2}(1+\varepsilon)^4A_t^n}])^{1/(1+\varepsilon)}(E[e^{\frac{1}{2}(1+\varepsilon)^2(2+\varepsilon)A_t^n}])^{\varepsilon/(1+\varepsilon)}.$$

Choose ε so small that $\frac{1}{2}(1+\varepsilon)^2(2+\varepsilon) < 1+\delta$. The first factor on the right side of the inequality equals 1, and the second factor is bounded since it is dominated by $(E[e^{(1+\delta)\langle N\rangle_T}])^{\varepsilon/(1+\varepsilon)}$. Hence, for all n and $t \in [t_1,t_2]$, $E(\eta_t^n)^{1+\varepsilon}$ is bounded by a constant, which shows that the sequence (η_t^n) is uniformly integrable. It follows that

$$E(\eta_{t_2}) = \lim_{n\to\infty} E(\eta_{t_2}^n) = 1. \tag{7.2.2}$$

The theorem follows upon taking $t_1 = 0$ and $t_2 = T$. □

It follows from Lemma 7.1.1 that for $t_1 < t_2$,

$$E(\eta_{t_2}|\mathscr{F}_{t_1}) = 1. \tag{7.2.3}$$

Theorem 7.2.2. *Suppose there exists a positive number α such that for every t in $[0,T]$ $(t+\alpha \le T)$,*

$$E\exp((1+\delta)[\langle N\rangle_{t+\alpha} - \langle N\rangle_t]) < \infty \tag{7.2.4}$$

for some $\delta > 0$. (The number δ may depend on t as well as on α.) Then $E(Y_T) = 1$.

PROOF. Let $\{t_j\}$ $(0 = t_0 < t_1 < \cdots < t_n = T)$ be a fixed partition of $[0,T]$ such that each $t_{j+1} - t_j < \alpha$. Note that condition (7.2.4) is valid with t, $t + \alpha$ replaced, respectively, by t_j, t_{j+1} $(j = 0, \ldots, n-2)$ and by t_{n-1}, T. In the proof of the last theorem replace $[t_1,t_2]$ by $[t_j,t_{j+1}]$ and η_t by $\eta_{t_j}^t$ $(t_j \le t \le t_{j+1})$. Then (7.2.3) can be rewritten as

$$E(\eta_{t_j}^{t_{j+1}} | \mathscr{F}_{t_j}) = 1 \quad \text{for every } j. \tag{7.2.5}$$

Now

$$E(\eta_0^{t_2} - 1) = E[\eta_0^{t_1} E\{(\eta_1^{t_2} - 1) | \mathscr{F}_{t_1}\}] = 0 \text{ .}$$

by (7.2.5) with $j = 1$. Hence $E(\eta_0^{t_2}) = 1$, and by repeating this argument, we get $E(\eta_0^T) = 1$, that is $E(Y_T) = 1$. □

Corollary 7.2.1. *In the original result of Girsanov (Theorem 7.1.3) condition (7.1.15) holds provided the following condition is satisfied. There exists $\alpha > 0$ such that for each t $(t + \alpha \le T)$*

$$E \exp\left((1 + \delta) \int_t^{t+\alpha} |f_s|^2 \, ds\right) < \infty \tag{7.2.6}$$

for some $\delta > 0$.

Corollary 7.2.2. *Suppose there exists $\varepsilon > 0$ and a positive constant C such that*

$$E \exp(\varepsilon |f_s|^2) \le C \tag{7.2.7}$$

for each $s \in [0,T]$. Then (7.1.15) is satisfied.

PROOF. Fix $\lambda > 1$ and choose a partition $\{t_j\}$ of $[0,T]$ such that λt_1, $\lambda(t_{j+1} - t_j)$ $(j = 1, \ldots, n-1; t_n = T)$ are all less than ε. By Jensen's inequality, for each j we have

$$\begin{aligned} E\left(\exp\left[\lambda \int_{t_j}^{t_{j+1}} |f_s|^2 \, ds\right]\right) &\le \frac{1}{t_{j+1} - t_j} \int_{t_j}^{t_{j+1}} E\left(\exp \varepsilon |f_s|^2\right) ds \\ &\le C. \end{aligned}$$

The conclusion now follows from Corollary 7.2.1. □

Finally, we state below a considerable improvement of the criteria derived above. For the proof, the reader is referred to the paper by Novikov cited in the bibliography. In Theorem 7.1.3 take $N = 1$.

Theorem 7.2.3 (Novikov). *Condition (7.1.15) in Theorem 7.1.3 holds provided*

$$E \exp\left(\frac{1}{2} \int_0^T f_s^2 \, ds\right) < \infty. \tag{7.2.8}$$

7.3 Stochastic Equations and Absolute Continuity of Induced Measures

The following stochastic processes are assumed given on $(\Omega,\mathscr{A},P)$:

$$(W_t,\mathscr{F}_t,P) \text{ is a Wiener martingale.} \tag{7.3.1}$$

$$\alpha = (\alpha_t) \text{ is } (\mathscr{F}_t)\text{-adapted and measurable;} \tag{7.3.2}$$

that is, the function $(t,\omega) \to \alpha_t(\omega)$ is $\mathscr{B} \times \mathscr{F}_t$-measurable, where $\mathscr{B}$ is the σ-field of Borel sets in $[0,T]$ and α satisfies the additional condition

$$P\left[\omega\colon \int_0^T \alpha_t^2(\omega)\,dt < \infty\right] = 1. \tag{7.3.3}$$

The process $\xi = (\xi_t)$ is defined by

$$\xi_t(\omega) = \int_0^t \alpha_s(\omega)\,ds + W_t(\omega) \tag{7.3.4}$$

$t \in [0,T]$ and $\omega \in \Omega$.

Observe that the process α is independent of the "future" of $W = (W_t)$ in the sense that for every s, $\sigma[\alpha_u, 0 \leq u \leq s] \perp\!\!\!\perp \sigma[W_v - W_u, v \geq u \geq s]$. This is a consequence of the $\mathscr{F}_t$-measurability of α_t and the fact that $(W_t,\mathscr{F}_t)$ is a Wiener martingale. As we have seen in Chapter 4, the latter property implies the independence of $\mathscr{F}_s$ and $\sigma[W_v - W_u, v \geq u \geq s]$.

Write

$$L(\omega) = L_T(\omega) = \exp\left[-\int_0^T \alpha_t(\omega)\,dW_t(\omega) - \frac{1}{2}\int_0^T \alpha_t^2(\omega)\,dt\right].$$

We know from Theorem 7.1.3 that if

$$E(L) = 1, \tag{7.3.5}$$

then $(\xi_t,\mathscr{F}_t,\tilde{P})$ is a Wiener martingale, where $\tilde{P}$ is given by $d\tilde{P} = L\,dP$.

Let $C = C[0,T]$ denote the Banach space, with sup norm, of real continuous functions x on $[0,T]$, and let us recall from Section 5.1 the definition of the σ-fields $\mathscr{B}_t(C)$, which we now denote by $\mathscr{B}_t$ for brevity. The process ξ given by (7.3.4) defines a map $\xi\colon \Omega \to C$ given by $\omega \to \xi(\omega)$, where $\xi(\omega)$ is the function $t \to \xi_t(\omega)$. This map is clearly $\mathscr{B}_T/\mathscr{F}_T$-measurable and induces on $(C,\mathscr{B}_T)$ the measure $P\xi^{-1}$ which will henceforth be denoted by P_ξ. Let $\mu = PW^{-1}$ be the standard Wiener measure. A consequence of (7.3.5) and Girsanov's theorem is that $P_\xi \equiv \mu[\mathscr{B}_T]$ since $\tilde{P} \equiv P[\mathscr{F}_T]$ and $\tilde{P}_{\xi^{-1}} = \mu$. The notation $P_\xi \equiv \mu[\mathscr{B}_T]$ means P_ξ and μ are mutually absolutely continuous relative to $\mathscr{B}_T$. In this section we investigate conditions under which $P_\xi \ll \mu$ or $\mu \ll P_\xi$. We shall, of course, not assume condition (7.3.5).

Theorem 7.3.1. *Let ξ be given by (7.3.4). Then condition (7.3.3), implies*

$$P_\xi \ll \mu[\mathscr{B}_T]. \tag{7.3.6}$$

PROOF. Define the $(\mathscr{F}_t)$ stopping times

$$\tau_n(\omega) = \begin{cases} \inf\left\{t: t \le T, \int_0^t \alpha_s^2(\omega)\,ds \ge n\right\} & \\ T & \text{if the above set is empty.} \end{cases}$$

Then (7.3.3) implies $\tau_n(\omega) \uparrow T$ (P-a.s.). Let

$$\xi_t^{(n)}(\omega) = \int_0^t \alpha_s^{(n)}(\omega)\,ds + W_t(\omega), \tag{7.3.7}$$

where $\alpha_t^{(n)}(\omega) = \alpha_t(\omega) I_{[0,\tau_n(\omega)]}(t)$. Since $\int_0^T [\alpha_t^{(n)}(\omega)]^2\,dt \le n$ (P-a.s.), it follows from Lemma 7.1.2 that $E\exp[-\int_0^T \alpha_t^{(n)}(\omega)\,dW_t - \frac{1}{2}\int_0^T (\alpha_t^{(n)}(\omega))^2\,dt] = 1$. Hence from Theorem 7.1.3 we have $P_{\xi^{(n)}} \equiv \mu$, where $\xi^{(n)}$ is the map from Ω into C given by (7.3.7). For any set $A \in \mathscr{B}_T$ such that $\mu(A) = 0$, we get

$$P(\{\omega: \xi(\omega) \in A\} \cap \{\omega: \tau_n(\omega) = T\}) = P(\{\omega: \xi^{(n)}(\omega) \in A\} \cap \{\omega: \tau_n(\omega) = T\}),$$

since on $\{\tau_n(\omega) = T\}$, $I_{[0,\tau_n(\omega)]}(t) = 1$ for all t in $[0,T]$. Hence

$$\begin{aligned} P(\{\omega: \xi(\omega) \in A\} \cap \{\omega: \tau_n(\omega) = T\}) &\le P(\{\omega: \xi^{(n)}(\omega) \in A\}) \\ &= P_{\xi^{(n)}}(A) = 0. \end{aligned}$$

Making n tend to infinity, we obtain $P_\xi(A) = 0$, since $\{\omega: \tau_n(\omega) = T\} = \{\omega: \int_0^T \alpha^2(t,\omega)\,dt \le n\}$ converges to $\{\omega: \int_0^T \alpha_t^2(\omega)\,dt < \infty\}$ which has P-probability 1 by (7.3.3). □

Remark 7.3.1. If (7.3.3.) is replaced by the stronger condition

$$\int_0^T \alpha_t^2(\omega)\,dt \le C \qquad (P\text{-a.s.}), \tag{7.3.3a}$$

where C is a constant, then we have $P_\xi \equiv \mu$. To see this, note that Lemma 7.1.2 now implies that (7.3.5) holds. The remark just preceding Theorem 7.3.1 establishes the conclusion.

Remark 7.3.2. We can use the remark following Lemma 7.1.5 to obtain an expression for the Radon-Nikodym derivative in Theorem 7.3.1

Let $\bar{\mathscr{B}}_t = \mathscr{B}_t \vee$ subsets of μ-null sets of $\mathscr{B}_T$ and $L_t(x) = (dP_\xi/d\mu)(x)[\bar{\mathscr{B}}_t]$, $x \in C$. There exists a right-continuous (and hence separable) version of the martingale $(L_t, \bar{\mathscr{B}}_t, \mu)$. Let $\sigma_n(x) = \inf\{t \le T: L_t(x) \ge n\}$, and $= T$ if this set is empty, and let $L_t'^n = L_{t \wedge \sigma_n}$. Since $(L_t'^n, \bar{\mathscr{B}}_t, \mu)$ is a separable, square-integrable martingale, it follows from the representation results of the last chapter that $(L_t'^n)$ is actually sample continuous. Since the paths of (L_t) are bounded on $[0,T]$, for almost all $x \in C$, there exists $N(x)$ such that, for every $n \ge N(x)$, $\sigma_n(x) = T$, and since $L_t'^n = L_t$ for $t \le \sigma_n$, we have that (L_t) is continuous on $[0,T]$ (a.s.) μ.

Now define the stopping times τ and $\{\tau_n\}$ as before except that Ω, $(\mathscr{F}_t)$ are now replaced by C, $(\bar{\mathscr{B}}_t)$. Write $N_t^n = \int_0^t (L_s^n)^{-1}\,dL_s^n$. Then N_t^n is a continuous square-integrable $(\overline{\mathscr{B}}_t,\mu)$-martingale. Again, by the result on the representation of functionals on Wiener space proved in the last chapter it follows that for all t,

$$N_t^n(x) = \int_0^t h_s^n(x)\,dx_s \qquad \text{(a.s.) } \mu \tag{7.3.8}$$

where the latter is an Ito integral with respect to the coordinate Wiener process (x_t). Also N_t^n has quadratic variation

$$A_t^n(x) = \int_0^t (h_s^n(x))^2\,ds. \tag{7.3.9}$$

From (7.1.11) we obtain

$$\frac{dP_\xi}{d\mu} = \begin{cases} exp\{N_t - \frac{1}{2}A_t\} & \text{if } t < \tau, \\ 0 & \text{if } t \geq \tau, \end{cases} \tag{7.3.10}$$

where for $t < \tau$ we have

$$N_{t\wedge\tau_n}(x) = \int_0^{t\wedge\tau_n} h_s^n(x)\,dx_s \tag{7.3.11}$$

and

$$A_{t\wedge\tau_n}(x) = \int_0^{t\wedge\tau_n} (h_s^n(x))^2\,ds.$$

Suppose now that the process ξ is given by (7.3.4). Let us further assume that, for each t, α_t is measurable with respect to $\mathscr{F}_t^\xi$. By Theorem 2.7.2 there exists a functional $\gamma: [0,T] \times C \to \mathbf{R}^1$ which is $\mathscr{B}[0,T] \times \mathscr{B}_T$-measurable and such that $\gamma(t,\cdot)$ is $\mathscr{B}_{t+}$-measurable for every t and such that for a.e. t, $\gamma(t,\xi(\omega)) = \alpha_t(\omega)$ (P-a.s.). Equation (7.3.4) then takes the form

$$\xi_t(\omega) = \int_0^t \gamma(s,\xi(\omega))\,ds + W_t(\omega). \tag{7.3.12}$$

In other words, ξ is the strong solution of the stochastic differential equation

$$d\xi_t = \gamma(t,\xi)\,dt + dW_t, \qquad \xi_0 = 0. \tag{7.3.13}$$

From Theorem 7.3.1 [with $\alpha_t(\omega) = \gamma(t,\xi(\omega))$] it follows that $P_\xi \ll \mu$ provided

$$P_\xi\left[x \in C: \int_0^T \gamma^2(t,x)\,dt < \infty\right] = 1. \tag{7.3.14}$$

Theorem 7.3.2. *Let ξ be a strong solution of (7.3.12) satisfying condition (7.3.14). Then $\mu \ll P_\xi$ if and only if*

$$\mu\left[x \in C: \int_0^T \gamma^2(t,x)\,dt < \infty\right] = 1, \tag{7.3.15}$$

or equivalently,

$$P\left[\omega: \int_0^T \gamma^2(t,W(\omega))\,dt < \infty\right] = 1. \tag{7.3.15a}$$

PROOF. (i) Assume (7.3.15). By Girsanov's theorem it is enough to verify (7.3.5). Define the stopping times

$$\sigma_n(x) = \begin{cases} \inf\left\{t \le T\colon \int_0^t \gamma^2(s,x)\,ds > n\right\}, \\ T & \text{if the above set is empty.} \end{cases} \tag{7.3.16}$$

Let $\tau_n(\omega) = \sigma_n(\xi(\omega))$. Letting $\alpha_s(\omega) = \gamma(s,\xi(\omega))I_{[0,\tau_n(\omega)]}(s)$ and defining

$$\xi_t^{(n)}(\omega) = \int_0^t \gamma(s,\xi(\omega))I_{[0,\tau_n(\omega)]}(s)\,ds + W_t(\omega), \tag{7.3.17}$$

it follows from Girsanov's theorem that $(\xi_t^{(n)},\mathscr{F}_t,\tilde{P}_n)$ is a Wiener martingale, where $d\tilde{P}_n = L^n\,dP$, and

$$\begin{aligned} L^n(\omega) &= \exp\left[-\int_0^T \gamma(t,\xi(\omega))I_{[0,\tau_n(\omega)]}(t)\,dW_t(\omega)\right. \\ &\qquad \left. - \frac{1}{2}\int_0^T \gamma^2(t,\xi(\omega))I_{[0,\tau_n(\omega)]}(t)\,dt\right] \\ &= \exp\left[-\int_0^T \gamma(t,\xi^n(\omega))I_{[0,\tau_n(\omega)]}(t)\,d\xi_t^{(n)}(\omega)\right. \\ &\qquad \left. + \frac{1}{2}\int_0^T \gamma^2(t,\xi^n(\omega))I_{[0,\tau_n(\omega)]}(t)\,dt\right], \end{aligned} \tag{7.3.18}$$

the last expression resulting from (7.3.17) and the fact that $\xi_t(\omega) = \xi_t^n(\omega)$ on $[\omega\colon 0 \le t \le \tau_n(\omega)]$. Note that Girsanov's theorem applies since $E(L^n) = 1$. Now

$$\begin{aligned} \int_{[L^n(\omega) > N]} L^n(\omega)\,dP(\omega) &= \tilde{P}_n([\omega\colon L^n(\omega) > N]) \\ &= \mu(E_N^n), \end{aligned} \tag{7.3.19}$$

where $E_N^n = [x \in C\colon Y_T^n(x) > N]$ ($N > 0$ is arbitrary),

$$Y_T^n(x) = \exp\left[-\int_0^T \gamma(t,x)I_{[0,\sigma_n(x)]}(t)\,dx_t + \frac{1}{2}\int_0^T \gamma^2(t,x)I_{[0,\sigma_n(x)]}(t)\,dt\right]. \tag{7.3.20}$$

The second equality in (7.3.19) follows because $\tilde{P}_n\xi^{n^{-1}} = \mu$. Setting $\gamma_n(t,x) = \gamma(t,x)I_{[0,\sigma_n(x)]}(t)$, we have

$$\begin{aligned} \mu(E_N^n) \le\ & \mu\left[x \in C\colon -\int_0^T \gamma_n(t,x)\,dx_t - \frac{1}{2}\int_0^T \gamma_n^2(t,x)\,dt > \frac{1}{2}\log N\right] \\ & + \mu\left[x \in C\colon \int_0^T \gamma_n^2(t,x)\,dt > \frac{1}{2}\log N\right]. \end{aligned} \tag{7.3.21}$$

The first term on the right-hand side of (7.3.21) is bounded by $N^{-\frac{1}{2}}$ and the second term does not exceed $\mu[x \in C\colon \int_0^T \gamma^2(t,x)\,dt > \frac{1}{2}\log N]$, which tends to 0 as $N \to \infty$ on account of (7.3.15). Hence (L^n) is a P-uniformly integrable sequence of random variables. Furthermore,

$$L^n(\omega) \to L(\omega) = \exp\left[-\int_0^T \gamma(t,\xi(\omega))\,dW_t(\omega)\right.$$
$$\left.-\frac{1}{2}\int_0^T \gamma^2(t,\xi(\omega))\,dt\right] \qquad (P\text{-a.s.}).$$

Thus we have $EL = 1$, which is (7.3.5). The conclusion follows from Girsanov's theorem (see remarks preceding Theorem 7.3.1).

(ii) If $\mu \ll P_\xi$, (7.3.15) follows at once from (7.3.14). □

The corollary below, is an immediate consequence of Theorems 7.3.1 and 7.3.2.

Corollary 7.3.1. *Suppose γ is a causal functional such that for every x in C,*

$$\int_0^T \gamma^2(t,x)\,dt < \infty. \tag{7.3.22}$$

If ξ is a strong solution of (7.3.13), then

$$P_\xi \equiv \mu. \tag{7.3.23}$$

Let us return to Theorem 7.3.1 for ξ satisfying Equation (7.3.12) and condition (7.3.14). Remark 7.3.2 following Theorem 7.3.1 can now be made more precise. Let

$$L_t(x) = \frac{dP_\xi}{d\mu}(x)[\mathscr{B}_t] \quad \text{and} \quad \Gamma = \left\{x \in C: \int_0^T \gamma^2(t,x)\,dt < \infty\right\}.$$

It is not assumed that $\mu(\Gamma) = 1$. Let σ_n be as defined in Theorem 7.3.2 and define

$$\sigma_\infty(x) = \begin{cases} \inf\left\{t \leq T: \int_0^t \gamma^2(s,x)\,ds = \infty\right\} & \\ T & \text{if the above set is empty.} \end{cases}$$

Note that if we set $\Gamma_n = \{x \in C: \int_0^T \gamma^2(s,x)\,ds \leq n\}$, it is easy to see that $t < \sigma_n(x)$ for all $t < T$ if and only if $\sigma_n(x) = T$, that is, if and only if $x \in \Gamma_n$. Hence if $A \in \mathscr{B}_T$,

$$\int_{A \cap [\sigma_n(x) = T]} L_T(x)\,d\mu(x) = P_\xi(A \cap [\sigma_n(x) = T])$$
$$= P[(\xi(\omega) \in A) \cap (\tau_n(\omega) = T)], \tag{7.3.24}$$

where $\tau_n(\omega) = \sigma_n(\xi(\omega))$. Hence from (7.3.17) the right-hand side of (7.3.24) becomes

$$P[(\xi^n(\omega) \in A) \cap (\tau_n(\omega) = T)] = P_{\xi^n}(A \cap [\sigma_n(x) = T]).$$

So

$$\int_{A \cap [\sigma_n(x)=T]} L_T(x)\, d\mu(x) = \int_{A \cap [\sigma_n(x)=T]} Y_T^n(x)\, d\mu(x). \tag{7.3.25}$$

Write $N_t^n(x) = \int_0^{t \wedge \sigma_n(x)} \gamma(s,x)\, dx_s$ and $A_t^n(x) = \int_0^{t \wedge \sigma_n(x)} \gamma^2(s,x)\, ds$. It is easy to see that processes $(N_t(x)),(A_t(x))$ are defined for $t \in [0,T]$ and $x \in \Gamma$, independent of n and such that $N_t(x) = N_t^n(x) = \int_0^t \gamma(s,x)\, dx_s$ for all t if $x \in \Gamma_n$. Similarly, $A_t(x) = A_t^n(x) = \int_0^t \gamma^2(s,x)\, ds$ for all t if $x \in \Gamma_n$. For all t in $[0,T]$ define

$$Y_t(x) = \begin{cases} \exp[N_t(x) - \frac{1}{2}A_t(x)] & \text{if } x \in \Gamma, \\ 0 & \text{if } x \in \Gamma^c. \end{cases} \tag{7.3.26}$$

Noting that $Y_T(x) = Y_T^n(x)$ on Γ_n, and writing (7.3.25) as

$$\int_{A \cap \Gamma_n} L_T(x)\, d\mu(x) = \int_{A \cap \Gamma_n} Y_T(x)\, d\mu(x),$$

and since $\Gamma = \bigcup_1^\infty \Gamma_n$ (Γ_n increasing), we obtain

$$\int_{A \cap \Gamma} L_T(x)\, d\mu(x) = \int_{A \cap \Gamma} Y_T(x)\, d\mu(x). \tag{7.3.27}$$

Next, since $\Gamma^c \subset \Gamma_n^c$ for every n,

$$\int_{\Gamma^c} L_T(x)\, d\mu(x) \leq \int_{\Gamma_n^c} L_T(x)\, d\mu(x) = P_\xi\left[x \in C: \int_0^T \gamma^2(s,x)\, ds > n\right].$$

Making $n \to \infty$, $\int_{\Gamma^c} L_T(x) du(x) = 0$ from (7.3.14). Thus we obtain

$$\int_A L_T(x)\, d\mu(x) = \int_A Y_T(x)\, d\mu(x) \tag{7.3.28}$$

for every $A \in \mathscr{B}_T$. We have thus established the following result.

Theorem 7.3.3. *Let ξ be a process given by (7.3.12) and let condition (7.3.14) be satisfied. Then $P_\xi \ll \mu$ and almost surely μ,*

$$\frac{dP_\xi}{d\mu}(x) = \begin{cases} \exp[N_T(x) - \frac{1}{2}A_T(x)] & \text{if } x \in \Gamma, \\ 0 & \text{if } x \notin \Gamma. \end{cases} \tag{7.3.29}$$

Remark 7.3.3. From (7.3.29) it follows immediately that (under the conditions of Theorem 7.3.3) $\mu \ll P_\xi$ if and only if $\mu(\Gamma^c) = 0$ (which is Theorem 7.3.2). In particular, if $P_\xi \equiv \mu,(dP_\xi/d\mu)(x)$ is given by (7.3.29), where $N_T(x) = \int_0^T \gamma(t,x) dx_t$ and $A_T(x) = \int_0^T \gamma^2(t,x) dt$.

7.4 Weak Solutions

Theorem 7.3.1 tells us that if (7.3.14) holds, a strong solution of (7.3.12) determines a measure on $(C,\mathscr{B}_T)$ which is absolutely continuous with respect to Wiener measure and is equivalent to it if the further condition (7.3.15) is satisfied. If we are concerned only with those properties of the solution of (7.3.12) which are described by its probability law, then the object of interest is the measure P_ξ. In other words, as was explained in Chapter 5, we should look for a weak solution of (7.3.12). Girsanov's theorem enables us to derive an easily verifiable criterion for a weak solution. We only consider weak solutions P_ξ satisfying (7.3.14).

Theorem 7.4.1. *Suppose that γ is the functional given in Equation (7.3.12). Let*

$$\mu\left[x \in C: \int_0^T \gamma^2(t,x)\,dt < \infty\right] = 1 \tag{7.4.1}$$

and

$$E_\mu \exp\left[\int_0^T \gamma(t,x)\,dx_t - \frac{1}{2}\int_0^T \gamma^2(t,x)\,dt\right] = 1, \tag{7.4.2}$$

E_μ denoting integration over C with respect to μ. Then Equation (7.3.12) has a weak solution.

PROOF. Observe that the coordinate process (x_t) on C is a $(\mathscr{B}_{t+},\mu)$ Wiener martingale. For $x \in C$, define

$$W_t(x) = x_t - \int_0^t \gamma(s,x)\,ds \quad \text{and} \quad dP = \rho\,d\mu, \tag{7.4.3}$$

where

$$\rho(x) = \exp\left[\int_0^T \gamma(t,x)\,dx_t - \frac{1}{2}\int_0^T \gamma^2(t,x)\,dt\right].$$

Then from (7.4.2) and Girsanov's theorem it follows that $(W_t,\mathscr{B}_{t+},P)$ is a Wiener martingale. Letting $\xi_t(x) = x_t$, we have

$$\xi_t(x) = \int_0^t \gamma(s,\xi(x))\,ds + W_t(x) \tag{7.4.4}$$

from (7.4.3). In order to conclude that the weak solution exists [with choice of the family $\{(C,\mathscr{B}_T,P),(\mathscr{B}_{t+}),\xi,W\}$] it is enough to see that since $\xi(x) = x$,

$$P\left[x: \int_0^T \gamma^2(s,\xi(x))^2\,ds < \infty\right] = P\left[x: \int_0^T \gamma^2(s,x)\,ds < \infty\right] = 1,$$

the last assertion being a consequence of (7.4.1) and $P \equiv \mu$. □

Remark 7.4.1. The weak solution whose existence has been shown above is P_ξ and is actually equivalent to μ.

We consider here weak solutions satisfying the requirement (7.4.1) [See Section 5.2(i).]

Theorem 7.4.2. *Let the nonanticipating functional γ satisfy (7.4.1). If μ^* is a weak solution of Equation (7.3.13), then we must have (7.4.2).*

PROOF. By the definition of weak solution given in Section 5.2 there is a process (ξ_t) defined on some probability space $(\Omega,\mathscr{A},P)$ which satisfies

(7.3.12) and such that $P_\xi = \mu^*$. Since P_ξ satisfies (7.3.14) by hypothesis, it follows that $P_\xi \ll \mu$. Condition (7.4.1) implies $\mu \ll P_\xi$. Hence $P_\xi \equiv \mu$, and by the remark following Theorem 7.3.3 we have (7.4.2). □

Theorem 7.4.3. *Let γ satisfy (7.4.1). Then a weak solution to Equation (7.3.13) exists if and only if (7.4.2) is satisfied.*

Corollary 7.4.1 (to Theorem 7.4.3). *Let $\int_0^T \gamma^2(t,x)\,dt < \infty$ for every $x \in C$. Then (7.4.2) is a necessary and sufficient condition for the existence of a weak solution. Furthermore, the weak solution is unique.*

7.5 Stochastic Equations Involving Vector-Valued Processes

The extension of Theorems 7.3.1 and 7.3.2 to vector-valued processes requires only obvious and minor modifications of the proofs in the one-dimensional case. We make the same assumptions as in Section 7.3 except that $(W_t,\mathscr{F}_t,P)$ is now a d-dimensional Wiener martingale and the process $\alpha = (\alpha_t)$ is also d-dimensional, that is, $\alpha\colon [0,T] \times \Omega \to \mathbf{R}^d$. Condition (7.3.3) is replaced by

$$P\left[\omega\colon \int_0^T |\alpha_t(\omega)|^2\,dt < \infty\right] = 1. \tag{7.5.1}$$

Theorem 7.5.1. *Let the process $\xi = (\xi_t)$ be defined by*

$$\xi_t(\omega) = \int_0^t \alpha_s(\omega)\,ds + W_t(\omega). \tag{7.5.2}$$

for all $t \in [0,T]$ (P-a.s.). Then condition (7.5.1) implies that $P_\xi \ll \mu[\mathscr{B}_T(C_d)]$, where μ is Wiener measure on $(C_d,\mathscr{B}_T(C_d))$ and $P_\xi = P\xi^{-1}$.

In the proof we take

$$L_T(\omega) = \exp\left[-\int_0^T (\alpha_t(\omega),dW_t(\omega)) - \frac{1}{2}\int_0^T |\alpha_t(\omega)|^2\,dt\right]$$

and

$$\tau_n(\omega) = \begin{cases} \inf\left\{t\colon t \le T,\ \int_0^t |\alpha_s(\omega)|^2\,ds > n\right\} & \\ T & \text{if the above set is empty.} \end{cases}$$

The proof of Theorem 7.3.1 then applies virtually without any change. □

Let us now assume that for each t,α_t is $\mathscr{F}_t^\xi$-measurable. By Theorem 2.7.2, for a.e. t, we have $\alpha_t(\omega) = \gamma(t,\xi(\omega))$, ($P$-a.s.), where $\gamma\colon [0,T] \times C_d \to \mathbf{R}^d$, is

$\mathscr{B}[0,T] \times \mathscr{B}_T(C_d)$-measurable and such that $\gamma(t,\cdot)$ is $\mathscr{B}_{t+}(C_d)$-measurable for each t. The following result is the d-dimensional analogue of Theorem 7.3.2.

Theorem 7.5.2. *Let* $\xi = (\xi_t)$ *be given by*

$$\xi_t(\omega) = \int_0^t \gamma(s,\xi(\omega))\, ds + W_t(\omega) \tag{7.5.3}$$

(a.s.) *for all* $t \in [0,T]$. *Also suppose that*

$$P_\xi\left[x \in C_d\colon \int_0^T |\gamma(t,x)|^2\, dt < \infty\right] = 1. \tag{7.5.4}$$

Then $\mu \ll P_\xi$ *if and only if*

$$\mu\left[x \in C_d\colon \int_0^T |\gamma(t,x)|^2\, dt < \infty\right] = 1. \tag{7.5.5}$$

The proof is exactly as in Theorem 7.3.2 if we define

$$\sigma_n(x) = \begin{cases} \inf\left\{t \leq T\colon \int_0^t |\gamma(s,x)|^2\, ds > n\right\} & \\ T & \text{if the above set is empty.} \end{cases}$$

Remark 7.5.1. Under assumptions (7.5.3) to (7.5.5), we have $P_\xi \equiv \mu$ and the Radon-Nikodym derivative $dP_\xi/d\mu$ is given by

$$\frac{dP_\xi}{d\mu}(x) = \exp\left[\int_0^T (\gamma(t,x), dx_t) - \frac{1}{2}\int_0^T |\gamma(t,x)|^2\, dt\right] \quad (\mu\text{-a.s.}). \tag{7.5.6}$$

The derivation of (7.5.6) follows the proof of Theorem 7.3.3.

7.6 Explosion Times and an Extension of Girsanov's Formula

The exponential supermartingale $(\xi_0^t(f),\mathscr{F}_t)$ $(t \geq 0)$ considered in Section 7.1 is a martingale if and only if $E\xi_0^t(f) = 1$ for every t. In connection with Girsanov's theorem where it is assumed that t lies in the finite interval $[0,T]$, we have seen that the condition $E\xi_0^T(f) = 1$ is very important. In particular, in the context of Theorem 7.1.3 (taking only real-valued processes for convenience) this condition is satisfied if f is bounded. The following formula is an immediate consequence of Theorem 7.1.3.

Suppose that $b \in C_b^1(\mathbf{R})$ [cf.(5.1.10)]. Let (X_t) be the unique solution for $t \in [0,T]$ of the equation

$$X_t = \int_0^t b(X_s)\, ds + W_t.$$

Then we have $\mathscr{F}_t^X = \mathscr{F}_t^W$. By Girsanov's theorem, $(X_t,\mathscr{F}_t^W,\tilde{P})$ is a Wiener process where $\tilde{P}$ is the probability measure on $(\Omega,\mathscr{F}_T^W)$ with $\tilde{P} \equiv P$ and Radon-Nikodym derivative given by

$$\frac{dP}{d\tilde{P}}(X) = \exp\left[\int_0^T b(X_s)\,dX_s - \frac{1}{2}\int_0^T b^2(X_s)\,ds\right].$$

Then for $A \in \mathscr{B}_T(C)$ we have

$$\begin{aligned} P[X \in A] &= \int_{[X\in A]} \exp\left[\int_0^T b(X_s)\,dX_s - \frac{1}{2}\int_0^T b^2(X_s)\,ds\right] d\tilde{P} \\ &= \int_{[W\in A]} \exp\left[\int_0^T b(W_s)\,dW_s - \frac{1}{2}\int_0^T b^2(W_s)\,ds\right] dP. \end{aligned}$$

The formula just derived will be called the *Girsanov* or the *Cameron-Martin-Girsanov formula.*

What we shall do in this section is to give a simple proof of an extension of this formula to cover the case of "explosion times" and thereby also furnish an example of an exponential supermartingale $(\xi_0^t(b),\mathscr{F}_t^W)$ which is not a martingale, that is, for which $E\xi_0^t(b) < 1$ for some $t > 0$.

The concept of a solution of a stochastic differential equation up to an explosion time is best introduced through the following theorem. It is not our purpose to undertake a complete study of questions of this sort. The reader is referred, for a more detailed discussion to the book of McKean, which also contains the extended Girsanov formula. The proof given there requires a knowledge of Weyl's lemma, whereas the one given in this section is based on a transparent application of the Girsanov formula already shown above.

Theorem 7.6.1. *The equation*

$$X_t = \int_0^t b(X_s)\,ds + W_t, \qquad X_0 = 0 \qquad \text{(a.s.)},$$

where $b \in C^1(\mathbf{R})$ *and* (W_t), $t \in \mathbf{R}_+$, *is a real-valued, Wiener process, has a solution in the following sense: There exists a unique pair* (τ^b,X), *where* $\tau^b[\tau^b > 0\text{(a.s.)}]$ *is a stopping time with respect to the family* $(\mathscr{F}_t^W)$ *and* $X_t(\omega)$ *is defined on* $[(t,\omega): 0 \leq t < \tau^b(\omega)]$ *such that*

1. $X_t(\omega) = \int_0^t b(X_s(\omega))\,ds + W_t(\omega)$, $0 \leq t < \tau^b(\omega)$, (a.s.).
2. *For almost all* ω, *the function* $t \to X_t(\omega)$ *is continuous on* $[0,\tau^b(\omega))$. *Moreover, for each* $t > 0$, X *defined on* $[0,t] \times [\omega: \tau^b(\omega) > t]$ *is* $\mathscr{B}[0,t] \times \mathscr{F}_t^W$*-measurable.*
3. $\limsup_{t\uparrow\tau^b(\omega)} |X_t(\omega)| = \infty$ on $[\omega: \tau^b(\omega) < \infty]$.

Proof. For each integer $N \geq 1$, let $b^N \in C^1(\mathbf{R})$ coincide with b on $[-N,N]$ and equal 0 on $(-\infty, -N-1] \cup [N+1, \infty)$. Let $X^N(t,\omega)$ be the unique, continuous (and hence progressively measurable) solution of

$$\xi_t = \int_0^t b^N(\xi_s)\,ds + W_t.$$

Define $\tau_N = \inf\{t: |X_t^N| \geq N\}$, and $= \infty$ if the set is empty. Then each τ_N is a stopping time for $(\mathscr{F}_t^W)$. Writing $\sigma_N = \tau_N \wedge \tau_{N+1}$, we have

$$\begin{aligned} X_{t\wedge\sigma_N}^{N+1} &= \int_0^t I_{[0,\sigma_N]}(s) b^{N+1}(X_s^{N+1})\,ds + W_{t\wedge\sigma_N} \\ &= \int_0^t I_{[0,\sigma_N]}(s) b(X_s^{N+1})\,ds + W_{t\wedge\sigma_N}, \\ X_{t\wedge\sigma_N}^{N} &= \int_0^t I_{[0,\sigma_N]}(s) b(X_s^{N})\,ds + W_{t\wedge\sigma_N} \end{aligned}$$

and so

$$I_{[0,\sigma_N]}(t)(X_{t\wedge\sigma_N}^{N+1} - X_{t\wedge\sigma_N}^{N}) = I_{[0,\sigma_N]}(t) \int_0^t I_{[0,\sigma_N]}(s)[b(X_s^{N+1}) - b(X_s^N)]\,ds.$$

Setting $\varphi(t) = E(I_{[0,\sigma_N]}(t)|X_t^{N+1} - X_t^N|^2)$, we note that φ is measurable on $[0,\infty)$ and that

$$\varphi(t) \leq tE \int_0^t I_{[0,\sigma_N]}(s)|b(X_s^{N+1}) - b(X_s^N)|^2\,ds \tag{7.6.1}$$

for all $t \in \mathbf{R}_+$. For $s \in [0,\sigma_N(\omega)]$ by the mean value theorem,

$$|b(X_s^{N+1}) - b(X_s^N)|^2 = |b'(c)|^2 |X_s^{N+1} - X_s^N|^2,$$

where c (depending on N and ω) lies between $-(N+1)$ and $(N+1)$. Denoting $\sup_{|x|\leq N+1} |b'(x)|$ by K_N and using (7.6.1), it follows that on each interval $[0,T]$, $\varphi(t) \leq T \cdot K_N^2 \int_0^t \varphi(s)\,ds$. Hence $\varphi(t) \equiv 0$ and therefore, $E(I_{[0,\sigma_N]}(t)|X_t^{N+1} - X_t^N|^2) = 0$ for all t. The continuity of the processes (X_t^N) and (X_t^{N+1}) then implies that $X_t^{N+1}(\omega) = X_t^N(\omega)$ on $[0,\sigma_N(\omega)]$, (a.s.). Now, if $\sigma_N(\omega) = \infty$, then $\tau_N(\omega) = \tau_{N+1}(\omega) = \infty$, and $\sigma_N(\omega) < \infty$ implies $|X_t^{N+1}(\omega)| = |X_t^N(\omega)| \leq N$ for $t \in [0,\sigma_N(\omega)]$. The definition of τ_{N+1} then shows that $\tau_{N+1}(\omega) > \sigma_N(\omega)$, from which we have $\sigma_N = \tau_N$ (a.s.). Hence, for a.a. ω, $\tau_N(\omega) \leq \tau_{N+1}(\omega)$ and $\tau_N(\omega) < \tau_{N+1}(\omega)$ if $\tau_N(\omega) < \infty$. We conclude that $\tau_N \uparrow \tau$ (a.s.) for some τ, which is a stopping time for $(\mathscr{F}_t^W)$. Define

$$X_t(\omega) = X_t^N(\omega) \quad \text{if } 0 \leq t < \tau_N(\omega) \qquad (N \geq 1).$$

Then X is defined consistently on $[(t,\omega): 0 \leq t < \tau(\omega)]$. We define τ^b of the theorem to be τ. Since $X_0^N = 0$ for all N, we have $\tau_N > 0$ (a.s.) and $\tau > 0$ (a.s.). For a.a. ω, if $0 \leq t < \tau(\omega)$, there exists N such that $t \leq \tau_N(\omega) < \tau(\omega)$. Then

$$X_u^N = \int_0^u b^N(X_s^N)\,ds + W_u \qquad [0 \leq u \leq \tau_N(\omega)].$$

In the integrand on the right-hand side, $|X_s^N| \leq N$ so that $b^N(X_s^N) = b(X_s^N)$ and $X_s^N = X_s$ $[0 \leq s \leq \tau_N(\omega)]$. Hence the above equation becomes

$$X_u = \int_0^u b(X_s)\,ds + W_u \quad \text{for all } u \in [0,\tau_N(\omega)].$$

Taking $u = t$, we obtain part 1.

The following argument establishes conclusion 2. Fix t and let $(s,\omega) \in [0,t] \times [\omega: \tau(\omega) > t]$. Then, since each X^N is progressively measurable, we have

$$X(s,\omega)I_{[0,t] \times [\omega: \tau_N(\omega) > t]}(s,\omega) = X^N(s,\omega)I_{[0,t] \times [\omega: \tau_N(\omega) > t]}(s,\omega)$$

and is hence $\mathscr{B}[0,t] \times \mathscr{F}_t^W$-measurable. Making $N \to \infty$, we have part 2.

Proof of part 3: If $\tau(\omega) < \infty$, then $\tau_N(\omega) < \infty$ for all N and $|X_{\tau_N(\omega)}(\omega)| = |X^N_{\tau_N(\omega)}(\omega)| = N \to \infty$ as $N \to \infty$. Hence $(\limsup)_{t \uparrow \tau(\omega)} |X_t(\omega)| = \infty$ on $[\omega: \tau(\omega) < \infty]$.

It remains now to prove uniqueness. Suppose that (τ',X') is another solution of part 1 having properties 2 and 3. Then for a.a. ω and all $u < \tau' \wedge \tau$, both $u \to X_u(\omega)$ and $u \to X'_u(\omega)$ solve

$$\eta_u = \int_0^u b(\eta_s)\,ds + W_u$$

for $0 \leq u < \tau \wedge \tau'$. Hence, using Proposition 5.1.1 we have

$$P[\omega: X_u(\omega) = X'_u(\omega), 0 \leq u < \tau(\omega) \wedge \tau'(\omega)] = 1.$$

Furthermore, we know that $(\limsup)_{t \uparrow \tau(\omega)} |X_t(\omega)| = \infty$ if $\tau(\omega) < \infty$. Combining these two facts, it is easy to see that we must have $\tau(\omega) = \tau'(\omega)$ (a.s.) and for a.a. ω, $X_t(\omega) = X'_t(\omega)$ on $[0,\tau(\omega))$. □

Let us now prove the extended Girsanov formula.

Theorem 7.6.2. *If $X = (X_t)$ is the solution of part 1 of Theorem 7.6.1, then for all $T > 0$ and $A \in \mathscr{B}_T(C)$,*

$$P[X \in A, \tau^b > T] = \int_{[W \in A]} \exp\left[\int_0^T b(W_s)\,dW_s - \frac{1}{2}\int_0^T b^2(W_s)\,ds\right] dP. \quad (7.6.2)$$

Proof. The set $B = [x \in C: x \in A, |x(s)| < N, 0 \leq s \leq T]$ belongs to $\mathscr{B}_T(C)$ if $A \in \mathscr{B}_T(C)$. Let b^N be the bounded function defined in the proof of Theorem 7.6.1. Also recall the definitions of X^N and τ_N. Then from the remark made above, it follows that

$$P[X^N \in B] = \int_{[W \in B]} \exp\left[\int_0^T b^N(W_s)\,dW_s - \frac{1}{2}\int_0^T (b^N(W_s))^2\,ds\right] dP.$$

$$= \int_{[W \in A,\ |W_s| < N,\ 0 \leq s \leq T]} \exp\left[\int_0^T b(W_s)\,dW_s - \frac{1}{2}\int_0^T b^2(W_s)\,ds\right] dP. \quad (7.6.3)$$

Now we have

$$\begin{aligned}[X^N \in B] &= [X^N \in A, |X^N_s| < N, 0 \leq s \leq T] \\ &= [X^N \in A, |X^N_s| < N, T < \tau_N] \\ &= [X \in A, |X_s| < N, T < \tau_N].\end{aligned}$$

Making $N \uparrow \infty$ on both sides of (7.6.3), we have (7.6.2) and the proof is complete. □

In Theorem 7.6.2 (as in Theorem 7.6.1) only the case $\sigma = 1$ has been considered. However, the method of proof employed above extends without difficulty to the case of a general σ.

Let us denote $\exp[\int_0^t b(W_s)\,dW_s - \frac{1}{2}\int_0^t b^2(W_s)\,ds]$ by $\xi_0^t(b)$. If we take $A = C$ in (7.6.2), we get

$$E\xi_0^T(b) = P[\tau^b > T]. \tag{7.6.4}$$

Since $P[\tau^b > T] \to P[\tau^b = \infty]$ as $T \uparrow \infty$, we see from (7.6.4) that if $P[\tau^b < \infty] > 0$, then for some T we must have

$$E\xi_0^T(b) < 1. \tag{7.6.5}$$

We shall now give an example of an explosion time τ^b which is finite with positive probability (p. 189). It will be necessary first to derive a sufficient condition for $P(\tau^b < \infty)$ to be positive. This condition is a part of Feller's test for explosions.

Define the function $q\colon \mathbf{R} \to \mathbf{R}$ as follows:

$$q(x) = \begin{cases} \int_0^x \exp\left[-2\int_0^u b(v)\,dv\right] du & \text{for } x \geq 0, \\ -\int_x^0 \exp\left[2\int_u^0 b(v)\,dv\right] du & \text{for } x < 0. \end{cases}$$

Notice that $q'' = -2bq'$ and $q(0) = 0$. Thus, if Ito's formula can be applied to the process $Y_t = q(X_t)$, its differential has no drift term:

$$dY_t = \hat{\sigma}(Y_s)\,dW_s,$$

where $\hat{\sigma} = q' \circ q^{-1}$.

Theorem 7.6.3. *If $q(-\infty)$ and $q(\infty)$ are both finite and*

$$\int_{-\infty}^0 [q(x) - q(-\infty)][q'(x)]^{-1}\,dx + \int_0^\infty [q(\infty) - q(x)][q'(x)]^{-1}\,dx < \infty, \tag{7.6.6}$$

then $P[\tau^b < \infty] > 0$.

Before giving the proof, it is useful to make the following observation.

Lemma 7.6.1.

$$\int_0^\infty \left[\int_0^x [q'(u)]^{-1}\,du\right] q'(x)\,dx < \infty \tag{7.6.7}$$

if and only if

$$q(\infty) < \infty \tag{7.6.8}$$

and

$$\int_0^\infty [q(\infty) - q(x)][q'(x)]^{-1}\,dx < \infty. \tag{7.6.9}$$

A similar statement holds for the situation at $-\infty$.

Proof. Setting $r(x) = \int_0^x [q'(u)]^{-1}\,du$, the integral in (7.6.7) may be written as $\int_0^\infty r(x)q'(x)\,dx$. If (7.6.7) holds with $r(\infty) = \infty$, then for some $a > 0$ we have $\int_0^\infty r(x)q'(x)\,dx \geq \int_a^\infty q'(x)\,dx$, which yields (7.6.8). If (7.6.7) is true with $r(\infty) < \infty$, since r is an increasing function, $r(1)\int_1^\infty q'(x)\,dx \leq \int_1^\infty r(x)q'(x)\,dx < \infty$, again giving (7.6.8). Thus (7.6.7) implies (7.6.8). Next, integrating by parts,

$$\begin{aligned}\int_0^x r(u)q'(u)\,du &= q(x)\int_0^x [q'(u)]^{-1}\,du - \int_0^x q(u)[q'(u)]^{-1}\,du \\ &= \int_0^x [q(x) - q(u)][q'(u)]^{-1}\,du. \end{aligned} \tag{7.6.10}$$

Hence, fixing $a > 0$ arbitrary and $x > a$, we have

$$\int_0^a [q(x) - q(u)][q'(u)]^{-1}\,du \leq \int_0^\infty r(u)q'(u)\,du < \infty,$$

and so

$$\int_0^a [q(\infty) - q(u)][q'(u)]^{-1}\,du \leq \int_0^\infty r(u)q'(u)\,du < \infty.$$

Suppose now that (7.6.8) and (7.6.9) are satisfied. Then for all $x > 0$,

$$\begin{aligned}\int_0^\infty [q(\infty) - q(u)][q'(u)]^{-1}\,du &> \int_0^x [q(x) - q(u)][q'(u)]^{-1}\,du \\ &= \int_0^x r(u)\,q'(u)\,du \quad \text{from (7.6.10)}.\end{aligned}$$

This shows $\int_0^\infty r(u)q'(u)\,du < \infty$, which is (7.6.7). □

In view of the lemma just proved, we may replace the assumptions of Theorem 7.6.3 by the equivalent condition (7.6.7).

Proof of Theorem 7.6.3. Let $f = f(q)$ be defined by $f = \sum_{n=0}^\infty f_n$, where $f_0 = 1$ and

$$f_n(q(x)) = 2\int_0^x \left[\int_0^u f_{n-1}(q(v))\hat{\sigma}^{-2}(q(v))\,dq(v)\right] dq(u)$$

for all x.

Then $f(q)$ satisfies the equation

$$\frac{1}{2}\hat{\sigma}^2 \frac{d^2 f}{dq^2} = f.$$

It can be verified that $f_n \leq f_1^n/n!$ for all $n \geq 0$, so that $1 + f_1 \leq f \leq e^{f_1}$. For $x > 0$,

$$f_1(q(x)) = 2\int_0^x \left(\int_0^u [q'(v)]^{-1}\,dv\right) q'(u)\,du.$$

From Lemma 7.6.1. and (7.6.6) we find that $f_1(q(\infty)) < \infty$. Analogously the lemma and (7.6.6) also yield the finiteness of $f_1(q(-\infty))$. It then follows that the function $f(q(x))$ is bounded.

Let τ_N, X_t^N, and X_t be defined as in Theorem 7.6.1. From

$$X_{t \wedge \tau_N} = \int_0^t b(X_s) I_{[0,\tau_N]}(s)\, ds + \int_0^t I_{[0,\tau_N]}(s)\, dW_s$$

Ito's formula yields

$$\begin{aligned} q(X_{t \wedge \tau_N}) &= q(X_0) + \int_0^t q'(X_{s \wedge \tau_N}) I_{[0,\tau_N]}(s)\, dW_s \\ &\quad + \int_0^t q'(X_{s \wedge \tau_N}) b(X_{s \wedge \tau_N}) I_{[0,\tau_N]}(s)\, ds \\ &\quad + \frac{1}{2}\int_0^t q''(X_{s \wedge \tau_N}) I_{[0,\tau_N]}(s)\, ds \\ &= q(X_0) + \int_0^t q'(X_{s \wedge \tau_N}) I_{[0,\tau_N]}(s)\, dW_s \end{aligned}$$

since $q''(x) = -2b(x)q'(x)$. Define $Y_t^N = q(X_{t \wedge \tau_N})$. Then

$$Y_t^N = Y_0^N + \int_0^t \hat{\sigma}(Y_s^N) I_{[0,\tau_N]}(s)\, dW_s.$$

Applying Ito's formula to $e^{-t}f(Y_t^N)$, we have

$$\begin{aligned} d[e^{-t}f(Y_t^N)] &= -e^{-t}f(Y_t^N)\, dt + e^{-t}f'(Y_t^N)\hat{\sigma}(Y_t^N) I_{[0,\tau_N]}(t)\, dW_t \\ &\quad + \tfrac{1}{2}e^{-t}f''(Y_t^N)\hat{\sigma}^2(Y_t^N) I_{[0,\tau_N]}(t)\, dt. \end{aligned}$$

Integrate the above relation between 0 and $t \wedge \tau_N$, where $t > 0$ is arbitrary. Then

$$\begin{aligned} e^{-t \wedge \tau_N}f(Y_{t \wedge \tau_N}^N) &= f(Y_0^N) - \int_0^{t \wedge \tau_N} e^{-s}f(Y_s^N)\, ds + \int_0^{t \wedge \tau_N} e^{-s}f'(Y_s^N)\hat{\sigma}(Y_s^N)\, dW_s \\ &\quad + \frac{1}{2}\int_0^{t \wedge \tau_N} e^{-s}f''(Y_s^N)\hat{\sigma}^2(Y_s^N)\, ds \\ &= f(Y_0^N) + \int_0^{t \wedge \tau_N} e^{-s}f'(Y_s^N)\hat{\sigma}(Y_s^N)\, dW_s \end{aligned}$$

by our choice of f. After verifying that

$$E\int_0^{t \wedge \tau_N} e^{-2s}[f'(Y_t^N)\hat{\sigma}(Y_t^N)]^2\, dt < \infty,$$

we get

$$E[e^{-t \wedge \tau_N}f(Y_{t \wedge \tau_N}^N)] = Ef(Y_0^N).$$

Note that $Ef(Y_0^N) = Ef(q(X_0)) = f(0)$ since $X_0 = 0$ (a.s.) and $f(0) = f_1(0) = 1$. Hence

$$E[e^{-t \wedge \tau_N}f(Y_{t \wedge \tau_N}^N)] = 1 \tag{7.6.11}$$

for all N. Write the term on the left-hand side as the sum of integrals over $[\tau^b = \infty]$ and $[\tau^b < \infty]$, respectively. Since, for $t > 0$ and all $N \geq 1$,

$$|f(Y^N_{t \wedge \tau_N})| \leq \sup_{x \in \mathbf{R}} |f(q(x))| = K < \infty$$

and $\tau_N \uparrow \tau^b$ (a.s.), it follows that

$$\limsup_{N \to \infty} \limsup_{t \to \infty} \left| \int_{[\tau^b = \infty]} e^{-t \wedge \tau_N} f(Y^N_{t \wedge \tau_N})\, dP \right| \leq K \int_{[\tau^b = \infty]} e^{-\tau^b}\, dP = 0$$

and

$$\limsup_{N \to \infty} \limsup_{t \to \infty} \left| \int_{[\tau^b < \infty]} e^{-t \wedge \tau_N} f(Y^N_{t \wedge \tau_N})\, dP \right| \leq K \int_{[\tau^b < \infty]} e^{-\tau^b}\, dP.$$

Thus finally from (7.6.11) we have

$$1 \leq K \int_{[\tau^b < \infty]} e^{-\tau^b}\, dP$$

which is impossible if $P[\tau^b < \infty] = 0$. Hence $P[\tau^b < \infty] > 0$. □

EXAMPLE 7.6.1. Let $b(x) = x^3$. Then there exists a T such that $E\xi_0^T(b) < 1$.
We have only to verify condition (7.6.6). For $x \geq 0$, $q(x) = \int_0^x e^{-\frac{1}{2}u^4}\, du$ and

$$\int_0^\infty [q(\infty) - q(x)][q'(x)]^{-1}\, dx = \int_0^\infty e^{\frac{1}{2}x^4} \left(\int_x^\infty e^{-\frac{1}{2}u^4}\, du \right) dx.$$

The finiteness of this integral follows from the simple bound

$$e^{\frac{1}{2}x^4} \int_x^\infty e^{-\frac{1}{2}u^4}\, du \leq \frac{1}{x^3} e^{\frac{1}{2}x^4} \int_x^\infty u^3 e^{-\frac{1}{2}u^4}\, du$$

$$= \frac{1}{2x^3} \qquad (x > 0).$$

The remaining term can be dealt with in a similar manner.

7.7 Nonexistence of a Strong Solution

We shall now present an example of a stochastic differential equation which does not have a strong solution but for which the weak solution exists and is unique. The example is due to Tsirel'son [53], but the form in which it is presented below was communicated in a letter to the author by A. N. Shiryaev.

EXAMPLE 7.7.1 (Tsirel'son). Let $(\Omega, \mathscr{F}, P)$ be a complete probability space on which is given a standard Wiener process (W_t), $0 \leq t \leq 1$. Suppose that $b(t,x)$ is a real-valued functional on $[0,1] \times C([0,1], \mathbf{R})$ defined as follows: choose and fix a sequence $\{t_k\}$ $(k = 0,-1,-2,\dots)$, contained in $[0,1]$ such that $t_0 = 1 > t_{-1} > t_{-2} > \cdots \to 0$.

Let

$$b(t,x) = \left\{\frac{x_{t_k} - x_{t_{k-1}}}{t_k - t_{k-1}}\right\} \text{ for } t_k \le t \le t_{k+1},\ k = -1,-2,\dots, \tag{7.7.1}$$

where $\{\alpha\}$ is the fractional part of the real number α. (Note that x_t is the value of the function x at t). Clearly, $b(t,x)$ satisfies the usual measurability conditions and is $(\mathscr{B}_t)$-adapted, and obviously,

$$\int_0^T b^2(t,x)\,dt < \infty \quad \text{for each } x.$$

Then the stochastic differential equation

$$\begin{aligned} d\xi_t &= b(t,\xi)\,dt + dW_t \qquad (0 < t < 1) \\ \xi_0 &= 0. \end{aligned} \tag{7.7.2}$$

has no solution ξ_t which is $\mathscr{F}_t^W$-measurable.

Proof. If $t_k \le t < t_{k+1}$ from (7.7.2),

$$\xi_t - \xi_{t_k} = \int_{t_k}^t \left\{\frac{\xi_{t_k} - \xi_{t_{k-1}}}{t_k - t_{k-1}}\right\} ds + W_t - W_{t_k}.$$

which upon setting

$$\eta_k = \frac{\xi_{t_k} - \xi_{t_{k-1}}}{t_k - t_{k-1}} \quad \text{and} \quad \varepsilon_{k+1} = \frac{W_{t_{k+1}} - W_{t_k}}{t_{k+1} - t_k}$$

becomes

$$\eta_{k+1} = \{\eta_k\} + \varepsilon_{k+1}. \tag{7.7.3}$$

Suppose that (7.7.2) has a solution ξ_t which is $\mathscr{F}_t^W$-measurable.

Then

$$\eta_k \text{ is } \mathscr{F}_{t_k}^W\text{-measurable,} \tag{7.7.4}$$

and the family of random variables

$$\{\eta_m\}\ (m = k, k-1, \dots) \text{ is independent of } \varepsilon_{k+1}. \tag{7.7.5}$$

From (7.7.5) we have

$$e^{2\pi i\eta_{k+1}} = e^{2\pi i\{\eta_k\}}e^{2\pi i\varepsilon_{k+1}} = e^{2\pi i\eta_k}e^{2\pi i\varepsilon_{k+1}}. \tag{7.7.6}$$

Writing $d_k = E(e^{2\pi i\eta_k})$, we obtain, from (7.7.5) and (7.7.6),

$$d_{k+1} = d_k E(e^{2\pi i\varepsilon_{k+1}}) = d_k e^{-2\pi^2/(t_{k+1}-t_k)}$$

Proceeding inductively, we have $(n = 0,1,\dots)$

$$\begin{aligned} d_{k+1} &= d_{k-n}\exp\left[-2\pi^2\left(\frac{1}{t_{k+1} - t_k} + \cdots + \frac{1}{t_{k-n+1} - t_{k-n}}\right)\right] \\ &\to 0 \quad \text{as } n \to \infty. \end{aligned}$$

Hence

$$d_k = 0, \qquad k = 0, -1, -2, \ldots.$$

Furthermore, since $e^{2\pi i \eta_{k+1}} = e^{2\pi i \eta_{k-n}} e^{2\pi i(\varepsilon_{k+1} + \cdots + \varepsilon_{k-n+1})}$,

$$E\left[e^{2\pi i \eta_{k+1}} \middle| \mathscr{F}^W_{[t_{k-n}, t_{k+1}]}\right] = d_{k-n} e^{2\pi i(\varepsilon_{k+1} + \cdots + \varepsilon_{k-n+1})},$$

where $\mathscr{F}^W_{[t_{k-n}, t_{k+1}]} = \sigma\{W_v - W_u, t_{k-n} \le u \le v \le t_{k+1}\}$. Noting that $d_{k-n} = 0$, we have

$$E\left[e^{2\pi i \eta_{k+1}} \middle| \mathscr{F}^W_{[t_{k-n}, t_{k+1}]}\right] = 0. \tag{7.7.7}$$

As $n \to \infty$, $\mathscr{F}^W_{[t_{k-n}, t_{k+1}]} \uparrow \mathscr{F}^W_{t_{k+1}}$, and from (7.7.4) and (7.7.7) it follows that

$$e^{2\pi i \eta_{k+1}} = E(e^{2\pi i \eta_{k+1}} | \mathscr{F}^W_{t_{k+1}}) = 0. \tag{7.7.8}$$

Since (7.7.8) is impossible, Equation (7.7.2) can have no $\mathscr{F}^W_t$-measurable solution ξ_t. □

If, in the definition of a strong solution given in Chapter 5, we take $\mathscr{F}_t = \mathscr{F}^W_t$, the counterexample just established shows that (7.7.2) has no strong solution, whereas from the discussion of Section 7.4 it follows that a weak solution exists and is unique.

8 The General Filtering Problem and the Stochastic Equation of the Optimal Filter (Part I)

8.1 The Filtering Problem and the Innovation Process

Before discussing the filtering problem, we prove a number of results in preparation for the martingale approach to the stochastic differential equation of the optimal filter which will be derived in the later sections of this chapter. Let us recall that $(\Omega,\mathscr{A},P)$ is a complete probability space and $(\mathscr{F}_t)$ $(t \in \mathbf{R}_+)$ is an increasing family of sub σ-fields of $\mathscr{A}$, and that it will be assumed that all P-null sets belong to $\mathscr{F}_0$. The following processes are given on Ω: (S_t) called the *signal* or *system process*; (Z_t), the *observation process*; and (B_t), the *noise process*. All three are related by the model

$$Z_t = S_t + B_t. \tag{8.1.1}$$

The signal (S_t) is not directly observed. Information concerning (S_t) is obtained by observations on (Z_t). In mathematical terms, to say that (Z_t) or any other process in this context is observable is equivalent to assuming that it is adapted to $(\mathscr{F}_t)$. It is important not to assume that $(\mathscr{F}_t)$ is right-continuous because the latter represents a σ-field dependent on observations. For example, if the model (8.1.1) is given in advance $\mathscr{F}_t^Z$ might be a permissible choice of $\mathscr{F}_t$. The property of right-continuity would then be difficult to verify in advance. Let us make the following assumptions about the three processes:

1. (Z_t) is $(\mathscr{F}_t)$-adapted.
2. $S_0 = 0$ and (S_t) has right-continuous sample functions and is of bounded variation (B.V.) in $[0,t]$ for every t.
3. $E\,\mathrm{Var}_\infty S < \infty$, for each t, where $\mathrm{Var}_t S$ is the total variation of S_u in $[0,t]$.

4. $B_0 = 0$. For every t, B_t is integrable and for $s < t$,

$$E(B_t - B_s|\mathscr{F}_s) = 0. \tag{8.1.2}$$

Note that it is not assumed that (B_t) is $(\mathscr{F}_t)$-adapted. If that were so, Equation (8.1.2) would imply that $(B_t,\mathscr{F}_t)$ is a martingale and (S_t) itself would be $(\mathscr{F}_t)$-adaptable, that is, observable.

5. (B_t) has right-continuous paths.

We shall also assume without restricting the generality that $Z_0 = S_0(=0)$. If this assumption is not made, since $B_0 = 0$, it would follow that S_0 is $\mathscr{F}_0$-measurable. Write $S = U - V$, where $U = (U_t)$ and $V = (V_t)$ are increasing processes, right-continuous and such that $U_0 = V_0 = 0$ (a.s.), $EU_t < \infty$, $EV_t < \infty$ for each t. From Theorem 3.1.4 (writing $\bar{S}$ instead of S^* for convenience), we have

$$E(U_\infty - U_t|\mathscr{F}_{t+}) = M_t - \bar{U}_t,$$

where (M_t) is a right-continuous $(\mathscr{F}_{t+})$-martingale and $(\bar{U}_t)$ is the uniquely determined, integrable increasing process which is $(\mathscr{F}_{t+})$-predictable. With $\bar{V}_t$ introduced in a similar fashion, let $\bar{S}_t = \bar{U}_t - \bar{V}_t$. $\{(\bar{U}_t)$ is the "dual predictable projection" in the terminology of Dellacherie and Meyer [42].$\}$

Definition 8.1.1 (Innovation Process). The process $\nu = (\nu_t)$,

$$\nu_t = Z_t - \bar{S}_t \tag{8.1.3}$$

is called the *innovation process.*

Theorem 8.1.1. *$(\nu_t,\mathscr{F}_t)$ is a martingale with right-continuous sample paths.*

PROOF. First note that the $(\mathscr{F}_{t+})$-predictability of $\bar{S}_t$ yields the fact that $(\bar{S}_t)$ is $(\mathscr{F}_t)$-adapted. Hence (ν_t) is $(\mathscr{F}_t)$-adapted. Furthermore if $s < t$, from (8.1.1) and assumption 5,

$$\begin{aligned} E[\nu_t - \nu_s|\mathscr{F}_s] &= E[(Z_t - Z_s) - (\bar{S}_t - \bar{S}_s)|\mathscr{F}_s] \\ &= E[(S_t - S_s)|\mathscr{F}_s] - E[\bar{S}_t - \bar{S}_s|\mathscr{F}_s] + E[B_t - B_s|\mathscr{F}_s] \\ &= 0 \end{aligned}$$

from (8.1.2) and the property of the dual projection mentioned in Theorem 3.1.4. The right-continuity is obvious. □

Since (ν_t) is a right-continuous martingale, it is known that

$$\text{prob}\lim \sum_{\Pi_n} (\nu_{t_{j+1}^n} - \nu_{t_j^n})^2 \text{ exists as } |\Pi_n| \to 0, \tag{8.1.4}$$

where $\Pi_n = \{t_j^n\}$ is a finite partition of $[0,t]$ and $|\Pi_n|$ is the norm of the partition Π_n. The limit in question is denoted by $[\nu,\nu]_t$. Even though we have not assumed in Theorem 8.1.1 that (B_t) is a martingale, we shall denote prob lim $\sum_{\Pi_n}(B_{t^n_{j+1}} - B_{t^n_j})^2$ also by $[B,B]_t$ provided the former exists. (We shall not enter into a characterization of the process $[v,v]_t$ in terms of the discontinuities and the continuous part of v. For this theory see [45]. By (8.13) and Theorem 8.1.1 it follows that B_t has left limits a.s.

Theorem 8.1.2. *Let assumptions 1 to 5 be satisfied. Then $[B,B]$ exists.*

(a) *If (S_t) is continuous, then $[B,B] = [v,v]$.*
(b) *If (S_t) and (B_t) are continuous, then $(v_t,\mathscr{F}_t)$ is a continuous martingale. Furthermore, if (B_t) is a martingale relative to some $(\mathscr{G}_t)$ with $\mathscr{G}_t \supseteq \mathscr{F}_t$. then $\langle B \rangle = \langle v \rangle$.*
(c) *If $(B_t,\mathscr{G}_t)$ is a Wiener martingale, then $(v_t,\mathscr{F}_t)$ is also a Wiener martingale.*

The result just stated as well as the proof which we give below is due to Meyer [42]. The proof is included here, although it makes use of facts pertaining to discontinuous martingales which have not been treated in this book.

Proof of Theorem 8.1.2. Write $\sigma_t = S_t - \bar{S}_t$. The sum in (8.1.4) is the sum of the three terms

(i) $\sum_{\Pi_n} (\sigma_{t_{j+1}} - \sigma_{t_j})^2$

(ii) $2\sum_{\Pi_n} (\sigma_{t_{j+1}} - \sigma_{t_j})(B_{t_{j+1}} - B_{t_j})$

(iii) $\sum_{\Pi_n} (B_{t_{j+1}} - B_{t_j})^2$

For any right-continuous process (M_t) with left limits, we adopt the notation ΔM_t to denote $M_t - M_{t-}$, the jump at t. Consider first, the sum in (i). For $\varepsilon > 0$ there is only a finite number, say m (depending on ω), of points of discontinuity, $s_1, \ldots ,s_m$, at which the height of the jump exceeds $\frac{1}{2}\varepsilon$. Hence at most m of the intervals of the partition Π_n contain the s_j. [Note that if $|\Pi_n|$ is sufficiently small, the points (s_j) will be contained in exactly m of these subintervals.] The part of the sum (i) ranging over the m intervals, say, $\sum'(\sigma_{t_{j+1}} - \sigma_{t_j})^2$, tends to

$$\sum_{s \le t,\, |\Delta\sigma_s| > \frac{1}{2}\varepsilon} \Delta\sigma_s^2 \quad \text{as } n \to \infty.$$

The remaining sum $\sum''$ is dealt with as follows. If $|\Pi_n|$ is made small enough, for each subinterval (t_t,t_{j+1}) of $\sum''$, we have $|\sigma_{t_{j+1}} - \sigma_{t_j}| \le \varepsilon$, so that we have $\sum'' \le \varepsilon \sum'' |\sigma_{t_{j+1}} - \sigma_{t_j}| \le \varepsilon \operatorname{Var}_t \sigma$. Hence the sum in (i) tends to $\sum_{s \le t} (\Delta\sigma_s)^2$ in probability.

In a similar way, considering the discontinuities of B where jumps $> \frac{1}{2}\varepsilon$, it follows that $\sum'(\sigma_{t_{j+1}} - \sigma_{t_j})(B_{t_{j+1}} - B_{t_j})$ ($\sum'$ is summation over just such discontinuities) converges to $\sum_{s \le t,\, |\Delta B_s| > \frac{1}{2}\varepsilon} (\Delta\sigma_s)(\Delta B_s)$. The remaining term [as in (i)] is dominated by $\varepsilon \operatorname{Var}_t \sigma$ and hence $\to 0$. Thus the sum in (ii) converges in probability to $\sum_{s \le t} (\Delta\sigma_s)(\Delta B_s)$. It then follows that the sum in (iii) has a limit in probability as $|\Pi_n| \to 0$, that is, $[B,B]_t$ exists.

If (S_t) is continuous, its dual predictable projection $(\bar{S}_t)$ is also continuous, and then it is clear that the limits of the sums in (i) and (ii) are both 0, so that we have $[B,B]_t = [\nu,\nu]_t$.

If (S_t) and (B_t) are continuous, then the martingale $(\nu_t, \mathscr{F}_t)$ is obviously continuous. The remaining assertions of the theorem follow immediately. □

We need two auxiliary results for future use.

Lemma 8.1.1. *Let (A_t) $(0 \le t \le T)$ with $A_0 = 0$ be an increasing process not necessarily adapted to $(\tilde{\mathscr{F}}_t)$, where $\tilde{\mathscr{F}}_t = \mathscr{F}_{t+}$. Suppose that $(\bar{A}_t)$ is the dual predictable projection of (A_t) relative to $(\tilde{\mathscr{F}}_t)$. If $E(A_T^2) < \infty$, then $E(\bar{A}_T^2) < \infty$.*

Proof. The potential generated by (A_t) is the essentially unique right-continuous version of $X_t = E(A_T - A_t | \tilde{\mathscr{F}}_t)$. Now $\bar{A}$ is defined by the fact that it is natural, increasing, and such that $X_t = E(\bar{A}_T - \bar{A}_t | \tilde{\mathscr{F}}_t)$. Let ζ_t be the right-continuous version of the martingale $E(A_T | \tilde{\mathscr{F}}_t)$. Then $X_t \le \zeta_t$ for each t and therefore also $X_{t-} \le \zeta_{t-}$. Hence from the formula of Theorem 2.4.7 we have

$$\begin{aligned} E(\bar{A}_T^2) &= E \int_0^T (X_t + X_{t-})\, d\bar{A}_t \\ &\le E \int_0^T (\zeta_t + \zeta_{t-})\, d\bar{A}_t \\ &= 2E \int_0^T \zeta_{t-}\, d\bar{A}_t \quad \text{(since } \bar{A} \text{ is natural).} \end{aligned}$$

Since ζ_{t-} is predictable, using [41], Chapter VII, Theorem 17, we have

$$\begin{aligned} E \int_0^T \zeta_{t-}\, d\bar{A}_t = E \int_0^T \zeta_{t-}\, dA_t &\le E\left[\left(\sup_s \zeta_s\right) A_T\right] \\ &\le \left\{E\left(\sup_s \zeta_s\right)^2 E(A_T^2)\right\}^{\frac{1}{2}} \\ &\le \{4E(\zeta_T^2)E(A_T^2)\}^{\frac{1}{2}} \le 2E(A_T^2) \quad \text{[by Doob's inequality (2.2.7)].} \end{aligned}$$

This proves the assertion. □

Lemma 8.1.2. *Let the integrable, increasing process (A_t) (not necessarily adapted to $\tilde{\mathscr{F}}_t$) be continuous, and let $\bar{A}_t$ be the dual predictable projection of A_t relative to $(\tilde{\mathscr{F}}_t)$. Suppose that $\bar{A}_T \le C < \infty$ (a.s.), where C is a constant. Then $E(A_T^2) < \infty$.*

Proof. The proof is similar to that of Lemma 8.1.1. Since (A_t) is continuous, the formula in Theorem 2.4.7 is applicable, and we have

$$E(A_T^2) = E \int_0^T (X_t + X_{t-})\, dA_t,$$

where X_t is the potential generated by (A_t). Hence $X_t = E(\bar{A}_T - \bar{A}_t|\tilde{\mathscr{F}}_t)$ by the definition of $(\bar{A}_t)$. Denoting by (ζ'_t), a right-continuous version of the martingale $E(\bar{A}_T|\tilde{\mathscr{F}}_t)$, we obtain $X_t \leq \zeta'_t$; so $X_{t-} \leq \zeta'_{t-}$ upon taking left limits. Hence

$$E(A_T^2) \leq E\int_0^T (\zeta'_t + \zeta'_{t-})\,dA_t.$$

Now for each $t, \zeta'_t = E(\bar{A}_T|\tilde{\mathscr{F}}_t) \leq C$, and so, taking left limits, ζ'_{t-} is also bounded by C. Hence $E(A_T^2) \leq 2CE(A_T) < \infty$. □

Assertion (c) of Theorem 8.1.2 establishes the basic property of the innovation process. The proof is very simple, but it makes use of Remark 2.5.5 which is stated without proof in Chapter 2. In view of the central role played by the innovation process we give below a complete but somewhat lengthy proof of (c) which is based on Theorem 2.5.2.

Theorem 8.1.3. *Let the following conditions be satisfied in the model* (*8.1.1*):

(i) S_t *is continuous* ($S_0 = 0$).
(ii) $E(\mathrm{Var}_T\, S) < \infty$.
(iii) $(B_t, \mathscr{G}_t)$ *is a Wiener martingale, where* $\mathscr{G}_t \supseteq \mathscr{F}_t$.

Then $(\nu_t, \mathscr{F}_t)$ *is a Wiener martingale, and* $\mathscr{F}_s \perp\!\!\!\perp \sigma[\nu_v - \nu_u, s \leq u \leq v \leq T]$ *for each* s.

Proof. Define the stopping times

$$\tau_N = \begin{cases} \inf\{t \leq T : \mathrm{Var}_t\, \bar{S} > N\} \\ T & \text{if the set is empty.} \end{cases}$$

Define $S_t^N = S_{t\wedge\tau_N}$ and $\bar{S}_t^N = \bar{S}_{t\wedge\tau_N}$. It is easy to see that S_t^N is of bounded variation in $[0,T]$ and that $(\bar{S}_t^N)$ is the dual predictable projection (relative to $\tilde{\mathscr{F}}_t$) of (S_t^N). Setting $\tilde{\mathscr{F}}_t^N = \tilde{\mathscr{F}}_{t\wedge\tau_N}$, we have

$$E[S_N^T - S_t^N|\tilde{\mathscr{F}}_t^N] = E[\bar{S}_T^N - \bar{S}_t^N|\tilde{\mathscr{F}}_t^N];$$

that is, $(\bar{S}_t^N)$ is the dual predictable projection of (S_t^N) relative to $(\tilde{\mathscr{F}}_t^N)$. Let us define $Z_t^N = Z_{t\wedge\tau_N}$, Z being given by (8.1.1). Then, defining (ν_t^N) by

$$\nu_t^N = Z_t^N - \bar{S}_t^N,$$

we see that $\nu_t^N = \nu_{t\wedge\tau_N}$. Since $(\nu_t, \mathscr{F}_t)$ is a continuous martingale by Theorem 8.1.1, it follows that $(\nu_t^N, \mathscr{F}_t^N)$ is a continuous martingale (where $\mathscr{F}_t^N = \mathscr{F}_{t\wedge\tau_N}$) and hence from Lemma 2.6.1 that $(\nu_t^N, \mathscr{F}_t)$ is a continuous martingale. Now, note that $S_t^N = U_t^N - V_t^N$ and $\bar{S}_t^N = \bar{U}_t^N - \bar{V}_t^N$, where $U, \bar{U}$, etc., are defined as in the proof of Theorem 8.1.1 and the superscript N has the obvious meaning. From the definition of τ_N it follows that $\mathrm{Var}_T\, \bar{S}^N \leq N$, and so $\bar{U}_T^N$ and $\bar{V}_T^N$ are both bounded by N. From condition (ii) and Lemma 8.1.2 we then have $E(\mathrm{Var}_T\, S^N)^2 < \infty$. Also, since (B_t) is a Wiener process, $B_t^N = B_{t\wedge\tau_N}$ is a continuous L^2-martingale relative to $(\mathscr{G}_t)$. Thus it follows that (ν_t^N) is square-integrable and by Theorem 2.5.2

$$\langle v^N \rangle_t = L^1\text{-lim } S^N_{\Pi_n} \tag{8.1.5}$$

as $|\Pi_n| \to 0$, where $S^N_{\Pi_n} = \sum_{\Pi_n} E[(v^N_{t_{j+1}} - v^N_{t_j})^2 | \mathscr{F}_{t_j}]$.

Now

$$\begin{aligned} S^N_{\Pi_n} &= \sum_{\Pi_n} E[(\sigma^N_{t_{j+1}} - \sigma^N_{t_j})^2 | \mathscr{F}_{t_j}] \\ &\quad + 2 \sum_{\Pi_n} E[(\sigma^N_{t_{j+1}} - \sigma^N_{t_j})(B^N_{t_{j+1}} - B^N_{t_j}) | \mathscr{F}_{t_j}] \\ &\quad + \sum_{\Pi_n} E[(B^N_{t_{j+1}} - B^N_{t_j})^2 | \mathscr{F}_{t_j}] = J^n_1 + 2J^n_2 + J^n_3, \end{aligned} \tag{8.1.6}$$

say. Note that N is fixed throughout the argument and it is n that tends to infinity and $|\Pi_n| \to 0$. Setting $\alpha_n = \sum_{\Pi_n} (\sigma^N_{t_{j+1}} - \sigma^N_{t_j})^2$, we have

$$0 \le \alpha_n \le \sup_{\Pi_n} |\sigma^N_{t_{j+1}} - \sigma^N_{t_j}| \cdot (\text{Var}_t\, S^N + \text{Var}_t\, \bar{S}^N) \to 0 \text{ (a.s.)}$$

since σ^N_t is continuous. Next, for all n, $0 \le \alpha_n \le [\text{Var}_t\, S^N + \text{Var}_t\, \bar{S}^N]^2$, and the right-hand side has finite expectation. Hence $E\alpha_n \to 0$ by the Lebesgue dominated convergence theorem and thus $E(J^n_1) \to 0$. For the second term,

$$|J^n_2| \le \left(\sum E[(\sigma^N_{t_{j+1}} - \sigma^N_{t_j})^2 | \mathscr{F}_{t_j}] \cdot \sum E[(B^N_{t_{j+1}} - B^N_{t_j})^2 | \mathscr{F}_{t_j}]\right)^{\frac{1}{2}}$$

and so, $E|J^n_2| \le [E(J^n_1)E(J^n_3)]^{\frac{1}{2}}$. Now, $E(J^n_3) = E[\sum E\{(B^N_{t_{j+1}} - B^N_{t_j})^2 | \mathscr{G}_{t_j}\}]$ and $[\sum E\{(B^N_{t_{j+1}} - B^N_{t_j})^2 | \mathscr{G}_{t_j}\}]$ converges to $\langle B^N \rangle_t$ in L^1. Hence $E(J^n_3)$ is bounded and it follows that $E|J^n_2| \to 0$. From (8.1.5) and (8.1.6) we have

$$\sum_{\Pi_n} E[(B^N_{t_{j+1}} - B^N_{t_j})^2 | \mathscr{F}_{t_j}] \xrightarrow{L^1} \langle v^N \rangle_t. \tag{8.1.7}$$

Since

$$\begin{aligned} E[(B^N_{t_{j+1}} - B^N_{t_j}) | \mathscr{G}_{t_j}] &= E[\langle B^N \rangle_{t_{j+1}} - \langle B^N \rangle_{t_j} | \mathscr{G}_{t_j}] \\ &= E[\langle B \rangle_{t_{j+1} \wedge \tau_N} - \langle B \rangle_{t_j \wedge \tau_N} | \mathscr{G}_{t_j}] \\ &= E[(t_{j+1} \wedge \tau_N - t_j \wedge \tau_N) | \mathscr{G}_{t_j}], \end{aligned} \tag{8.1.8}$$

the random variable on the left-hand side of (8.1.7) equals

$$\sum_{\Pi_n} E[(t_{j+1} \wedge \tau_N - t_j \wedge \tau_N) | \mathscr{F}_{t_j}]. \tag{8.1.9}$$

Denoting the latter sum by $S_{n,N}$ and the set $[\tau_N > t]$ by Λ_N it is easily seen that $S_{n,N}(\omega) = t$ for $\omega \in \Lambda_N$. Hence

$$\int_{\Lambda_N} |S_{n,N} - \langle v^N \rangle_t| \, dP \le E|S_{n,N} - \langle v^N \rangle_t| \to 0$$

as $n \to \infty$ from (8.1.7) and (8.1.9). But the left-hand side equals $\int_{\Lambda_N} |t - \langle v^N \rangle_t| \, dP$

which, therefore, equals 0 for all N. Now,

$$P[(|\langle v^N\rangle_t - t| > \varepsilon) \cap \Lambda_N] \leq \frac{1}{\varepsilon}\int_{\Lambda_N} |\langle v^N\rangle_t - t|\, dP = 0$$

and

$$P[(|\langle v^N\rangle_t - t| > \varepsilon) \cap \Lambda_N^c] \leq P(\Lambda_N^c) = P(\tau_N \leq t).$$

Recalling that $\langle v^N\rangle_t = \langle v\rangle_{t\wedge\tau_N}$, we get

$$P[|\langle v\rangle_{t\wedge\tau_N} - t| > \varepsilon] \leq P(\tau_N \leq t) \to 0$$

as $N \to \infty$. Since $v_{t\wedge\tau_N} \to v_t$ (a.s.), it follows that $\langle v\rangle_t = t$. The conclusion then is a consequence of Theorem 4.4.1. □

Consider the model (8.1.1) where now (Z_t), (S_t), and (B_t) are d-dimensional vector-valued processes. The dual predictable projection of S_t is $\bar{S}_t = (\bar{S}_t^i)$ $(i = 1, \ldots, n)$, where $\bar{S}_t^i$ is the dual predictable projection of the ith component (S_t^i) of (S_t). The innovation process $v_t = Z_t - \bar{S}_t$ is now d-dimensional. The vector-valued version of Theorem 8.1.3 is proved in an exactly similar manner as in the one-dimensional case. Only the definition of τ_N has to be appropriately modified. We let

$$\tau_N = \begin{cases} \inf\{t\colon |\mathrm{Var}_t\, \bar{S}| > n\} & \\ T & \text{if the above set is empty,} \end{cases}$$

where $|\mathrm{Var}_t\, \bar{S}|^2 = \sum_{i=1}^d (\mathrm{Var}_t\, \bar{S}^i)^2$.

Theorem 8.1.4. *For the vector-valued model (8.1.1) assume that the following conditions are satisfied:*

(i) (S_t) *is continuous,* $S_0 = 0$.
(ii) $E(\mathrm{Var}_T\, S^i) < \infty$ *for* $i = 1, \ldots, d$.
(iii) $(B_t, \mathscr{G}_t)$ *is a d-dimensional Wiener martingale.*

Then $(v_t, \mathscr{F}_t)$ *is a d-dimensional Wiener martingale and*

$$\mathscr{F}_s \perp\!\!\!\perp \sigma[v_v - v_u, s \leq u \leq v \leq T].$$

For convenience of applications, from now on we consider a finite time interval $[0,T]$.

Remark 8.1.1. Suppose $U_t = \int_0^t h_s^+\, ds$, where $h_s^+ \geq 0$ is progressively measurable and such that $\int_0^T E(h_s^+)\, ds < \infty$.

Then $\bar{U}_t$, the dual predictable projection of U_t relative to $(\mathscr{F}_{t+})$ is given by $\int_0^t E(h_s^+ | \mathscr{F}_s)\, ds$.

PROOF. Set $A_t = \int_0^t E(h_s^+ | \mathscr{F}_s)\, ds$. Clearly (A_t) is $(\mathscr{F}_t)$-adapted, continuous and hence predictable. Furthermore, if $s < s' < t$, since $\mathscr{F}_{s+} \subseteq \mathscr{F}_u$ for all $u \geq s'$, we have

$$\begin{aligned}E[U_t - U_{s'}|\mathscr{F}_{s+}] &= \int_{s'}^{t} E(h_u^+|\mathscr{F}_{s+})\,ds \\ &= \int_{s'}^{t} E\{E(h_u^+|\mathscr{F}_u)|\mathscr{F}_{s+}\}\,ds \\ &= E\left(\int_{s'}^{t} E(h_u^+|\mathscr{F}_u)\,du|\mathscr{F}_{s+}\right) \\ &= E(A_t - A_{s'}|\mathscr{F}_{s+}). \end{aligned}\tag{8.1.10}$$

Since $A_{s'} \leq A_T$ and $U_{s'} \leq U_T$ for all $s' > s$ and all the random variables are integrable, it is permissible by the monotone convergence theorem to make $s' \to s$ ($s' > s$) inside the conditional expectations in (8.1.10). Because of the continuity of (A_t) and (U_t), we obtain

$$E(U_t - U_s|\mathscr{F}_{s+}) = E(A_t - A_s|\mathscr{F}_{s+}).\tag{8.1.11}$$

From Equation (8.1.11) we can write

$$E(U_T - U_t|\mathscr{F}_{t+}) = M_t - A_t,$$

where M_t is the right-continuous version of the martingale $E(A_T|\mathscr{F}_{t+})$ and (A_t) is predictable and obviously natural. But such a decomposition of the potential $E(U_T - U_t|\mathscr{F}_{t+})$ is unique. Hence $A_t = \bar{U}_t$ for all t, (a.s.). □

Remark 8.1.2. Absolutely continuous signal. Assume that (S_t) is of the form

$$S_t = \int_0^t h_s\,ds,$$

h_s progressively measurable and $\int_0^T E(|h_s|)\,ds < \infty$. Then by Remark 8.1.1 it follows that

$$\bar{S}_t = \int_0^t E(h_s|\mathscr{F}_s)\,ds.$$

Corollary 8.1.1 (to Theorem 8.1.4). *Suppose that (S_t) is absolutely continuous,*

$$S_t = \int_0^t h_s\,ds,$$

where h_s is a progressively measurable, d-dimensional process such that

$$\int_0^T E|h_s|\,ds < \infty.\tag{8.1.12}$$

Then $(\nu_t, \mathscr{F}_t)$ is a d-dimensional Wiener martingale.

PROOF. Since $S_t^i = \int_0^t h_s^i\,ds$ $(i = 1, \ldots, d)$, $\mathrm{Var}_T S^i = \int_0^T |h_s^i|\,ds$. Hence

$$E(\mathrm{Var}_T S^i) = E\left(\int_0^T |h_s^i|\,ds\right) \leq E\left(\int_0^T |h_s|\,ds\right) < \infty$$

from (8.1.12). The conclusion follows immediately from Theorem 8.1.4. □

Remark 8.1.3. When S_t is absolutely continuous and (B_t) is a Wiener process, we have in model (8.1.1) an example of practical importance. In fact, this is the case which will almost exclusively be considered in this book.

The model obtained by formally differentiating (8.1.1) is more natural but lacks rigor:

$$\dot{Z}_t = \dot{S}_t + \dot{B}_t. \tag{8.1.13}$$

Here $\dot{B}_t$ is Gaussian "white noise," that is, the "derivative" of the Wiener process (B_t). The best estimate of $\dot{S}_t$ is $\hat{\dot{S}}_t = E(\dot{S}_t|\mathscr{F}_t)$ and the "innovation process" of (8.1.13) is given by $I_t = \dot{Z}_t - \hat{\dot{S}}_t$. Integrating both sides, we have

$$\int_0^t I_u\,du = Z_t - \int_0^t \hat{\dot{S}}_u\,du.$$

The right-hand side is the innovation process v_t, for by Remark 8.1.2, if $S_t = \int_0^t \dot{S}_u\,du$, then the dual predictable projection $\overline{S}_t = \int_0^t \hat{\dot{S}}_u\,du$. This remark will also help explain why the definition of (v_t) uses $\overline{S}_t$ instead of $\hat{S}_t = E(S_t|\mathscr{F}_t)$. Note also the fact that if $t' > t$, $\overline{S}_{t'} = \overline{S}_t + \int_t^{t'} \hat{\dot{S}}_u\,du$, whereas $\hat{S}_{t'}$ cannot be expressed as the sum of $\hat{S}_t$ and the new information from t to t'.

A generalization, due to Meyer [42], of Theorem 8.1.1 (or of Corollary 8.1.1 to Theorem 8.1.4, $d = 1$) is the following.

Theorem 8.1.5. *Suppose that*

(i) $\int_0^t |h_s|\,ds < \infty$ (a.s.) *for every* $t \in \mathbf{R}_+$.
(ii) $E(|h_t|\,\mathscr{F}_t) < \infty$ (a.s.) *for each* t:
(iii) $(B_t,\mathscr{G}_t)$ *is a Wiener process.*

Then

$$E\int_0^t |\hat{h}_s|\,ds < \infty \quad \textit{for every } t, \tag{8.1.14}$$

and

$$v_t = Z_t - \int_0^t \hat{h}_s\,ds$$

is a Wiener martingale with respect to $(\mathscr{F}_t)$.

PROOF. First let us prove (8.1.14). Write $h_t^* = E(|h_t||\mathscr{F}_t)$ and set $f_t^n = I_{[h_t^* \le n]} \operatorname{sgn} \hat{h}_t$, where $\operatorname{sgn} x = 1$ if $x \ge 0$, and $= -1$ if $x < 0$. The process f_t^n is bounded and predictable; so the stochastic integral with respect to Z_t exists, and we have

$$\int_0^t f_s^n\,dZ_s = \int_0^t f_s^n h_s\,ds + \int_0^t f_s^n\,dB_s.$$

Let us write this as

$$Z_t^n = S_t^n + B_t^n.$$

Then it is easy to see that Z_t^n is $\mathscr{F}_t$-measurable, that $(B_t^n,\mathscr{G}_t)$ is a continuous L^2-martingale, and that $\overline{S}_t^n = \int_0^t f_s^n \hat{h}_s\,ds$, which exists and is finite and positive because it equals

$$\int_0^t I_{[h_s^* \le n]} \operatorname{sgn} \hat{h}_s \cdot \hat{h}_s \, ds = \int_0^t I_{[h_s^* \le n]} |\hat{h}_s| \, ds$$
$$\le \int_0^t I_{[h_s^* \le n]} h_s^* \, ds.$$

From Theorem 8.1.1 the innovation process

$$v_t^n = Z_t^n - \bar{S}_t^n = S_t^n - \bar{S}_t^n + B_t^n$$

is a continuous $(\mathscr{F}_t)$-martingale with

$$\langle v^n \rangle_t = \langle B^n \rangle_t \le \int_0^t (f_s^n)^2 \, ds \le t.$$

Therefore, as $n \to \infty$, $E(v_t^n)^2$ remains bounded, $B_t^n \to \int_0^t \operatorname{sgn} \hat{h}_s \, dB_s$ in L^2, and $S_t^n \to \int_0^t \operatorname{sgn} \hat{h}_s \cdot h_s \, ds$, which is finite (a.s.). Note also that $\bar{S}_t^n = \int_0^t I_{[h_s^* \le n]} |\hat{h}_s| \, ds$ increases to a limit $\bar{S}_t = \int_0^t |\hat{h}_s| \, ds$, finite or infinite, as $n \to \infty$. For a subsequence (n') we have $B_t^{n'} \to \int_0^t \operatorname{sgn} \hat{h}_s \cdot h_s \, ds$ (a.s.). Hence $v_t^{n'}$ tends to a limit, which has to be finite by Fatou's lemma since $E[v_t^{n'}]^2$ is bounded. Thus $\bar{S}_t$ has to be finite (a.s.). Let us now define the stopping time sequence

$$\tau_n = \begin{cases} \inf\left\{t \colon \int_0^t |\hat{h}_s| \, ds \ge n\right\} & \\ \infty & \text{if the event never occurs.} \end{cases}$$

Let $S_t^n = S_{t \wedge \tau_n}$, $B_t^n = B_{t \wedge \tau_n}$ and $Z_t^n = Z_{t \wedge \tau_n}$. Then $\bar{S}_t^n = \int_0^t \hat{h}_s I_{[0,\tau_n]}(s) \, ds$ and $(B_t^n, \mathscr{G}_t)$ is a continuous, L^2-martingale with $\langle B^n \rangle_t = t \wedge \tau_n$. From Theorem 8.1.1 we conclude that the innovation process $v_t^n = Z_t^n - \bar{S}_t^n$ is a continuous, L^2-martingale with $\langle v^n \rangle_t = \langle B^n \rangle_t \le t$. Hence $E[v_t^n]^2$ is bounded for every t, and so (for each t) the sequence (v_t^n) is uniformly integrable. Setting $v_t = Z_t - \int_0^t \hat{h}_s \, ds$, we see that $v_t^n = v_{t \wedge \tau_n} \to v_t$ (a.s.). It follows that $(v_t, \mathscr{F}_t)$ is a continuous L^2-martingale and $\langle v \rangle_t = \lim_{n \to \infty} \langle v \rangle_{t \wedge \tau_n} = \lim_{n \to \infty} \langle v_t^n \rangle = \lim_{n \to \infty} \langle B^n \rangle_t = t$. This completes the proof. □

The Nonlinear Filtering Problem

The general problem of nonlinear filtering or estimation may be described as follows.

We are interested in the estimation of a signal or system process (X_t), $(0 \le t \le T)$ which cannot be observed directly. Instead we have an observation process (Z_t) which is related to (X_t) through the model (8.1.1). The state space, that is, the space in which (X_t) assumes its values can be quite general, for example, a complete, separable metric space. The observation process will be assumed to be n-dimensional. The least squares estimate of $f(X_t)$ (where f is a suitable real-valued function) based on the σ-field $\mathscr{F}_t^Z$ of observations up to time t is given by the conditional expectation

$E[f(X_t)|\mathscr{F}_t^Z]$. This estimate depends, in the general case, nonlinearly on the observations and is called the nonlinear *filter*. A Bayes formula for the conditional expectation has been given in [32] in the case when (X_t) and the observation "noise" are independent, but it is directly useful only when t is fixed. When the observations are being received continuously—as in most engineering and technological applications—one requires an estimate which can be continuously revised to take into account the new data. The Bayes formula does not lend itself to recursive calculations of this sort for the estimate at time t cannot be effectively used in computing the estimate at time $t + h$. A practical method which is also mathematically more appealing is to derive a stochastic differential equation for the filter and exploit the power of the Ito stochastic calculus. (It is with this in mind that the essential tools of the latter were developed in Chapters 3 to 5).

The Bayes formula can also be used in deriving the basic stochastic differential equation [33].

The martingale approach to filtering theory, which is both more general and elegant has been adopted in the works of Liptser and Shiryaev, Fujisaki, Kunita, and Kallianpur, and—for the case of linear filtering independently, by Balakrishnan ([40], [14], [2]). The innovation process makes its appearance in this theory and plays much the same role as in the prediction theory of stationary stochastic processes. (We return to this point later in connection with the linear theory of Chapter 10). For the case of an absolutely continuous signal, the discovery of the innovation Wiener process is due to many workers in the field (see the references in [24]) but its systematic use in filtering theory seems to have been made first by Kailath [25].

The totality of stochastic differential equations obtained for $E[f(X_t)|\mathscr{F}_t^Z]$ for each f belonging to a suitably rich class of functions may be looked upon [for example, in the case when (X_t) and (Z_t) are jointly a Markov diffusion process] as yielding a measure-valued stochastic differential equation satisfied by the conditional probability distribution or the conditional density function (when it exists) of X_t given $\mathscr{F}_t^Z$.

The crux of the martingale argument used in filtering problems is contained in the following simple result. The σ-fields $\mathscr{F}_t$, $\tilde{\mathscr{F}}_t$, and $\mathscr{G}_t$ have the same meaning as before.

Theorem 8.1.6. *Let*

$$\xi_t = \xi_0 + L_t + M_t, \qquad t \in [0,T] \tag{8.1.15}$$

be a d-dimensional semimartingale relative to the family $(\mathscr{G}_t)$. The following conditions are assumed to hold:

(i) *The process (L_t) is right-continuous and of bounded variation in $[0,T]$, $L_0 = 0$, and $(\mathscr{G}_t)$-adapted.*
(ii) $E|\mathrm{Var}_T L|^2 < \infty$.
(iii) *$(M_t,\mathscr{G}_t)$ is a right-continuous L^2-martingale with $M_0 = 0$.*
(iv) $E|\xi_0|^2 < \infty$.

Then $E(\xi_t|\mathscr{F}_t)$ *is an* L^2*-semimartingale relative to* $(\mathscr{F}_t)$ *and has the following form*:

$$E(\xi_t|\mathscr{F}_t) = E(\xi_0|\mathscr{F}_0) + \bar{L}_t + \bar{M}_t, \tag{8.1.16}$$

where

(a) $(\bar{L}_t)$ *is the dual predictable projection of* (L_t) *relative to* $(\tilde{\mathscr{F}}_t)$, *with* $\bar{L}_0 = 0$.
(b) $E|\text{Var}_T \bar{L}|^2 < \infty$.
(c) $(\bar{M}_t, \mathscr{F}_t)$ *is a right-continuous* L^2*-martingale.*

PROOF. For $(\bar{L}_t)$ given by (a) we have $E|\text{Var}_T \bar{L}|^2 < \infty$ by condition (ii) and Lemma 8.1.1. Define $\eta_t = E(\xi_t|\mathscr{F}_t) - \bar{L}_t - E(\xi_0|\mathscr{F}_0)$. Then for $s < t$,

$$E(\eta_t - \eta_s|\mathscr{F}_s) = E(\xi_t - \xi_s|\mathscr{F}_s) - E(\bar{L}_t - \bar{L}_s|\mathscr{F}_s).$$

The first term on the right-hand side becomes

$$\begin{aligned} E[E(\xi_t - \xi_s|\mathscr{G}_s)|\mathscr{F}_s] &= E[\{E(L_t - L_s|\mathscr{G}_s) + E(M_t - M_s|\mathscr{G}_s)\}|\mathscr{F}_s] \\ &= E(L_t - L_s|\mathscr{F}_s). \end{aligned}$$

Hence

$$E(\eta_t - \eta_s|\mathscr{F}_s) = E(L_t - L_s|\mathscr{F}_s) - E(\bar{L}_t - \bar{L}_s|\mathscr{F}_s) = 0.$$

Moreover, $E|\eta_t|^2 < \infty$ in view of the following facts. $E|E(\xi_t|\mathscr{F}_t)|^2 \leq E|\xi_t|^2$, whose finiteness is ensured by the conditions of the theorem and $E|\bar{L}_t|^2 < \infty$ which follows from (b). Since η_t is clearly $\mathscr{F}_t$-measurable, it follows that $(\eta_t, \mathscr{F}_t)$ is an L^2-martingale. Define $\bar{M}_t$ to be a right-continuous version of η_t. Then we have (8.1.16). □

If X_t is the solution of a stochastic differential equation and f is a C^2-function, we have seen that an application of Ito's formula yields the fact that $\xi_t = f(X_t)$ is a semimartingale. If the conditions of Theorem 8.1.6 are met, then $E[f(X_t)|\mathscr{F}_t]$ is a right-continuous semimartingale with respect to the observation σ-fields. Now, if every right-continuous L^2-martingale relative to $\mathscr{F}_t$ (in particular, $\bar{M}_t$), can be represented as a stochastic integral with respect to a Wiener process "living" on the σ-fields $\mathscr{F}_t$, then we have, in (8.1.16), the beginning of a stochastic differential equation for $E[f(X_t)|\mathscr{F}_t]$. Thus the question of a stochastic integral representation for L^2-martingales is of importance for filtering problems. This, in a nutshell, is the approach we adopt in this chapter. The stochastic integral representation problem will be solved for separable L^2-martingales with respect to $\mathscr{F}_t^Z$, where (Z_t) is the observation process in the model considered in the next section.

For an important class of problems involving discontinuous processes (which lie outside the scope of this book) it is of interest to seek a stochastic integral representation with respect to some other process such as the martingale $N(t) - \lambda t$ associated with the Poisson process with parameter λ. In general, if (ζ_t) is a continuous L^2-martingale with respect to $(\mathscr{F}_t)$, then it is *not* possible to represent every separable L^2-$(\mathscr{F}_t)$ martingale M_t as a stochastic integral with respect to ζ_t. This is shown by the following example which was described to the author by Marc Yor who attributes it to H. Kunita.

Let $W_t = (W_t^1, W_t^2)$ be a Wiener process in $\mathbf{R}^2$, and let $\zeta_t = \int_0^t W_s^1 \, dW_s^2$. Choose $\mathscr{F}_t = \mathscr{F}_t^\zeta$. Then $\langle\zeta\rangle_t$ is $\mathscr{F}_t^\zeta$-measurable, and since $\langle\zeta\rangle_t = \int_0^t (W_s^1)^2 \, ds$, it follows that $(W_t^1)^2$ is $\mathscr{F}_t^\zeta$-measurable. Now, $M_t = (W_t^1)^2 - t$ is an L^2-martingale relative to $\mathscr{F}_t^\zeta$ and equals $2\int_0^t W_s^1 \, dW_s^1$. Suppose M_t can be represented as a stochastic integral $\int_0^t H_s \, d\zeta_s$, where (H_s) is predictable and $E\int_0^t H_s^2 \, d\langle\zeta\rangle_s < \infty$. From

$$M_t = 2\int_0^t W_s^1 \, dW_s^1 = \int_0^t H_s \, d\zeta_s = \int_0^t H_s W_s^1 \, dW_s^2,$$

we have

$$\begin{aligned} 0 &= E\left[2\int_0^t W_s^1 \, dW_s^1 - \int_0^t H_s W_s^1 \, dW_s^2\right]^2 \\ &= 4E\int_0^t (W_s^1)^2 \, ds + E\int_0^t H_s^2 (W_s^1)^2 \, ds, \end{aligned}$$

which is impossible.

8.2 Observation Process Model with Absolutely Continuous (S_t)

The signal process $h_t(\omega)$ and the observation process $Z_t(\omega)(t \in [0,T])$ will be assumed to be N-dimensional processes defined on a complete probability space $(\Omega,\mathscr{A},P)$ and further related as follows:

$$Z_t(\omega) = \int_0^t h_u(\omega) \, du + W_t(\omega), \tag{8.2.1}$$

where

$$W_t \text{ is an } N\text{-dimensional Wiener process} \tag{8.2.2}$$

and

$$h_t(\omega) \text{ is } (t,\omega)\text{-measurable}$$

and such that

$$\int_0^T E|h_t|^2 \, dt < \infty. \tag{8.2.3}$$

Here $|\cdot|$ denotes the norm of the N vector.

We make the following assumption concerning the signal and noise processes, which seems natural and takes into account possible applications to stochastic control.

For each s ($s \in [0,T]$), the σ-fields

$$\mathscr{F}_s^{h,W} \quad \text{and} \quad N_s^T = \sigma\{W_v - W_u, \, s \le u < v \le T\} \tag{8.2.4}$$

are independent.

The σ-fields $\mathscr{F}_t^Z$ $(0 \le t \le T)$ will be called the *observation σ-fields. In other words, the σ-field $\mathscr{F}_t$ introduced in Section 8.1 is now chosen to be* $\mathscr{F}_t^Z$. The family $(\mathscr{F}_t)$ represents all the statistical data or information concerning (h_t). Note that (8.2.4) implies that $\mathscr{F}_s \perp\!\!\!\perp \sigma[W_v - W_u, s \leq u \leq v \leq T]$ for each s.

Remark 8.2.1. Recall that $\mathscr{F}_s^{h,W}$ in (8.2.4) is the σ-field generated by the set of random variables $\{h_u, W_u, 0 \leq u \leq s\}$ which is augmented by adding to it all A sets of P-measure zero.

Remark 8.2.2. From condition (8.2.3), Theorem 2.7.1, and Proposition 2.1.4, it follows that there exists an N-vector function $H(t,\omega)$ such that for a.e. t (Lebesgue measure in $[0,T]$), $E(h_t|\mathscr{F}_t)(\omega) = H(t,\omega)$ (a.s.). The same remark will apply to conditional expectations of other processes to be encountered later.

The observation process models in many applications can be reduced to the "standard form" (8.2.1). In a naturally arising model which is considered below, the signal process, which we now denote by $X_t(\omega)$, $t \in [0,T]$, and the observation process $Z_t(\omega)$ are given by the following conditions:

1. $X_t(\omega)$ takes values in a complete, separable metric space S.
2. For each s,

$$\mathscr{F}_s^{X,W} \perp\!\!\!\perp \sigma\{W_v - W_u, s \leq u \leq v \leq T\},$$

   where (W_t) is the Wiener process as in (8.2.2). Z_t is defined by (8.2.1), where $h_t(\omega)$ satisfies (8.2.3).
3. For each $s < T$, $h_s(\omega)$ is $\mathscr{F}_s^{X,W}$-measurable.

Then conditions 2 and 3 imply (8.2.4). Thus the three processes (Z_t, h_t, W_t) satisfy conditions (8.2.1) to (8.2.4). For convenience, let us write $\mathscr{G}_t$ for $\mathscr{F}_t^{X,W}$.

Observation Process Determined by a Stochastic Differential Equation

An important form of the filtering problem is the following where the observation process, say, (Y_t) is determined by the stochastic differential equation

$$\begin{aligned} dY_t = {} & a(t, X_s, s \leq t, Y_s, s \leq t)\,dt \\ & + b(t, X_s, s \leq t, Y_s, s \leq t)\,dW_t. \end{aligned} \tag{8.2.5}$$

The coefficients a and b will be dependent, in general, on the "past" values of the signal and noise as well as on their values at the instant t. The notation in (8.2.5) will now be made precise.

Let S be a complete, separable metric space, C the space of all continuous mappings from $[-T,0]$ to $\mathbf{R}^N$ with the usual uniform topology and D the space of all right-continuous mappings with left-hand limits from $[-T,0]$ to S endowed with the Skorokhod topology (see [52]). We now make the following assumptions. (See Chapter 5.)

(A.1) The sample paths $X_t(\omega)$ of the signal process take values in S and are continuous with left-hand limits.

(A.2) W_t is an M-dimensional Wiener process (M assumed larger than N) which satisfies assumption 2 above.

For each sample path (X_t) define $(\pi_t X)(u) = X_{t+u}$ if $-t \leq u \leq 0$, and $= X_0$ if $-T \leq u \leq -t$. Then $\pi_t X \in D$ for each $t \in [0,T]$.

(A.3) Let $a(t,g,f)$ be a functional on $[0,T] \times D \times C$ taking values in $\mathbf{R}^N$ and $b(t,g,f)$ an $N \times M$-matrix-valued functional on $[0,T] \times D \times C$ satisfying the following conditions:

$$a(t,g,f) \text{ and } b(t,g,f) \text{ are Borel measurable in } [0,T] \times D \times C. \quad (8.2.6)$$

There exists a bounded measure Γ on $[-T,0]$ and a positive constant K such that,

$$|a(t,g,f) - a(t,g,\tilde{f})|^2 + |b(t,g,f) - b(t,g,\tilde{f})|^2 \leq K \int_{-T}^{0} |f(s) - \tilde{f}(s)|^2 \, d\Gamma(s) \quad (8.2.7)$$

and

$$|a(t,g,f)|^2 + |b(t,g,f)|^2 \leq K\left[1 + \int_{-T}^{0} |f(s)|^2 \, d\Gamma(s) + |L(t,g)|^2\right], \quad (8.2.8)$$

where $L(t,g)$ is a Borel measurable real-valued functional in $[0,T] \times D$ such that

$$\int_0^T E|L(t,\pi_t X|^2 \, dt < \infty.$$

In (8.2.7) $\tilde{f}$ denotes an element of C and $|b|$ denotes $\sum_{i,j} (b^{ij})^2$, the norm of the $N \times M$ matrix $b = (b^{ij})$.

The precise meaning of the stochastic differential equation (8.2.5) is the stochastic integral equation

$$Y_t = \eta + \int_0^t a(s,\pi_s X,\pi_s Y) \, ds + \int_0^t b(s,\pi_s X,\pi_s Y) \, dW_s, \quad (8.2.9)$$

where η is taken to be a constant.

Theorem 8.2.1. *Under assumptions (A.1) to (A.3) the stochastic Equation (8.2.9) has a unique solution (Y_t) such that Y_t is square-integrable and $\mathscr{G}_t$-measurable.*

PROOF. We use the standard method of successive approximations. Set $Y_0(t) = \eta$ and

$$Y_t^n = \eta + \int_0^t a(s,\pi_s X,\pi_s Y^{n-1}) \, ds + \int_0^t b(s,\pi_s X,\pi_s Y^{n-1}) \, dW_s$$

for $n \geq 1$. Then a direct calculation which makes use of (8.2.7) and (8.2.8) shows that $\rho_n(t) \equiv \sup_{s \leq t} E|Y_s^n - Y_s^{n-1}|^2$ satisfies the inequality

$$\rho_{n+1}(t) \leq 2K(T+1)\|\Gamma\| \int_0^t \rho_n(s)\,ds$$

$$\leq (2K(T+1)\|\Gamma\|)^n \frac{\rho_1(t)}{n!},$$

where

$$\rho_1(t) \leq 2K\left\{(T+1)(1+|\eta|^2)\|\Gamma\| + \int_0^t E|L(s,\pi_s X)|^2\,ds\right\} < \infty.$$

Here $\|\Gamma\|$ denotes the total mass of the measure Γ. Hence Y_t^n converges to a continuous process Y_t which satisfies Equation (8.2.9) by a standard argument. It is easy to see that Y_t is $\mathscr{G}_t$-measurable and square-integrable. The uniqueness of the solution can be proved similarly. □

Let us introduce the following additional conditions.

(A.4) $b(t,g,f)$ does not depend on $g \in D$.
(A.5) The $N \times N$ matrix $c(t,f) = b(t,f)b^*(t,f)$ is positive definite ($bb^* > 0$) for all t and f, b^* being the transpose of b.
(A.6)

$$\int_0^T E|h'(t,\pi_t X,\pi_t Y)|^2\,dt < \infty,$$

where

$$h'(t,g,f) = c^{-\frac{1}{2}}(t,f)a(t,g,f).$$

Theorem 8.2.2. *Assume (A.1) to (A.6). Let*

$$W_t' = \int_0^t c^{-\frac{1}{2}}(s,\pi_s Y)b(t,\pi_s Y)\,dW_s$$

and

$$Z_t' = \int_0^t h'(s,\pi_s X,\pi_s Y)\,ds + W_t'.$$

Then W_t' is an N-dimensional Wiener process adapted to $(\mathscr{G}_t)$ such that $\sigma\{W_v' - W_u', s \leq u \leq v \leq T\} \perp\!\!\!\perp \mathscr{G}_s$. Furthermore the three processes W_t', h_t', and Z_t' satisfy conditions (8.2.1) to (8.2.4).

Proof. Since $W_t - W_s \perp\!\!\!\perp \mathscr{G}_s$ by assumption (A.2), W_t is a $\mathscr{G}_t$-martingale. From Theorem 8.2.1 $c^{-\frac{1}{2}}(s,\pi_s Y)b(s,\pi_s Y)$ is $\mathscr{G}_s$-measurable so that the process W_t' is also a $\mathscr{G}_t$-martingale. Furthermore we have, writing $c_u = c(u,\pi_u Y)$ and $b_u = b(u,\pi_u Y)$,

$$E[(W_t'^i - W_s'^i)(W_t'^j - W_s'^j)|\mathscr{G}_s] = E\left[\int_s^t (c_u^{-\frac{1}{2}} b_u b_u^* c_u^{-\frac{1}{2}})_{ij}\,du \Big|\mathscr{G}_s\right]$$

$$= (t-s)\delta_{ij}.$$

Hence we obtain the first assertion. Since $\{h'_u,\ u \le s\}$ is $\mathscr{G}_s$-measurable, condition (8.2.4) is satisfied for (h'_t, W'_t). Condition (8.2.3) for h'_t is obvious from (A.6) thus proving the theorem. □

From the definition of the innovation process given in Section 8.1 and Remark 2 of the same section, it follows that

$$\nu_t = Z_t - \int_0^t \hat{h}_s \, ds$$

is the innovation process corresponding to the model (8.2.1).

The following result is simply a restatement of the Corollary 8.1.1 to Theorem 8.1.4 since condition (8.2.3) implies (8.1.17).

Theorem 8.2.3. *Under assumptions (8.2.1) to (8.2.4), $(\nu_t, \mathscr{F}_t, P)$ is an N-dimensional Wiener martingale. Furthermore for each s,*

$$\mathscr{F}_s \perp\!\!\!\perp \sigma[\nu_v - \nu_u,\ s \le u \le v \le T]. \tag{8.2.10}$$

8.3 Stochastic Integral Representation of a Separable Martingale on $(\Omega, \mathscr{F}_t, P)$

Theorem 8.3.1. *Under conditions (8.2.1) to (8.2.4), every separable, square-integrable martingale $(Y_t, \mathscr{F}_t, P)$ is sample-continuous and has the representation*

$$Y_t - E(Y_0) = \int_0^t (\Phi_s, d\nu_s) \equiv \sum_{i=1}^N \int_0^t \Phi_s^i \, d\nu_s^i, \tag{8.3.1}$$

where

$$\int_0^T E|\Phi_s|^2 \, ds < \infty \tag{8.3.2}$$

and $\Phi_s = (\Phi_s^1, \ldots, \Phi_s^N)$ is jointly measurable and adapted to $(\mathscr{F}_s)$.

Proof. For $\Phi(t,\omega)$ any (t,ω)-measurable N vector process adapted to $(\mathscr{F}_t)$ and which satisfies the condition

$$\int_0^T E|\Phi_t|^2 \, dt < \infty,$$

let us define

$$\alpha_s^t(\Phi) = \exp\left[\int_s^t (\Phi_u, d\nu_u) - \frac{1}{2}\int_s^t |\Phi_u|^2 \, du\right].$$

Let

$$T_n = \begin{cases} \inf\left\{t : 0 \le t \le T,\ \alpha_0^t(-\hat{h}) > n \quad \text{or} \quad \int_0^t |\hat{h}_u|^2 \, du > n\right\} \\ T \quad \text{if the above set is empty.} \end{cases}$$

We have here taken $\Phi_t = -\hat{h}_t = -E(h_t|\mathscr{F}_t)$, (h_t) being the process in (8.2.1) the model of the observation process (Z_t).

For each n, let $d\tilde{P}_n = \alpha_0^{T_n}(-\hat{h})\,dP$. Then $\tilde{P}_n$ is a probability measure. Recall one useful fact before proceeding further. Write $\tilde{E}$ for expectation with respect to $\tilde{P}_n$ and α_0^t for $\alpha_0^{t\wedge T_n}(-\hat{h}) = \alpha_0^t(-\hat{h}I_{[0,T_n]})$. From Lemma 7.1.3 it then follows that a measurable process (Y_t) is an $(\mathscr{F}_t,\tilde{P}_n)$-martingale if and only if $(\alpha_0^t Y_t)$ is an $(\mathscr{F}_t,P)$-martingale.

By Girsanov's theorem the process

$$Z_t^n = v_t + \int_0^{t\wedge T_n} \hat{h}_s\,ds$$

is an N-vector standard $(\mathscr{F}_t,\tilde{P}_n)$ Wiener martingale. Set $\mathscr{F}_t^n = \sigma\{Z_s^n : s \le t\}$. It is well known that every separable, square-integrable $(\mathscr{F}_t^n,\tilde{P}_n)$-martingale $\tilde{Y}_t$ is continuous and is represented as $\tilde{Y}_t = \int_0^t (\Phi_s^n, dZ_s^n)^\dagger$. Note that $Z_s = Z_s^n$ holds for $s < T_n$ and apply Doob's optional sampling theorem. Then we see that $\tilde{Y}_{t\wedge T_n} = \int_0^{t\wedge T_n} (\Phi_s^n, dZ_s)$ holds and it is an $(\mathscr{F}_{t\wedge T_n}^n,\tilde{P}_n)$-martingale, where

$$\mathscr{F}_{t\wedge T_n}^n = \{B \in \mathscr{F}_t^n : B \cap \{T_n \le s\} \in \mathscr{F}_s^n \text{ for all } 0 \le s \le T\}. \tag{8.3.3}$$

On the other hand, it is easily seen that

$$\mathscr{F}_{t\wedge T_n}^n = \sigma\{Z_{s\wedge T_n}^n : 0 \le s \le t\} = \sigma\{Z_{s\wedge T_n} : 0 \le s \le t\} = \mathscr{F}_{t\wedge T_n},$$

where the σ-field $\mathscr{F}_{t\wedge T_n}$ is defined by (8.3.3) replacing $\mathscr{F}_t^n$ by $\mathscr{F}_t$. We have thus seen that if Y_t is a separable square-integrable $(\mathscr{F}_{t\wedge T_n},\tilde{P}_n)$-martingale, it is represented as

$$\tilde{Y}_t = \tilde{Y}_{t\wedge T_n} = \int_0^{t\wedge T_n} (\Phi_s^n, dZ_s). \tag{8.3.4}$$

Suppose now that Y_t is a separable square-integrable $(\mathscr{F}_t,P)$-martingale and let $\tilde{Y}_t = (\alpha_0^t)^{-1} Y_t$. Then $\tilde{Y}_{t\wedge T_n}$ is a separable square-integrable $(\mathscr{F}_{t\wedge T_n},\tilde{P}_n)$-martingale from Lemma 7.1.3, so that it has the representation (8.3.4). Consequently,

$$\tilde{Y}_{t\wedge T_n} = \int_0^{t\wedge T_n} (\Phi_s^n, dv_s) + \int_0^{t\wedge T_n} (\Phi_s^n, \hat{h}_s)\,ds. \tag{8.3.5}$$

Ito's formula applied to $Y_{t\wedge T_n} = \alpha_0^{t\wedge T_n} \tilde{Y}_{t\wedge T_n}$ enables us to write

$$Y_{t\wedge T_n} = \int_0^{t\wedge T_n} (\Phi_s^n, dv_s) + \int_0^{t\wedge T_n} \Psi_s^n\,ds.$$

The second term of the right-hand side vanishes (a.s.) since it is an $(\mathscr{F}_{t\wedge T_n},P)$-martingale with bounded variation. We have thus obtained

$$Y_{t\wedge T_n} = \int_0^{t\wedge T_n} (\Phi_s^n, dv_s), \tag{8.3.6}$$

where

$$\int_0^t E|I_{[0,T_n]}(s)\Phi_s^n|^2\,ds = E(Y_{t\wedge T_n}^2) \le E(Y_T^2) < \infty.$$

† We may and do assume that $\tilde{Y}_0 = 0$. The same remark applies to other martingales.

The uniqueness of the representation yields $\Phi_s^n = \Phi_s^m$ if $s < T_n$ and $n < m$, that is, there exists Φ_s, jointly measurable and $(\mathscr{F}_s)$-adapted such that $\Phi_s = \Phi_s^n$ for $s < T_n$. It follows that $\int_0^T E|\Phi_s|^2\,ds < \infty$ and the proof is complete. □

8.4 A Stochastic Equation for the General Nonlinear Filtering Problem

Assume (8.2.1) to (8.2.3) and conditions 1 to 3 of Section 8.2. Let f be a real measurable function on S [the state space of the signal process (X_t)] such that

$$E[f(X_t)]^2 < \infty \quad \text{for all } t \text{ in } [0,T]. \tag{8.4.1}$$

The function f is said to belong to the class $\mathscr{D}$ if there exists a jointly measurable, real process $B_t[f](\omega)$ adapted to $(\mathscr{G}_t)$ and having the following properties:

1. Almost all trajectories of the process $(B_t[f])$ are right-continuous and of bounded variation over the interval $[0,T]$ with $B_0[f] = 0$.
2. $E(\operatorname{Var} B[f])^2 < \infty$, where $\operatorname{Var} B[f](\omega)$ is the total variation of the trajectory $B_t[f](\omega)(0 \leq t \leq T)$.
3. The process $M_t(f) = f(X_t) - E[f(X_0)] - B_t[f]$ is a $(\mathscr{G}_t,P)$-martingale which we take to be right-continuous. Note that from (8.4.1) and conditions 8.4.1 and 2 it follows that $(M_t(f))$ is a square-integrable martingale.

The following result is an immediate consequence of Theorems 8.1.6 and 8.3.1.

Theorem 8.4.1. *Let (8.4.1) and the conditions 1 to 3 hold. Then for every f in $\mathscr{D}$ there exists a jointly measurable process $(\bar{B}_t[f])$ adapted to the family $(\mathscr{F}_t)$ such that almost all its trajectories are right-continuous and of bounded variation over the interval $[0,T]$ with $\bar{B}_0[f] = 0$. Furthermore,*

$$E(\operatorname{Var} \bar{B}[f])^2 < \infty, \tag{8.4.2}$$

and the conditional expection process [writing E^t for $E(\ |\mathscr{F}_t)$]$\{E^t f(X_t)\}$ is a square-integrable semimartingale with respect to $(\mathscr{F}_t)$ and is of the form

$$E^t[f(X_t)] = E(f(X_0)] + \int_0^t (\Phi_s, dv_s) + \bar{B}_t(f). \tag{8.4.3}$$

We are now in a position to derive the basic result of this section which describes the time evolution of the optimal nonlinear filter. The problem that now remains is to discover the particular form that Φ_s takes in (8.4.3).

First, we need the following lemmas.

Lemma 8.4.1. *Let $(M_t(f),\mathscr{G}_t,P)$ be the square-integrable martingale in condition 3. Then there exists a unique continuous process $\langle M(f),W^i\rangle$ $(i = 1,\ldots,N)$*

adapted to $\mathscr{G}_t$ such that almost all sample functions are of bounded variation and $M_t(f)W_t^i - \langle M(f),W^i\rangle_t$ are $\mathscr{G}_t$-martingales. Furthermore, each $\langle M(f),W^i\rangle_t$ has the following properties. It is absolutely continuous with respect to Lebesgue measure in $[0,T]$. There exists a modification of the Radon-Nikodym derivative which is (t,ω)-measurable and adapted to $(\mathscr{G}_t)$ and which we shall denote by $\tilde{D}_t^i f(\omega)$. Then using the vector notation $\tilde{D}_t f = (\tilde{D}_t^1 f, \ldots, \tilde{D}_t^N f)$,

$$\langle M(f),W\rangle_t = \int_0^t \tilde{D}_s f\, ds \qquad \text{(a.s.)}, \tag{8.4.4}$$

where

$$\int_0^T E|\tilde{D}_s f|^2\, ds < \infty. \tag{8.4.5}$$

If the processes (X_t) and (W_t) are independent, then (a.s.)

$$\langle M(f),W\rangle_t = 0. \tag{8.4.6}$$

Proof. The first assertion of the lemma is a consequence of Theorem 2.6.3, while (8.4.4) and (8.4.5) follow from Section 6.8 upon noting that $(M_t(f),\mathscr{G}_t)$ and $(W_t,\mathscr{G}_t)$ are both square-integrable martingales, $M_t(f)$ is right-continuous, and $\langle W^i,W^j\rangle_t = t\delta_{ij}$. To show (8.4.6), it suffices to prove the equivalent assertion that $(M_t(f)W_t,\mathscr{G}_t)$ is a martingale. For $s < t$, a direct calculation shows

$$E[(M_t(f) - M_s(f))(W_t - W_s)|\mathscr{G}_s] = E[M_t(f)W_t|\mathscr{G}_s] - M_s(f)W_s.$$

On the other hand, the assumption of the complete independence of (X_t) and (W_t) implies that $W_t - W_s$ and $\mathscr{G}_s \vee \sigma\{M_t(f) - M_s(f)\}$ are independent. Hence

$$E[(M_t(f) - M_s(f))(W_t - W_s)|\mathscr{G}_s] = E[M_t(f) - M_s(f)|\mathscr{G}_s]E[W_t - W_s] = 0$$

(a.s.), thus proving (8.4.6). □

Lemma 8.4.2. *Let $Y(\omega) = \int_0^T f(s,\omega)\, dv_s(\omega)$, where f is measurable, f_s is $\mathscr{F}_s^Z$-measurable and $EY^2 = \int_0^T Ef^2(s)\, ds < \infty$. Let (B_t) be a right-continuous stochastic process whose sample functions are of bounded variation in $[0,T]$ with* $\operatorname{Var} B(\omega)$ *denoting the total variation of $B_t(\omega)$. Write $\gamma(\omega) = \operatorname{Var} B(\omega)$.*

Let $\Pi_n = \{t_j^n\}$ be a partition of $[0,T]$.

Write

$$\Delta_j Y = \int_{t_{j-1}}^{t_j} f(s)\, dv_s \quad \text{and} \quad \eta_n(\omega) = \max_{\Pi_n} |\Delta_j Y(\omega)|.$$

Let

$$\zeta_{\Pi_n} = \sum_{\Pi_n} (\Delta_j B)(\Delta_j Y), \qquad \Delta_j B = B_{t_{j+1}} - B_{t_{j-1}}.$$

Suppose that

$$E\gamma^2 < \infty.$$

Then

$$\lim_{|\Pi_n|\to 0} E|\zeta_{\Pi_n}| = 0.$$

PROOF. Let $Y_j = \int_0^{t_j} f(s)\,dv_s$, $Y_0 = 0$ $(t_0 = 0)$, $t_n = T$, $Y_n = Y$. For $j \le n$,

$$\begin{aligned} E[Y_n|Y_1,\ldots,Y_j] &= E\left[Y_j + \int_{t_j}^T f(s)\,dv_s \middle| Y_1,\ldots,Y_j\right] \\ &= Y_j + E\left\{E\left[\int_{t_j}^T f(s)\,dv_s \middle| \mathscr{F}_{t_j}^Z\right] \middle| Y_1,\ldots,Y_j\right\} \\ &= Y_j. \end{aligned}$$

Hence from Doob's inequality (2.2.7) we get

$$E(\max|Y_j|)^2 \le 4E|Y_n|^2 = 4\int_0^T E[f^2(s)]\,ds = \kappa,$$

say. Thus

$$\eta_n = \max_j |\Delta_j Y| = \max_j |Y_j - Y_{j-1}| \le 2\max_j |Y_j|,$$

so that

$$\int_{[\eta_n \ge A]} \eta_n\,dP \le \frac{E(\eta_n^2)}{A} \le \frac{K}{A}, \qquad (K = 4\kappa). \tag{8.4.7}$$

Next, let $\varepsilon > 0$ be arbitrary. Choose $C = C_\varepsilon$ positive such that

$$\int_{[\gamma^2 > C^2]} \gamma^2\,dP < \frac{\varepsilon^2}{4K}.$$

Now

$$\begin{aligned} |\zeta_{\Pi_n}| &\le \sum_{\Pi_n} |\Delta_j B|\,|\Delta_j Y| \\ &\le \max_j |\Delta_j Y| \cdot \sum_{\Pi_n} |\Delta_j B| \\ &\le \eta_n \sup_{\Pi}\left(\sum_{\Pi} |\Delta_j B|\right) \\ &= \eta_n \operatorname{Var}(B) = \eta_n \gamma. \end{aligned}$$

$$\int_{[|\zeta_{\Pi_n}| \ge N]} |\zeta_{\Pi_n}|\,dP \le \int_{[\eta_n \ge N/C] \cap [\gamma \le C]} \eta_n \gamma\,dP + \int_{[\gamma > C]} \eta_n \gamma\,dP. \tag{8.4.8}$$

The first term on the right-hand side of (8.4.8) is dominated by

$$C\int_{[\eta_n \ge N/C]} \eta_n\,dP < \frac{\varepsilon}{2} \quad \text{for } N \ge N(\varepsilon),$$

independent of n. This is done by taking $A = N/C$ in (8.4.7) and $N(\varepsilon)$ such that $KC^2/N < \varepsilon/2$. The second term on the right-hand side of (8.4.8)

$$= \int \eta_n \gamma I_{[\gamma > C]} \, dP$$
$$\leq \left\{ E(\eta_n^2) \int_{[\gamma^2 > C^2]} \gamma^2 \, dP \right\}^{\frac{1}{2}} \leq K^{\frac{1}{2}} \left(\frac{\varepsilon^2}{4K} \right)^{\frac{1}{2}} = \frac{\varepsilon}{2}.$$

Thus there exists $N(\varepsilon)$ such that $N \geq N(\varepsilon)$ implies

$$\int_{[|\zeta_{\Pi_n}| \geq N]} |\zeta_{\Pi_n}| \, dP < \varepsilon \quad \text{for all } n,$$

that is, $\{\zeta_{\Pi_n}\}$ is P-uniformly integrable. Since by assumption $\operatorname{Var} B < \infty$ (P-a.s.) and

$$\eta_n = \max_{\Pi_n} \left| \int_{t_{j-1}}^{t_j} f(s) \, dv_s \right| \to 0 \qquad (P\text{-a.s.}) \text{ as } |\Pi_n| \to 0$$

if we choose $\int_0^t f(s) \, dv(s)$ to be sample continuous, it follows that $\zeta_{\Pi_n} \to 0$ (P-a.s.) ($|\Pi_n| \to 0$). The uniform integrability of $\{\zeta_{\Pi_n}\}$ then implies

$$E|\zeta_{\Pi_n}| \to 0. \qquad \square$$

Theorem 8.4.2. *Assume (8.2.1) to (8.2.3), conditions 1 to 3 of Section 8.2 and the conditions of Theorem 8.4.1. If $f \in \mathscr{D}$ satisfies*

$$\int_0^T E|f(X_t)h_t|^2 \, dt < \infty,$$

then $E^t[f(X_t)]$ satisfies the stochastic equation

$$E^t[f(X_t)] = E[f(X_0)] + \bar{B}_t(f)$$
$$+ \int_0^t ([E^s(f(X_s)h_s) - E^s(f(X_s))E^s(h_s) + E^s(\tilde{D}_s f)], dv_s). \quad (8.4.9)$$

Proof. The problem is to calculate the integrand in the stochastic integral in (8.4.3). Equation (8.4.9) is equivalent to saying that $\bar{M}_t(f)$ equals the stochastic integral on the right-hand side which we denote by $M_t^*(f)$. The method of proof is to show $E[\bar{M}_t(f)Y_t] = E[M_t^*(f)Y_t]$ for all Y_t such that

$$Y_t = \int_0^t (\Phi_s, dv_s) \qquad (\Phi_s \text{ bounded})$$

since such Y_t are dense in $L^2(\mathscr{F}_t, P)$ (up to constants) by virtue of Theorem 8.3.1. (Note that the Φ_s here is not to be confused with the same symbol in Theorem 8.4.1.) Since f will be fixed throughout the argument, we suppress it and write B_t, M_t, $\bar{B}_t$, $\bar{M}_t$, and M_t^*. Also for ease of writing we shall consider the scalar case (that is, $N = 1$). We calculate $E(M_t Y_t)$ and $E[(\bar{M}_t - M_t)Y_t]$ separately. The details of each calculation are given in several steps and need no additional explanation.

(I) $E(M_t Y_t) = E([f(X_t) - f(X_0) - B_t]Y_t)$

$$= E\left[M_t \int_0^t \Phi_s \, dW_s\right] + E\left[f(X_t)\int_0^t g_s \, ds\right] - E\left[B_t \int_0^t g_s \, ds\right],$$

where we set $g_s = \Phi_s(h_s - \hat{h}_s)$.

Consider the right-hand side of I:

$$\text{First term} = E\left(\int_0^t \Phi_s \, d\langle M,W\rangle_s\right)$$

$$= E\int_0^t \Phi_s \tilde{D}_s f \, ds \quad \text{(by Lemma 8.4.1)}$$

$$= E\left[Y_t \int_0^t E^s(\tilde{D}_s f)\, dv_s\right].$$

$$\text{Second term} = E\int_0^t f(X_t) g_s \, ds$$

$$= \int_0^t (E\{[f(X_t) - f(X_s)]g_s\} + E\{f(X_s)g_s\})\, ds$$

$$= \int_0^t E[f(X_s)g_s]\, ds + \int_0^t E\{(M_t - M_s + B_t - B_s)g_s\}\, ds.$$

Now

$$\begin{aligned} E\{(M_t - M_s + B_t - B_s)g_s\} &= E(g_s E\{M_t - M_s + B_t - B_s|\mathscr{G}_s\}) \\ &= E[g_s E(B_t|\mathscr{G}_s)] - E(g_s B_s) \\ &= E(B_t g_s) - E(B_s g_s). \end{aligned}$$

Hence

$$E\int_0^t f(X_t) g_s \, ds = E\int_0^t f(X_s) g_s \, ds + E\left(B_t \int_0^t g_s \, ds\right) - E\int_0^t B_s g_s \, ds.$$

and so

$$E(M_t Y_t) = E\left(Y_t\left[\int_0^t E^s(\tilde{D}_s f)\, dv_s\right]\right) + E\left(\int_0^t f(X_s) g_s \, ds\right) - E\left(\int_0^t B_s g_s \, ds\right).$$

The second term on the right equals

$$E\int_0^t (\Phi_s E^s[f(X_s)(h_s - \hat{h}_s)])\, dv_s = E\left(Y_t \int_0^t E^s[f(X_s)(h_s - \hat{h}_s)]\, dv_s\right).$$

Hence

$$E(M_t Y_t) = E(M_t^* Y_t) - E\left(\int_0^t B_s g_s \, ds\right). \tag{8.4.10}$$

(II) $$\begin{aligned} E[(\bar{M}_t - M_t)Y_t] &= E([E^t(f(X_t)) - f(X_t) + \bar{B}_t - B_t]Y_t) \\ &= E[(\bar{B}_t - B_t)Y_t]. \end{aligned} \tag{8.4.11}$$

Now let $\Pi = \{t_j\}$ be a finite partition of $[0,t]$. Actually Π stands for a sequence Π_n for which $|\Pi_n| = \max_j (t^n_{j+1} - t^n_j) \to 0$ as $n \to \infty$. Set $\Delta_j B = B_{t_{j+1}} - B_{t_j}$, and similarly define $\Delta_j \bar{B}$, etc. We have

$$E\{(B_t - \bar{B}_t)Y_t\} = E\left\{\sum_\Pi (\Delta_j B - \Delta_j \bar{B})Y_t\right\}$$
$$= \sum_\Pi E\{(\Delta_j B - \Delta_j \bar{B})Y_{t_j}\} + \sum_\Pi E\{(\Delta_j B - \Delta_j \bar{B})(Y_t - Y_{t_j})\}.$$

The first term $= \sum_\Pi E\{Y_{t_j} E[(\Delta_j B - \Delta_j \bar{B})|\mathscr{F}_{t_j}]\} = 0.$

$$Y_t - Y_{t_j} = \int_{t_j}^t (\Phi_s, dW_s) + \int_{t_j}^t g_s\, ds.$$

Thus

$$E\{(B_t - \bar{B}_t)Y_t\} = \sum_\Pi E\left\{(\Delta_j B - \Delta_j \bar{B}) \int_{t_j}^{t_{j+1}} \Phi_s\, dW_s\right\}$$
$$+ \sum_\Pi E\left\{(\Delta_j B - \Delta_j \bar{B}) \int_{t_{j+1}}^t \Phi_s\, dW_s\right\}$$
$$+ \sum_\Pi E\left[(\Delta_j B - \Delta_j \bar{B}) \int_{t_j}^t g_s\, ds\right]. \tag{8.4.12}$$

Consider the second term on the right-hand side of (8.4.12):

$$E\left[(\Delta_j B - \Delta_j \bar{B}) \int_{t_{j+1}}^t \Phi_s\, dW_s\right] = E\left[(\Delta_j B - \Delta_j \bar{B}) E\left(\int_{t_{j+1}}^t \Phi_s\, dW_s \middle| \mathscr{G}_{t_{j+1}}\right)\right] = 0$$

since $\int_0^t (\Phi_s, dW_s)$ is an $L^2 - \mathscr{G}_t$ martingale.
In the third term,

$$E(\Delta_j B - \Delta_j \bar{B}) \int_{t_j}^t g_s\, ds = E\left[(\Delta_j B - \Delta_j \bar{B}) \int_{t_j}^{t_{j+1}} g_s\, ds\right]$$
$$+ E\left[\Delta_j B \int_{t_{j+1}}^t g_s\, ds\right] - E\left[\Delta_j \bar{B} \int_{t_{j+1}}^t g_s\, ds\right]$$

and

$$E\left[\Delta_j \bar{B} \int_{t_{j+1}}^t g_s\, ds\right] = E \int_{t_{j+1}}^t E[\Delta_j \bar{B}(\Phi_s[h_s - \hat{h}_s])|\mathscr{F}_s]\, ds$$
$$= E \int_{t_{j+1}}^t \Delta_j \bar{B}(\Phi_s E^s(h_s - \hat{h}_s))\, ds = 0.$$

Hence

$$E[(\bar{M}_t - M_t)Y_t] = E\left(\sum_\Pi (\Delta_j B - \Delta_j \bar{B})\, \Delta_j Y\right) + E\left\{\sum_\Pi (\Delta_j B) \int_{t_{j+1}}^t g_s\, ds\right\}.$$

The conditions $E(\operatorname{Var} B)^2 < \infty$, $E(\operatorname{Var} \bar{B})^2 < \infty$ (see Theorem 8.4.1) imply that $E\left|\sum_{\Pi_n}(\Delta_j B)(\Delta_j Y)\right| \to 0$ and $E\left|\sum_{\Pi_n}(\Delta_j \bar{B})(\Delta_j Y)\right| \to 0$. Next we show that

$$E\left\{\sum_{\Pi_n}(\Delta_j B)\int_{t_{j+1}}^t g_s\,ds\right\} \to E\left(\int_0^t B_s g_s\,ds\right).$$

For this write

$$D = \sum_{\Pi} \Delta_j B \int_{t_{j+1}}^t g_s\,ds - \int_0^t B_s g_s\,ds,$$

and let $a_k = \int_{t_k}^{t_{k+1}} g_s\,ds$. Then

$$\begin{aligned}
D &= \sum_{\Pi}(\Delta_j B)\{a_{j+1} + \cdots + a_{n-1}\} \\
&= B_{t_1}(a_1 + \cdots + a_{n-1}) + [B_{t_2} - B_{t_1}](a_2 + \cdots + a_{n-1}) \\
&\quad + (B_{t_3} - B_{t_2})(a_3 + \cdots + a_{n-1}) + \cdots + (B_{t_{n-2}} - B_{t_{n-3}})(a_{n-2} + a_{n-1}) \\
&\quad + (B_{t_{n-1}} - B_{t_{n-2}})a_{n-1} \\
&= B_{t_1}a_1 + B_{t_2}a_2 + B_{t_3}a_3 + \cdots + B_{t_{n-2}}a_{n-2} + B_{t_{n-1}}a_{n-1}.
\end{aligned}$$

So

$$\begin{aligned}
D &= \sum B_{t_j}\int_{t_j}^{t_{j+1}} g_s\,ds - \sum \int_{t_j}^{t_{j+1}} B_s g_s\,ds \\
&= \sum \int_{t_j}^{t_{j+1}} (B_{t_j} - B_s) g_s\,ds
\end{aligned}$$

Since for $s \in [t_j, t_{j+1}]$,

$$\begin{aligned}
|B_{t_j} - B_s| &= |U_{t_j} - U_s + V_s - V_{t_j}| \le U_{t_{j+1}} - U_{t_j} + V_{t_{j+1}} - V_{t_j}, \\
|D| &\le \left[\max_{\Pi_n} \int_{\Delta_j} |g_s|\,ds\right] \sum_{\Pi_n} \Delta_j |B| \\
&\le \max_{\Pi_n} \int_{\Delta_j} |g_s|\,ds(\operatorname{Var} B) \to 0 \qquad \text{(a.s.) as } |\Pi_n| \to 0.
\end{aligned}$$

Also $|D| \le \int_0^T |g_s|\,ds(\operatorname{Var} B)$ and

$$E\left[\int_0^T |g_s|\,ds(\operatorname{Var} B)\right] \le \left\{E\int_0^T |g_s|^2\,ds\,TE(\operatorname{Var} B)^2\right\}^{\frac{1}{2}} < \infty.$$

Hence

$$E(D_{\Pi_n}) \to 0 \quad \text{as } |\Pi_n| \to 0$$

(writing D_{Π_n} for D) and so

$$E\left[\sum_{\Pi_n}(\Delta_j B)\int_{t_{j+1}}^t g_s\,ds\right] \to E\int_0^t B_s g_s\,ds.$$

Thus finally we have

$$E[(\bar{M}_t - M_t)Y_t] = E\left[\int_0^t B_s g_s \, ds\right].$$

Comparing this with (8.4.10), we at once get

$$E(\bar{M}_t Y_t) = E(M_t^* Y_t)$$

concluding the proof of the theorem. □

An important special case of the above theorem, suggested by applications in which signal and observation processes are governed by stochastic differential equations, is the following: Assume that the process $B_t[f](\omega)$ is given by $\int_0^t (\tilde{A}_s f)(\omega)\, ds$, where $(\tilde{A}_t f)(\omega)$ is real-valued, jointly measurable, $\mathscr{G}_t$-adapted, and such that

$$\int_0^T E|\tilde{A}_t f|^2 \, dt < \infty.$$

In the present context we write $\mathscr{D}(\tilde{A})$ in place of $\mathscr{D}$.

Theorem 8.4.3. *Assume the conditions of Theorem 8.4.1. If $f \in \mathscr{D}(\tilde{A})$ satisfies*

$$\int_0^T E|f(X_t)h_t|^2 < \infty,$$

then $E^t f(X_t)$ satisfies the following stochastic differential equation

$$E^t[f(X_t)] = E[f(X_0)] + \int_0^t E^s[\tilde{A}_s f]\, ds$$
$$+ \int_0^t (E^s(f(X_s)h_s) - E^s f(X_s)E^s(h_s) + E^s(\tilde{D}_s f), dv_s). \quad (8.4.13)$$

Although Theorem 8.4.2 was established for a real-valued observation process (Z_t), the extension to vector-valued processes introduces no new difficulties. However, an independent, somewhat simpler proof of Theorem 8.4.3 is given below.

PROOF OF THEOREM 8.4.3. The idea of the proof is the same as that of Theorem 8.4.1. Equation (8.4.13) is equivalent to stating that $\bar{M}_t(f)$ equals $M_t^*(f)$, and as before we calculate $E[\bar{M}_t(f)Y_t]$ and $E[M_t^*(f)Y_t]$ and show they are equal for all Y_t such that

$$Y_t = \int_0^t (\Phi_s, dv_s) \qquad (\Phi_s \text{ bounded}), \quad (8.4.14)$$

since such Y_t are dense in $L^2(\mathscr{F}_t, P)$ (up to constants). We show this again by calculating $E[(M_t(f) - \bar{M}_t(f))Y_t]$ and $E[M_t(f)Y_t]$ separately. In what follows we again write $M_t, \bar{M}_t$, and M_t^*, suppressing f as it is fixed throughout the argument. A simple calculation yields

$$E[(\bar{M}_t - M_t)Y_t] = E\left[\int_0^t (Y_t - Y_s)\tilde{A}_s f\, ds\right]. \tag{8.4.15}$$

From (8.4.14), writing $Y_t = \int_0^t (\Phi_s, dW_s) + \int_0^t (\Phi_s, h_s - \hat{h}_s)\, ds$, the right-hand side of (8.4.15) is reduced to the form

$$\int_0^t E\left[\tilde{A}_s f \int_s^t (\Phi_u, dW_u)\right] ds + E\left[\int_0^t \tilde{A}_s f \left(\int_s^t (\Phi_u, h_u - \hat{h}_u)\, du\right) ds\right]. \tag{8.4.16}$$

The integrand in the first term of (8.4.16) is zero because $\tilde{A}_s f$ is $\mathscr{G}_s$-measurable and $E(\int_s^t (\Phi_u, dW_u) | \mathscr{G}_s) = 0$. The latter fact follows since $\int_0^t (\Phi_s, dW_s)$ is a $\mathscr{G}_t$-martingale. The quantity inside the brackets in the second term of (8.4.16) becomes (after an integration by parts) $\int_0^t [\int_0^s \tilde{A}_u f\, du](\Phi_s, h_s - \hat{h}_s)\, ds$. Hence the right-hand side of (8.4.15) equals

$$E\left[\int_0^t \left[\int_0^s A_u f\, du\right](\Phi_s, h_s - \hat{h}_s)\, ds\right]. \tag{8.4.17}$$

On the other hand, it is easy to verify that

$$\begin{aligned} E(M_t Y_t) = {} & E\left[M_t \int_0^t (\Phi_s, dW_s)\right] + E\left[\int_0^t f(X_s)(\Phi_s, h_s - \hat{h}_s)\, ds\right] \\ & - E\left[\int_0^t \left(\int_0^s \tilde{A}_u f\, du\right)(\Phi_s, h_s - \hat{h}_s)\, ds\right]. \end{aligned} \tag{8.4.18}$$

Consider the right-hand side of (8.4.18). From Lemma 8.4.1 and the properties of stochastic integrals, the first term is equal to

$$\begin{aligned} E\left[\int_0^t (\Phi_s, \tilde{D}_s f)\, ds\right] &= E\left[\int_0^t (\Phi_s, E(\tilde{D}_s f | \mathscr{F}_s))\, ds\right] \\ &= E\left[Y_t \int_0^t (E^s(\tilde{D}_s f), dv_s)\right]. \end{aligned} \tag{8.4.19}$$

The second term equals

$$E\left[\int_0^t (\Phi_s, E^s[f(X_s)(h_s - \hat{h}_s)])\, ds\right] = E\left[Y_t \int_0^t (E^s[f(X_s)(h_s - \hat{h}_s)], dv_s)\right]. \tag{8.4.20}$$

From (8.4.15) and (8.4.17) to (8.4.20) we immediately obtain $E[\bar{M}_t Y_t] = E[M_t^* Y_t]$. The proof is complete. □

Let us now consider the observation process (Y_t) introduced in Section 8.2. In order to derive a stochastic differential equation for the optimal filter in such problems, we first need a result which permits us to work with the σ-fields $\mathscr{F}_t^{Z'}$ [generated by (Z_t') of Theorem 8.2.2] without losing information. Write $\mathscr{F}_t' = \mathscr{F}_t^{Z'}$ for brevity.

Lemma 8.4.3. *$\mathscr{F}_t = \mathscr{F}'_t$ for all $0 \le t \le T$.*

PROOF. Since Y_t and Z'_t are related by

$$Y_t = \eta + \int_0^t c_s^{\frac{1}{2}}\, dZ'_s \quad \text{or} \quad Z'_t = \int_0^t c_s^{-\frac{1}{2}}\, dY_s \tag{8.4.21}$$

and $c_s^{-\frac{1}{2}}$ is $\mathscr{F}_s$-measurable, it is clear that $\mathscr{F}'_t \subseteq \mathscr{F}_t$. For the proof of the converse relation, we proceed as in Theorem 8.4.1. Set $\beta_t = v'_t$ and $\Phi_t = -\hat{h}'_t$. Then $Z'^n_t = v'_t + \int_0^{t\wedge T_n} \hat{h}'_s\, ds$ is an $(\mathscr{F}'_t,\tilde{P}_n)$-standard Wiener process. The solution Y^n_t of the stochastic differential equation

$$Y^n_t = \eta + \int_0^t c^{\frac{1}{2}}(s,\pi_s Y^n)\, dZ'^n_s$$

considered in $(\mathscr{F}'_t,\tilde{P}_n)$ exists uniquely and $\sigma\{Y^n_s; s \le t\} \subseteq \sigma\{Z'^n_s; s \le t\}$ by Theorem 8.2.1, since $c^{\frac{1}{2}}(t,f)$ is Lipschitz continuous in the sense of (8.4.7) and (8.4.8) at least locally. On the other hand, since $Z'^n_t = Z'_t$ holds for $t < T_n$, $Y_t = Y^n_t$ holds for $t < T_n$ by the uniqueness of the stochastic equation. This implies $\mathscr{F}'_{t\wedge T_n} \supseteq \mathscr{F}_{t\wedge T_n}$. Since $T_n \uparrow T$ as $n \to \infty$, we get $\mathscr{F}'_t \supseteq \mathscr{F}_t$. □

We have thus reduced the filtering problem for the observation process Y_t to that for the new observation process Z'_t. Theorem 8.4.3 is then modified as follows.

First let us define

$$v'_t = Z'_t - \int_0^t \hat{h}'_s\, ds,$$

where $\hat{h}'_s = E(h'_s|\mathscr{F}'_s)$. Then $(v'_t,\mathscr{F}'_t,P)$ is a Wiener martingale.

Theorem 8.4.4. *Assume (8.2.1), (8.2.7) to (8.2.9), and (A.4) to (A.6) of Section 8.2. If f belongs to $\mathscr{D}(\tilde{A})$ and satisfies $\int_0^T E|f(X_t)h'_t|^2\, dt < \infty$, then $E^t[f(X_t)] = E[f(X_t)|\mathscr{F}_t]$ satisfies the following stochastic differential equation:*

$$\begin{aligned} E^t[f(X_t)] &= E[f(X_0)] + \int_0^t E^s[\tilde{A}_s f]\, ds \\ &\quad + \int_0^t (c_s^{-\frac{1}{2}}[E^s(f(X_s)a_s) - E^s(f(X_s))E^s(a_s) \\ &\quad + b_s E^s(\tilde{D}_s f)], dv'_s), \end{aligned} \tag{8.4.22}$$

where v'_t is the standard Wiener process defined above.

PROOF. The only difference between (8.4.13) and (8.4.22) is in the term corresponding to $\tilde{D}_s f$. In Equation (8.4.22) the M column vector $\tilde{D}_s f = (\tilde{D}^1_s f, \ldots, \tilde{D}^M_s f)$ is defined as the Radon-Nikodym derivative of $\langle M(f),W'\rangle_t$ with respect to t, which is related to that of $\langle M(f),W\rangle_t$ in the following form

$$\langle M(f),W\rangle_t = \int_0^t c_s^{-\frac{1}{2}} b_s d\langle M(f),W'\rangle_s.$$

Therefore $\tilde{D}_s f$ in (8.4.13) corresponds to $c_s^{-\frac{1}{2}} b_s \tilde{D}$ in (8.4.22). □

8.5 Applications

Let us consider the case when the system and observation processes are solutions of a stochastic differential equation of the type considered by Fleming and Nisio [12].

$$dX_t = A(t,\pi_t X,\pi_t Y)\,dt + B(t,\pi_t X,\pi_t Y)\,dW_t, \tag{8.5.1}$$

$$dY_t = a(t,\pi_t X,\pi_t Y)\,dt + b(t,\pi_t Y)\,dW_t. \tag{8.5.2}$$

Here W_t is an M-vector standard Wiener process, a and A are N- and $(M-N)$-vector functionals $(M > N)$, respectively, and b and B are $N \times M$ and $(M-N) \times M$-matrix functionals, respectively. We assume similar Lipschitz conditions as (8.2.7) and (8.2.8) for both of (a,b) and (A,B). For the initial random variables, we assume that $Y_0 = 0$ and that

$$X_0 \text{ is independent of } \mathscr{F}_T^W. \tag{8.5.3}$$

Then the system of equations (8.5.1) and (8.5.2) has an unique solution (X_t, Y_t) which is measurable with respect to $\mathscr{F}_t^{X_0,W}$. If additional conditions (A.5) and (A.6) are imposed on a and b, Theorem 8.4.4 can be applied to this case. We shall obtain explicit representations of $\tilde{A}_t$ and $\tilde{D}$. The results are that if f is a C^2-class function on R^{M-N}, it belongs to $\mathscr{D}(\tilde{A})$ and

$$\tilde{A}_t f(\omega) = \sum_{i=1}^{M-N} A^i(t,\pi_t X,\pi_t Y) f_{x^i}(X_t) + \frac{1}{2}\sum_{i,j=1}^{M-N} (BB^*)^{ij}(t,\pi_t X,\pi_t Y) f_{x^i x^j}(X_t) \tag{8.5.4}$$

and

$$\tilde{D}_t^i f(\omega) = \sum_{j=1}^{M-N} B^{ji}(t,\pi_t X,\pi_t Y) f_{x^j}(X_t). \tag{8.5.5}$$

To prove this, apply Ito's formula to $f(X_t)$. We get

$$f(X_t) - f(X_0) = \sum_{i,j} \int_0^t B^{ij}(s,\pi_s X,\pi_s Y) f_{x^i}(X_s) dW_s^j + \int_0^t \tilde{A}_s f(\omega)\,ds,$$

where the integrand in the last term is the right-hand side of (8.5.4). Since the first term on the right-hand side is a $\mathscr{G}_t$-martingale, we see that

$$M_t(f) = f(X_t) - f(X_0) - \int_0^t \tilde{A}_s f\,ds$$

is a $\mathscr{G}_t$-martingale. This proves (8.5.4). The proof of (8.5.5) is immediate from

$$\begin{aligned}\langle M(f),W^i\rangle_t &= \sum_{k,j} \int_0^t B^{kj}(x,\pi_s X,\pi_s Y) f_{x^k}(X_s) d\langle W^j,W^i\rangle_s \\ &= \sum_k \int_0^t B^{ki}(s,\pi_s X,\pi_s Y) f_{x^k}(X_s)\,ds\end{aligned}$$

Using the vector notation, $c_s^{-\frac{1}{2}}b_s\tilde{D}_s f = c_s^{-\frac{1}{2}}bB^*f'$, where B^* is the transpose of B and $f' = (f'_{x^1}, \ldots, f'_{x^N})$. Therefore, the term involving $\tilde{D}_s f$ disappears if and only if $bB^* \equiv 0$.

Remark 8.5.1. There are many other ways of choosing $N \times N$ matrices $c^{\frac{1}{2}}$ and $c^{-\frac{1}{2}}$ in the discussion of this section. In fact, in the case where the dimension M is equal to N, it is more natural to replace $c^{\frac{1}{2}}$ and $c^{-\frac{1}{2}}$ by b and b^{-1}, respectively. More generally, if we choose an $N \times N$ matrix d with the Lipschitz condition (8.2.8) such that $c = dd^*$, then all the steps are valid replacing $c^{\frac{1}{2}}$ and $c^{-\frac{1}{2}}$ by d and d^{-1}, respectively. It should be noted that condition (A.6) does not depend on the choice of such d. Although the innovation process (v_t') is changed by such a replacement, the expression (8.4.22) does not depend on the replacement. In fact, the last member of (8.4.22) is equal to

$$\int_0^t (E^s[f(X_s)a_s] - E^s[f(X_s)]E^s[a_s] + b_s E^s[\tilde{D}_s f], c_s^{-1}\, dY_s - c_s^{-1}E^s[a_s]\, ds). \tag{8.5.6}$$

8.6 The Case of Markov Processes

The material of this section is taken from [14]. We consider the special case where the pair (X_t,Y_t) of the signal process X_t and the observation process Y_t is Markov with respect to $(\mathscr{G}_t,P)$. In this context the stochastic differential equation has a more definite meaning. We show in this section that $\tilde{A}_t f(\omega)$ is replaced by $A_t f(X_t,Y_t)$, where A_t is the generator of the process (X_t,Y_t), and that with additional conditions $\tilde{D}^i f(t,\omega)$ is replaced by $D^i f(t,X_t,Y_t)$, where $D^i f(t,X,Y)$ is a measurable function in $[0,T] \times S \times R^N$. D^i may be regarded as a first-order linear differential operator if $\eta_t = (X_t,Y_t)$ is a diffusion Markov process.

Let us first investigate conditions for $\eta_t = (X_t,Y_t)$ to be Markov.

Lemma 8.6.1. *Suppose that (X_t) is a Markov process which is completely independent of (W_t). Assume further that coefficients $a(t,g,f)$ and $b(t,g,f)$ of (8.2.7) depend only on the values $f(0)$ and $g(0)$, that (A.4) of Section 82 holds and that the measure Γ is concentrated at the point $\{0\}$. Then $((X_t,Y_t),\mathscr{G}_t,P)$ is a Markov process.*

Since the proof uses arguments which are standard in the theory of Markov processes, we shall state only the outline. First observe that the latter condition of the lemma states that $a(t,\pi_t X,\pi_t Y) = a(t,X_t,Y_t)$ and $b(t,\pi_t Y) = b(t,Y_t)$. Hence equation (8.2.9) becomes

$$Y_t = Y_s + \int_s^t a(u,X_u,Y_u)\, du + \int_s^t b(u,Y_u)\, dW_u. \tag{8.6.1}$$

Now let $\{Q_\omega\}$ be the regular conditional distribution relative to $(\mathscr{G}_s,P)$, that is, $Q_\omega(A)$, $A \in \mathscr{A}$ is $\mathscr{G}_s$-measurable for each A, a probability measure for each ω and that $Q_\omega(A) = P(A|\mathscr{G}_s)$ (a.s.). Since $\mathscr{F}_T^X$ and $\sigma\{W_v - W_u : s \leq u < v \leq T\}$ are independent relative to Q_ω for a.e. ω, $(Y_t)(t > s)$ may be considered as an observation process related to the signal process (X_t,Q_ω), $(t > s)$, by the formula (8.6.1). Then the uniqueness of the solution of the above stochastic differential equation and the Markov property of X_t proves that the joint distribution of (X_t,Y_t), $t > s$, relative to the measure Q_ω depends only on the initial value (X_s,Y_s) [together with the transition probability function of X_t and coefficients $a(s,X,Y)$ and $b(s,Y)$]. This shows the Markov property of $((X_t,Y_t),\mathscr{G}_t,P)$. □

Remark 8.6.1. The conditions of Lemma 8.6.1 are not necessary for the Markov property of (X_t,Y_t). For instance, in the case of the example considered in Section 8.5, if we assume that all coefficients a, b, A, B depend only on t and the values $f(0)$ and $g(0)$, then the process (X_t,Y_t) is Markov as is well known. However the process X_t is generally not Markov.

Let $P(s,\eta;t,B)$ $(\eta = (x, y))$ be the transition probability function of (η_t) assumed jointly measurable in (s,η,t) and let

$$P_s^t f(\eta) = \int P(s,\eta;t,d\eta')f(\eta').$$

A family of linear operators A_t, $t \in [0,T]$ defined in the space of real valued measurable functions on $S \times R^N$ is called an *extended generator* if

$$P_s^t f(\eta) - f(\eta) = \int_s^t P_s^u A_u f(\eta)\,du \tag{8.6.2}$$

is satisfied for all $0 \leq s < t \leq T$. (See Chapter 5.) We denote by $\mathscr{D}(A)$ the set of all f depending only on the first variable x and satisfying (8.6.2) together with

$$E|f(\eta_t)|^2 < +\infty \quad \text{for each } t \text{ in } [0,T], \tag{8.6.3}$$

$$\int_0^T E|A_t f(\eta_t)|^2\,dt < \infty. \tag{8.6.4}$$

Set

$$M_t(f) = f(X_t) - f(X_0) - \int_0^t A_s f(\eta_s)\,ds. \tag{8.6.5}$$

Lemma 8.6.2. *$M_t(f)$ defined by (8.6.5) is a square-integrable $(\mathscr{G}_t,P)$-martingale.*

PROOF. Since $f \in \mathscr{D}(A)$, we write $f(X_t)$ for $f(\eta_t)$. Note, however, that $A_u f(\eta)$ need not involve only x. From this observation, the fact that $(\eta_t,\mathscr{G}_t,P)$ is Markov, and (8.6.2) to (8.6.4), we see that Proposition 5.5.1 applies, and the lemma is proved.

Lemma 8.6.2 shows that $\mathscr{D}(A) \subset \mathscr{D}(\tilde{A})$ and that $A_t f(\eta_t) = \tilde{A}_t f(\omega)$, where $\tilde{A}_t$ is the operator defined earlier. In order to derive the property of the operators D^i, it is necessary to quote rather deep results concerning additive functionals in the theory of Markov processes. The following terminology and results are due to Motoo-Watanabe and Meyer (See [14]). It is well known that the space-time process (t,X_t,Y_t) is stationary Markov. We assume that

$$(t,X_t,Y_t) \text{ is a Hunt process with Meyer's hypothesis (L)} \tag{8.6.6}$$

and

$$M_t(f) \text{ is } \mathscr{F}_t^{X,W}\text{-measurable.} \tag{8.6.7}$$

By the above two conditions, each ith component W_t^i of W_t together with $M_t(f)$ is an additive functional of the process η_t. Then the process $\langle M(f),W^i\rangle_t$ introduced in Lemma 8.4.1 is again an additive functional that is absolutely continuous with respect to t for a.a. ω. Then there exists a jointly measurable function $D^if(t,x,y)$ such that

$$\langle M(f),W^i\rangle_t = \int_0^t D^if(s,X_s,Y_s)\,ds \qquad \text{(a.s.).} \tag{8.6.8}$$

[Such $D^if(s,X,Y)$ is determined uniquely (a.s.) relative to a suitable measure called *canonical*.]

Theorem 8.4.4 yields the following result for the Markov process case.

Theorem 8.6.1. *Let $\eta_t = (X_t,Y_t)$, where Y_t is given by (8.6.1) be a Markov process. Assume condition (8.4.1) and let the coefficients a and b satisfy (8.2.7) to (8.2.9) and (A.5) and (A.6) of Section 8.2. If f belongs to $\mathscr{D}(A)$ and $\int_0^T E|f(X_t)h_t'|^2\,dt < \infty$, then $E^tf(X_t)$ satisfies the stochastic differential equation (8.4.22), where $\tilde{A}_t$ and $\tilde{D}_t$ are replaced by A_t the generator of (X_t,Y_t) and by the operator D_t whose components D_t^i are defined by 8.6.8.*

If, as in Lemma 8.6.1, (X_t) and (W_t) are completely independent, it follows from (8.4.6) of Lemma 8.4.1 that the term involving D_t in the stochastic differential equation of Theorem 8.6.1 disappears leading us to the case treated often in the literature. (See [32].)

Example 8.6.1. If the coefficients a, b, A, and B depend only on $(t,f(0),g(0))$ in Fleming-Nisio's case (see Section 8.5), the operators A_t and D are given by

$$A_tf(x,y) = \sum_{i=1}^{M-N} A_i(t,x,y)f_{x^i}(x) + \frac{1}{2}\sum_{i,j=1}^{M-N} (B^*B)_{ij}(t,x,y)f_{x^ix^j}(x)$$

and

$$D^if(t,x,y) = \sum_{j=1}^{M-N} B_{ji}(t,x,y)f_{x^j}(x),$$

where f is a C^2-class function in $\mathbf{R}^{M-N}$.

It is possible and sometimes convenient to regard the pair of processes (X_t) and (Y_t) as a Markov process under a more general setting by enlarging the state space. Let $X^t = (X_s, s \le t)$ and $Y^t = (Y_s, s \le t)$. More precisely, set $\eta_t = (\pi_tX,\pi_tY)$, where π_t has been introduced in Section 8.2. Then we can prove that $(\eta_t,\mathscr{G}_t,P)$ is a Markov process by a similar argument as in Lemma

8.6.1, making use of the uniqueness of the solution of (8.2.9). The state space of the Markov process (η_t) is the function space $D \times C$ (see Section 8.2). Letting (A_t), $t \in [0,T]$, defined on the space of real-valued measurable functions on $D \times C$ be the extended generator family of the transition operators (P_s^t), if $\mathscr{D}(A)$ is defined as before, we have that for $f \in \mathscr{D}(A)$, $(M_t(f),\mathscr{G}_t,P)$ is a martingale, where

$$M_t(f) = f(\eta_t) - f(\eta_0) - \int_0^t (A_s f)(\eta_s)\,ds.$$

Remark 8.6.2. The case when $((X_t,Y^t),\mathscr{G}_t,P)$ is Markov is discussed in problems of stochastic control based on a partially observable process. Such a case occurs if the coefficients a, b, A, and B in (8.5.1)–(8.5.2) depend on t, f, and $g(0)$. It is obvious that the discussion of this section can be applied to these cases.

Gaussian Solutions of Stochastic Equations 9

Gaussian processes play an important role in the theory of linear filtering to be discussed in the next chapter. In the general stochastic filtering model it has been seen that the observation process and the innovation (Wiener) process are connected by an equation of the kind studied in Chapter 8. When the observation process is Gaussian, we have an example of the equation which will now be considered. The theory of stochastic equations whose solutions are Gaussian processes is an instructive special case of the general theory of functional stochastic differential equations because it is subsumed in the theory of nonanticipative representations of equivalent Gaussian measures and is identical with the latter if one of the measures is Wiener measure. We shall therefore present it in more detail than is strictly necessary for the purpose of solving linear filtering problems.

We begin by establishing the existence of a nonanticipative representation. The techniques employed are essentially operator-theoretic and are based on the well-known factorization theorem due to Gohberg and Krein [17]. Once such a representation has been explicitly derived in the case of a Gaussian measure equivalent to Wiener measure we apply it to obtain the main results.

9.1 The Gohberg-Krein Factorization Theorem

It will take us too far afield to give a proof of this theorem, but we introduce the notation and ideas which lead up to the form of the theorem useful to us.

Let H be a separable Hilbert space. A family of orthoprojectors $\pi = \{P\}$ is called a *chain* if for any distinct $P_1, P_2 \in \pi$, either $P_1 < P_2$ or $P_2 < P_1$, where $P_1 < P_2$ means $P_1 H \subset P_2 H$, that is, $P_1 P_2 = P_2 P_1 = P_1$. We shall write $P_1 \leq P_2$ if either $P_1 < P_2$ or $P_1 = P_2$. A chain π is said to be *bordered* if $0, I \in \pi$. The *closure* of a chain π is the set of all operators which are the strong limits of sequences in π. The closure of a chain is again a chain and if a chain coincides with its closure, it is said to be *closed*. A pair (P^-, P^+) of orthoprojectors in a closed chain π with $P^- < P^+$ is called a *gap* of π if for any $P \in \pi$ either $P \leq P^-$ or $P \geq P^+$, and the dimension of $P^+ - P^-$, that is, $\dim[P^+ H \ominus P^- H]$, is called the *dimension* of the gap (P^-, P^+). A chain is said to be *maximal* if it cannot be enlarged, or, equivalently, if it is bordered, closed, and its gaps (if any) are one-dimensional. A chain π is called an *eigenchain* of a bounded linear operator A on H if $PAP = AP$ for all $P \in \pi$.

Let π be a closed chain. A partition ζ of π is a chain consisting of a finite number of elements $\{P_0 < P_1 < \cdots < P_n\}$ of π such that $P_0 = \min_{P \in \pi} P$ and $P_n = \max_{P \in \pi} P$. Let $F(P)$ be an operator function defined on π and having as its values bounded linear operators on H. For a partition $\zeta = \{P_0 < P_1 < \cdots < P_n\}$ of π, define

$$S(\zeta) = \sum_{j=1}^{n} F(P_{j-1}) \Delta P_j, \qquad \Delta P_j = P_j - P_{j-1}.$$

An operator A is called the *limit in norm* of $S(\zeta)$, denoted by

$$A = (m) \int_{\pi} F(P)\, dP, \tag{9.1.1}$$

if for any $\varepsilon > 0$ there exists a partition $\zeta(\varepsilon)$ of π such that, for every partition $\zeta \supset \zeta(\varepsilon)$, $\|S(\zeta) - A\| < \varepsilon$. If the limit of $S(\zeta)$ exists, we shall say that the integral (9.1.1) converges. The integral

$$B = (m) \int_{\pi} dP\, F(P)$$

is defined analogously.

The *dual* $\pi^\perp$ of a chain π is a chain consisting of all orthoprojectors of the form $P^\perp = I - P$, $P \in \pi$. If π is an eigenchain of an operator A, then the dual chain $\pi^\perp$ is an eigenchain of the adjoint operator A^*.

By a *special factorization* of an operator A *along a chain* π we mean the representation of A in the form

$$A = (I + X_+) D (I + X_-), \tag{9.1.2}$$

where X_+ and X_- are Volterra operators (that is, completely continuous operators with the one-point spectrum $\lambda = 0$) having π and $\pi^\perp$ as eigenchains respectively, D commutes with all $P \in \pi$, and $D - I$ is completely continuous.

The factors $I + X_+$ and $I + X_-$ are invertible, and if A is invertible, so is the factor D. If an invertible operator A admits a special factorization relative to a maximal chain π, then the factorization is unique, and from the uniqueness it follows that if a self-adjoint invertible operator $A = A^*$ has such a factorization, then $X_+^* = X_-$ and $D^* = D$.

The following theorem is a special case of Theorems 6.1 and 6.2, Chapter IV, Gohberg-Krein [17]. We denote by $\mathscr{S}_2$ the class of all Hilbert-Schmidt operators on H.

Theorem 9.1.1. *Let π be a maximal chain. Then, for every operator $T \in \mathscr{S}_2$ such that each of the operators $I - PTP$, $P \in \pi$, is invertible, the integrals*

$$X_+ = (m)\int_\pi (I - PTP)^{-1}PT\,dP$$

and

$$X_- = (m)\int_\pi dP\,TP(I - PTP)^{-1} \tag{9.1.3}$$

converge in norm, and the operator $A = (I - T)^{-1}$ has a special factorization (9.1.2) along π with $X_+, X_-, D - I \in \mathscr{S}_2$ and

$$D = I + \sum_j (P_j^+ - P_j^-)[(I - P_j^+TP_j^+)^{-1} - I](P_j^+ - P_j^-), \tag{9.1.4}$$

where $\{(P_j^-,P_j^+)\}$ is the set of all gaps in the chain π.

For the convenience of the reader it is perhaps worth pointing out that the deduction of this result from the above-mentioned theorems of Gohberg and Krein is based on the fact that if $T \in \mathscr{S}_2$, then the integral $(m)\int_\pi PT\,dP$ converges in uniform norm (in fact, in Hilbert-Schmidt norm) and belongs to $\mathscr{S}_2$. The verification is simple and is a part of the proof of Theorem 10.1, Chapter I of [17].

Lemma 9.1.1. *Let $T \in \mathscr{S}_2$. If $I - T$ is self-adjoint, positive, and invertible, then for any orthoprojector P, $I - PTP$ is invertible.*

Proof. The proof is immediate as can be seen from the following inequality.

$$\begin{aligned}\langle (I - PTP)f,f\rangle &= \langle (I - P)f,f\rangle + \langle (I - T)Pf,Pf\rangle \\ &= \|(I - P)f\|^2 + \|(I - T)^{1/2}Pf\|^2 \\ &\geq \|(I - P)f\|^2 + c^2\|Pf\|^2 \geq c_1^2\|f\|^2,\end{aligned}$$

where c is some positive constant, $\langle\cdot,\cdot\rangle$ and $\|\cdot\|$ are the inner product and the norm of H, and $c_1^2 = \min(1,c^2)$. □

Lemma 9.1.2. *If a self-adjoint, positive, invertible operator A has a special factorization (9.1.2), then the factor D is self-adjoint, positive, and invertible.*

Proof. We need only to prove the positive definiteness of D. Since $I + X_+$ and $I + X_-$ are invertible,

$$\langle Df,f\rangle = \langle (I + X_+)^{-1}A(I + X_-)^{-1}f,f\rangle.$$

Set $(I + X_-)^{-1}f = g$. Then

$$\begin{aligned}\langle Df,f\rangle &= \langle (I + X_+)^{-1}Ag, (I + X_-)g\rangle \\ &= \langle (I + X_-)^*(I + X_+)^{-1}Ag,g\rangle \\ &= \langle (I + X_+)(I + X_+)^{-1}Ag,g\rangle \\ &= \langle Ag,g\rangle.\end{aligned}$$

□

Lemma 9.1.3. *If V is a Volterra operator $\in \mathscr{S}_2$ with π as an eigenchain, then the operator $W = (I + V)^{-1} - I$ is also Volterra $\in \mathscr{S}_2$ and has π as an eigenchain.*

PROOF. That $W \in \mathscr{S}_2$ follows immediately from the relation

$$W + V + VW = 0. \tag{9.1.5}$$

Since V and $I + W$ are permutable, we have (cf. Riesz-Nagy, *Functional Analysis*, p. 426 see Notes).

$$r_{V+VW} \leq r_V \cdot r_{I+W}$$

where r_A denotes the spectral radius of A, that is, the radius of the smallest closed disk centered at 0 which contains all the spectrum of A. By assumption $r_V = 0$; so $r_{V+VW} = 0$, that is, $V + VW$ is Volterra. Hence, from Equation (9.1.5), W is Volterra. Since $I + W$ is the resolvent at 1 of the operator $-V$,

$$W = \sum_{n=1}^{\infty} (-1)^n V^n,$$

the right-hand side converging in norm. It is readily verified that, for $P \in \pi$, $PV^nP = V^nP$ for any n. Hence we have $PWP = WP$. □

From Theorem 9.1.1 and Lemmas 9.1.1 to 9.1.3, we have the following theorem.

Theorem 9.1.2. *Let $S = I - T$ with $T \in \mathscr{S}_2$ be a self-adjoint positive and invertible operator. Then S and S^{-1} have the following factorizations along any maximal chain $\pi = \{P\}$:*

$$S = (I + W_-)D^{-1}(I + W_+) \quad \text{and} \quad S^{-1} = (I + X_+)D(I + X_-),$$

where

(a) *W_+, W_-, X_+, X_- are Volterra operators in $\mathscr{S}_2$, X_+, X_- are given by (9.1.3), the integral converging in norm, and $I + W_+ = (I + X_+)^{-1}$, $I + W_- = (I + X_-)^{-1}$.*
(b) *W_+, X_+ have π and W_-, X_- have $\pi^\perp$ as eigenchains.*
(c) *$W_+^* = W_-$ and $X_+^* = X_-$.*
(d) *D is a self-adjoint, positive, and invertible operator given by (9.1.4).*
(e) *$D - I \in \mathscr{S}_2$, $DP = PD$ for all $P \in \pi$.*

Chains of Orthoprojectors Associated with a Gaussian Process

Let $\{X_t, t \in [a,b]\}$ be a Gaussian process defined on a probability space $(\Omega,\mathscr{A},Q)$ with $E_Q X_t \equiv 0$ and covariance function $\Gamma_Q(s,t)$, where $[a,b]$ is taken to be either a finite closed or an infinite interval. For the sake of simplicity we assume that $[0,1] \subset [a,b]$. We assume throughout that $\{X(t)\}$ is continuous in quadratic mean (q.m.).

For $0 \le t \le 1$, let $L(X;t)$ be the closed linear subspace spanned by $\{X_\tau, 0 \le \tau \le t\}$ of $L^2(\Omega,\mathscr{A},Q)$, and let $\tilde{P}(t)$ be the orthoprojector defined on $L(X;1)$ with range $L(X;t)$. We are interested in a maximal chain containing the chain $\{\tilde{P}(t), 0 \le t \le 1\}$ (or $\{\tilde{P}(t), 0 < t \le 1\}$).

Let

$$L(X;t+) = \bigcap_{s>t} L(X;s)$$

and

$$L(X;t-) = \text{smallest closed linear space containing all } L(X;s),\ s < t.$$

Obviously $L(X;0) \subseteq L(X;0+)$ and $L(X;t-) \subseteq L(X;t) \subseteq L(X;t+)$ for $t > 0$. It is also easy to verify that $L(X;t-) = L(X;t)$ for all $0 < t \leqslant 1$. Since by the assumption on (X_t) the Hilbert space $L(X;1)$ is separable, the set of discontinuities $D = \{t \in [0,1]: L(X;t) \ne L(X;t+)\}$ is at most countable. Let $\tilde{P}(t_j+)$ be the orthoprojector with range $L(X;t_j+)$ for $t_j \in D$. The closure of the chain $\{\tilde{P}(t), 0 \le t \le 1\}$ consists of $\{\tilde{P}(t), 0 < t \le 1\}$ and $\{\tilde{P}(t_j+), t_j \in D\}$. If $D \ne \varnothing$, it has gaps $(\tilde{P}(t_j), \tilde{P}(t_j+))$, $t_j \in D$. If the dimension of the gap $(\tilde{P}(t_j), \tilde{P}(t_j+))$ is $n_j > 1$, we write the space $(\tilde{P}(t_j+) - \tilde{P}(t_j))L(X;1)$ as the orthogonal sum of one-dimensional subspaces $L(j,i)$:

$$(\tilde{P}(t_j+) - \tilde{P}(t_j))L(X,1) = \sum_{i=1}^{n_j} \oplus L(j,i).$$

Let $\tilde{Q}(j,k)$ be the orthoprojector with range $\sum_{i=1}^{k} \oplus L(j,i)$. Now consider the family of orthoprojectors π consisting of 0, $\{P(t), 0 \le t \le 1\}$, $\{\tilde{P}(t_j+), t_j \in D\}$ and $\{\tilde{P}(t_j) + \tilde{Q}(j,k), k = 1, \dots, n_j - 1; t_j \in D\}$. It is clear that $\tilde{\pi}$ is a maximal chain.

Remark 9.1.1. A maximal chain containing the chain $\{\tilde{P}(t), 0 \le t \le 1\}$ is obviously not unique, in general. If the dimension of $(\tilde{P}(t_j), \tilde{P}(t_j+))$ is greater than 1, we may take different orthoprojectors $\tilde{Q}(j,k)$.

Remark 9.1.2. The gap $(\tilde{P}(0),\tilde{P}(0+))$, if it exists and $\tilde{P}(0) \ne 0$, is special. Instead of filling in the gap $(\tilde{P}(0),\tilde{P}(0+))$, we may insert any set of orthoprojectors $\{\tilde{Q}(j), j = 0,1,\dots,n\}$ such that $\tilde{Q}(0) = 0$, $\tilde{Q}(n) = \tilde{P}(0+)$ and $\dim(\tilde{Q}(j) - \tilde{Q}(j-1)) = 1$. The maximal chain thus obtained will suffice for our purposes. In other words, we need only a maximal chain $\tilde{\pi}$ containing $\{\tilde{P}(t), 0 < t \le 1\}$.

The space $L(X;0+)$ is of particular interest. It can be trivial, n-dimensional $(1 \le n < \infty)$, or even infinite-dimensional.

Let ϕ denote the congruence (isometric isomorphism) from $L(X;1)$ onto the reproducing kernel Hilbert space $(RKHS)$ $H = H(\Gamma_Q)$ with reproducing kernel $\Gamma_Q(s,t)$, $0 \le s, t \le 1$, such that $\phi X_t = \Gamma_Q(\cdot,t)$, $0 \le t \le 1$. We note the following relation between subspaces of $L(X;1)$ and of H.

Let

$$F(t) = \{f \in H: f(s) = 0, 0 \le s \le t\}$$

and

$$M(t) = H \ominus F(t), \quad [\text{orthogonal complement of } F(t)].$$

Lemma 9.1.4. $\phi[L(X;t)] = M(t)$.

PROOF. Let $f \in H$ and $\xi = \phi^{-1}f$. Then, for all $s \in [0,t]$,

$$0 = f(s) = \langle f(\cdot), \Gamma(\cdot,s)\rangle \Leftrightarrow (\xi, X_s) = 0,$$

where $\langle \cdot,\cdot \rangle$ and $(\cdot,\cdot)$ denote, respectively, the inner products of H and $L(X;1)$. Hence $\xi \perp L(X;t)$, and thus $f \in F(t)$ if and only if $\phi^{-1}f \in L(X;1) \ominus L(X;t)$. This is equivalent to the assertion of the lemma. □

Let $P(t)$ denote the orthoprojector on H with range $M(t)$. Then $P(t) = \phi\tilde{P}(t)\phi^{-1}$. If $\tilde{\pi} = \{\tilde{P}\}$ is a maximal chain containing $\{\tilde{P}(t), 0 \leq t \leq 1\}$, then, obviously, the chain $\pi = \{\phi\tilde{P}\phi^{-1}\}$ is maximal and contains $\{P(t), 0 < t \leq 1\}$. We shall consistently use the following notation. If A is any linear operator on H, $\tilde{A}$ denotes the operator on $L(X;1)$ given by $\tilde{A} = \phi^{-1}A\phi$.

9.2 Nonanticipative Representations of Equivalent Gaussian Processes

Let $\mathscr{F}^{X,0}$ denote the σ-field generated by the random variables $\{X_t, 0 \leq t \leq 1\}$. Let P be another probability measure on $(\Omega, \mathscr{F}^{X,0})$ such that $\{X_t, 0 \leq t \leq 1, P\}$ is a q.m. continuous Gaussian process with $E_P X(t) = 0$ and covariance function $\Gamma_P(s,t)$. [$\{X_t, 0 \leq t \leq 1, Q\}$ is, by assumption, a q.m. continuous Gaussian process with $E_Q X_t = 0$ and covariance function $\Gamma_Q(s,t)$.] Assume that P and Q are equivalent, that is, mutually absolutely continuous relative to $\mathscr{F}^{X,0}$.

A nonanticipative representation of a Gaussian process with respect to another is defined as follows.

Definition 9.2.1. The process $\{X_t, 0 \leq t \leq 1, P\}$ has a nonanticipative representation with respect to $\{X_t, 0 \leq t \leq 1, Q\}$ if, on $(\Omega, \mathscr{F}^{X,0})$ there is a Gaussian process $\{Y_t, 0 \leq t \leq 1, Q\}$, having zero mean and Γ_P for its covariance, with the following property:

$$Y_t \in L(X;t) \quad \text{for each } t \in [0,1]. \tag{9.2.1}$$

We shall now prove the following theorem.

Theorem 9.2.1. *Every Gaussian process $\{X_t, 0 \leq t \leq 1, P\}$ which is equivalent to a given Gaussian process $\{X_t, 0 \leq t \leq 1, Q\}$ has a nonanticipative representation with respect to $\{X_t, 0 \leq t \leq 1, Q\}$. The processes are assumed to be q.m. continuous.*

The proof is based on Theorem 9.1.2 and on the following necessary and sufficient conditions for equivalence of P and Q (cf. [31]).

Theorem 9.2.2. *Gaussian measures P and Q are equivalent if and only if Γ_P defines an operator S on the RKHS $H(\Gamma_Q)$ with the following properties:*

(a) $\Gamma_P(\cdot,t) = S\Gamma_Q(\cdot,t)$ *for* $0 \le t \le 1$.
(b) S *is a bounded, self-adjoint, positive operator.*
(c) $T = I - S \in \mathscr{S}_2$.
(d) $1 \notin \sigma(T)$, *the spectrum of* T.

PROOF OF THEOREM 9.2.1. Consider a maximal chain π in $H(\Gamma_Q)$ described in the preceding section. Applying Theorem 9.1.2 to the operator S defined in Theorem 9.2.2, we have

$$S = (I + W_-)\Delta(I + W_+),$$

where $\Delta = D^{-1}$. The operator Δ is self-adjoint and positive. Since D commutes with all $P \in \pi$, we have $\Delta^{\frac{1}{2}}P = P\Delta^{\frac{1}{2}}$ for all $P \in \pi$. If we write

$$F = \Delta^{\frac{1}{2}}(I + W_+),$$

then

$$S = F^*F,$$

because $F^* = (I + W_+^*)\Delta^{\frac{1}{2}} = (I + W_-)\Delta^{\frac{1}{2}}$.

Consider now the operator $\tilde{F}$ on $L(X;1)$ corresponding to F:

$$\tilde{F} = \tilde{\Delta}^{\frac{1}{2}}(I + \tilde{W}_+). \tag{9.2.2}$$

Define

$$Y_t = \tilde{F}X_t, \qquad t \in [0,1].$$

Since $\tilde{F}$ is a linear operator on $L(X;1)$, $\{Y(t), 0 \le t \le 1,Q\}$ is Gaussian. Furthermore, $E_Q Y(t) \equiv 0$ and

$$\begin{aligned} E_Q Y_s Y_t &= (\tilde{F}X_s, \tilde{F}X_t) \\ &= \langle F\Gamma_Q(\cdot,s), F\Gamma_Q(\cdot,t)\rangle \\ &= \langle S\Gamma_Q(\cdot,s), \Gamma_Q(\cdot,t)\rangle \\ &= \langle \Gamma_P(\cdot,s), \Gamma_Q(\cdot,t)\rangle \\ &= \Gamma_P(s,t). \end{aligned}$$

Thus $\{Y_t, 0 \le t \le 1,Q\}$ is a Gaussian process with zero mean and covariance function $\Gamma_P(s,t)$. We have also for each $t \in [0,1]$,

$$\begin{aligned} Y_t &= \tilde{\Delta}^{\frac{1}{2}}(I + \tilde{W}_+)X_t \\ &= \tilde{\Delta}^{\frac{1}{2}}(I + \tilde{W}_+)\tilde{P}(t)X_t \\ &= \tilde{\Delta}^{\frac{1}{2}}\tilde{P}(t)(I + \tilde{W}_+)\tilde{P}(t)X_t \\ &= \tilde{P}(t)\tilde{\Delta}^{\frac{1}{2}}(I + \tilde{W}_+)\tilde{P}(t)X_t \\ &= \tilde{P}(t)Y_t. \end{aligned}$$

This shows that $Y_t \in L(X;t)$. The proof of Theorem 9.2.1 is complete. $\square$

9.3 Nonanticipative Representation of a Gaussian Process Equivalent to a Wiener Process

Suppose that $(X_t, t \in [0,1], Q)$ is a (standard) Wiener process. Then it is easy to see that $L(X; t+) = L(X; t)$ for all $t \in [0,1)$, that is, $D = \varnothing$ and that the chain $\tilde{\pi} = \{\tilde{P}(t), 0 \leq t \leq 1\}$ is maximal. We have

$$\begin{aligned} \Gamma_Q(s,t) &= EX_sX_t = \min(s,t) \\ &= \int_0^1 \chi(s,u)\chi(t,u)\,du, \end{aligned} \tag{9.3.1}$$

where $\chi(t,u) = 1$ if $0 \leq u \leq t$, $= 0$ if $t < u \leq 1$. Relation (9.3.1) defines an isometric isomorphism from $H(\Gamma_Q)$ onto $L^2[0,1]$ which sends $\Gamma_Q(\cdot,t)$ to $\chi(t,\cdot)$, and any element $f \in H(\Gamma_Q)$ is represented in the form $f(t) = \int_0^t \hat{f}(u)\,du$ with $\hat{f} \in L^2[0,1]$. Correspondingly, there is an isometric isomorphism θ from $L(X;1)$ onto $L^2[0,1]$ such that $\theta X_t = \chi(t,u)$ and, for any $\xi \in L(X;1)$ we have

$$\xi = \int_0^1 \hat{f}(u)\,dX_u \tag{9.3.2}$$

and

$$\theta[L(X;t)] = \{\hat{f} \in L^2[0,1]: \hat{f}(s) = 0 \text{ a.e. for } t < s \leq 1\}.$$

The chain $\pi = \{P(t) = \theta\tilde{P}(t)\theta^{-1}, 0 \leq t \leq 1\}$ is maximal and $P(t) \in \pi$ is characterized by

$$P(t)\hat{f}(u) = \hat{f}(u)\chi(t,u). \tag{9.3.3}$$

We denote by A the operator on $L^2[0,1]$ corresponding to an operator $\tilde{A}$ on $L(X;1)$ (instead of A on $H(\Gamma_Q)$).

Lemma 9.3.1. *Let $K(u,v)$ be the kernel in $L^2([0,1] \times [0,1])$ corresponding to a Hilbert-Schmidt operator K on $L^2[0,1]$. If the chain π is an eigenchain for K, then*

$$K(u,v) = 0 \quad \text{a.e. } \textit{for } u > v.$$

PROOF. Since, by assumption,

$$KP(t)\hat{f} = P(t)KP(t)\hat{f}, \qquad 0 \leq t \leq 1,$$

for $\hat{f} \in L^2[0,1]$ it follows from (9.3.3) that

$$\int_0^t K(u,v)\hat{f}(v)dv = 0 \quad \text{a.e. for } u > t. \tag{9.3.4}$$

Hence we have

$$\int_0^1 \int_0^1 K(u,v)(1 - \chi(t,u))\chi(t,v)\hat{g}(u)\hat{f}(v)\,du\,dv = 0$$

for all $0 \leq t \leq 1$ and for any $\hat{f},\hat{g} \in L^2[0,1]$. Let $\hat{f}(u) = \chi((a,b],u) = \chi(b,u) - \chi(a,u)$ and $\hat{g}(v) = \chi((c,d],v) = \chi(d,v) - \chi(c,v)$. Then, for either $b \leq c$ or $a \leq d$,

$$\int_0^1 \int_0^1 (1 - \chi(u,v))K(u,v)\chi((a,b],u)\chi((c,d],v)\,du\,dv = 0.$$

Since the family $\{\chi((a,b],u)\chi((c,d],v)$ with $b \le c$ or $a \ge d\}$ spans $L^2([0,1] \times [0,1]$, Lebesgue measure), we have

$$(1 - \chi(u,v))K(u,v) = 0 \quad \text{a.e. for } u,v \ge 0$$

that is,

$$K(u,v) = 0 \quad \text{a.e. for } u > v \ge 0. \qquad \square$$

Note that since there are no gaps in the chain $\tilde{\pi}$, $D = \Delta = I$ in the factorization theorem. Hence

$$\theta Y_t = (I + W_+)\theta X_t = (I + W_+)\chi(t,\cdot).$$

Applying Lemma 9.3.1 to the Volterra operator W_+, we have

$$(I + W_+)\chi(t,\cdot) = \chi(t,\cdot) + \int_0^1 W_+(\cdot,v)\chi(t,v)\,dv$$
$$= \chi(t,\cdot) + \int_0^t W_+(\cdot,v)\,dv,$$

where $W_+(u,v)$ is the Volterra kernel [that is, $W_+(u,v) = 0$ a.e. for $u > v$] in $L^2([0,1] \times [0,1])$ corresponding to W_+. Hence

$$\theta Y_t(u) = \chi(t,u) + \int_0^t W_+(u,v)\,dv \quad \text{for } 0 \le u \le 1.$$

Taking $\xi = Y(t)$ in (9.3.2), we obtain the desired representation

$$Y_t = \int_0^1 \chi(t,u)\,dX_u + \int_0^1 \int_0^t W_+(u,v)\,dv\,dX_u$$
$$= X_t + \int_0^t \left\{ \int_0^v W_+(u,v)\,dX_u \right\} dv, \tag{9.3.5}$$

where $W_+(u,v)$ is a Volterra kernel belonging to $L^2([0,1] \times [0,1])$. A derivation of (9.3.5) using martingale methods is also available [19]. However, the operator-theoretic approach has been presented above because it yields the more general theorem of Section 9.2. It would be of great interest to develop comparable functional analytic techniques powerful enough to deal with the non-Gaussian case.

9.4 Gaussian Solutions of Stochastic Equations

Let $\gamma(t,x), (t \in [0,T], x \in C)$ be a causal or nonanticipative functional as defined in Chapter 5. Suppose that there exists a complete probability space $(\Omega,\mathscr{A},P)$, and increasing family $(\mathscr{F}_t)(0 \le t \le T)$ of sub σ-fields of $\mathscr{A}$, and processes $\xi = (\xi_t)$ and $W = (W_t)$ defined on Ω, adapted to $(\mathscr{F}_t)$ and satisfying the following conditions:

1. $(W_t,\mathscr{F}_t,P)$ is a Wiener martingale.
2. For all t in $[0,T]$,

$$\xi_t(\omega) = \int_0^t \gamma(s,\xi(\omega))\,ds + W_t(\omega), \qquad (P\text{-a.s.}).$$

3. $P[\omega \in \Omega: \int_0^T \gamma^2(s,\xi(\omega))\, ds < \infty] = 1$.
4. The induced measure $\mu_\xi = P\xi^{-1}$ on $(C,\mathscr{B}_T(C))$ is Gaussian.

Because of condition 3, Theorem 7.3.1 implies that $\mu_\xi \ll \mu$. Hence we must have $\mu_\xi \equiv \mu$ since both measures are Gaussian (see Notes), and so

$$\mu\left[x \in C: \int_0^T \gamma^2(s,x)\, ds < \infty\right] = 1. \tag{9.4.1}$$

From condition 2 we see that (W_t) *is* $(\mathscr{F}_t^\xi)$-adapted so that $\mathscr{F}_t^W \subseteq \mathscr{F}_t^\xi \subseteq \mathscr{F}_t$. Let us now fix our attention on the probability space $(\Omega,\mathscr{F}_T^\xi,P)$ on which is defined the standard Wiener process $(W_t(\omega))$. The Gaussian process (ξ_t) is defined on the same probability space and has a continuous covariance function which we denote by R. The mean function will be assumed, without loss of generality to be zero. Applying the theory discussed in Section 9.3, we obtain a nonanticipative representation $(\tilde{\xi}_t)$ of (ξ_t) given by [see (9.3.5)]

$$\tilde{\xi}_t(\omega) = W_t(\omega) + \int_0^t \left[\int_0^v G(u,v)\, dW_u(\omega)\right] dv \tag{9.4.2}$$

[for all t, (P-a.s.)], a (t,ω)-measurable version of the right-side stochastic integral being chosen to ensure the (t,ω)-measurability of $\tilde{\xi}_t(\omega)$.

We now indicate how (9.4.2) can be inverted to obtain $W_t(\omega)$ explicitly in terms of the process $(\tilde{\xi}_t)$. If $f \in L^2[0,T]$ the existence of the stochastic integral $\int_0^T f(t)\, d\tilde{\xi}_t$ follows in the usual manner. Suppose (f_n) is a sequence of step functions [with $f_n(t) = c_j$ for $t_j \le t < t_{j+1}$] such that $f_n \to f$ in L^2. Define

$$\int_0^T f_n(t)\, d\tilde{\xi}_t = \sum f_n(t_j)[\tilde{\xi}_{t_{j+1}} - \tilde{\xi}_{t_j}].$$

From (9.4.2) $\int_0^T f_n(t)\, d\tilde{\xi}_t = \int_0^T f_n(t)\, dW_t + \int_0^T f_n(t)\phi_t\, dt$, where we introduce the temporary notation $\phi_t(\omega) = \int_0^t G(u,t)\, dW_u(\omega)$. Since $\int_0^T E(\phi_t)^2\, dt < \infty$, it is easy to see that both terms on the right-hand side of the above relation have $\int_0^T f(t)\, dW_t$ and $\int_0^T f(t)\phi_t\, dt$ for their respective $L^2(P)$ limits as $n \to \infty$. Hence $\int_0^T f_n(t)\, d\tilde{\xi}_t$ converges in the $L^2(P)$ sense to a limit which, following the usual definition, we denote by $\int_0^T f(t)\, d\tilde{\xi}_t$ and which has all the familiar properties of a stochastic integral. It can, in fact, be shown quite easily that every element $\zeta \in L_P(\tilde{\xi};T)$ is of the form $\int_0^T f(t)\, d\tilde{\xi}_t$ for a uniquely determined $f \in L^2[0,T]$. Here $L_P(\tilde{\xi};T)$ is the closure in $L^2(\Omega,F_T^\xi,P)$ of the linear manifold of finite linear combinations of $\tilde{\xi}_t (t \in [0,T])$.

Furthermore, we have

$$\int_0^T f(t)\, d\tilde{\xi}_t = \int_0^T f(t)\, dW_t + \int_0^T f(t)\phi_t\, dt.$$

This relation justifies our formally writing $d\tilde{\xi}_t = dW_t + \phi_t\, dt$ in computations involving integrals with respect to $(\tilde{\xi}_t)$.

Let us now turn to the question of inverting (9.4.2). Denote by $K(u,v)$ the resolvent kernel of G [see Equation (9.4.4)]. Note that $K(u,v) = 0$ a.e. for

$u > v$. Then we have the relations

$$G(s,t) + K(s,t) + \int_s^t K(s,u)G(u,t)\,du = 0,$$

$$G(s,t) + K(s,t) + \int_s^t G(s,u)K(u,t)\,du = 0 \qquad (s < t).$$

Using these, Equation (9.4.2), and the easily verified relation

$$\int_0^s\left(\int_0^s a(u,v)\,dW_u\right)dv = \int_0^s\left(\int_0^s a(u,v)\,dv\right)dW_u$$

for a square-integrable, (u,v)-measurable kernel $a(u,v)$, we obtain

$$\begin{aligned}
\tilde{\xi}_t + \int_0^t\left[\int_0^v K(u,v)\,d\tilde{\xi}_u\right]dv &= W_t + \int_0^t\left[\int_0^v G(u,v)\,dW_u\right]dv \\
&\quad + \int_0^t\left[\int_0^v K(u,v)\,dW_u + \int_0^v K(u,v)\phi_u\,du\right]dv \\
&= W_t + \int_0^t\left[\int_0^v G(u,v)\,dW_u\right]dv \\
&\quad + \int_0^t\left(\int_0^v K(u,v)\,dW_u\right)dv \\
&\quad + \int_0^t\left[\int_0^v K(u,v)\left(\int_0^u G(s,u)\,dW_s\right)du\right]dv \\
&= W_t + \int_0^t\left[\int_0^v G(u,v)\,dW_u\right]dv \\
&\quad + \int_0^t\left[\int_0^v\left\{K(s,v) + \int_s^v K(u,v)G(s,u)\,du\right\}dW_s\right]dv \\
&= W_t + \int_0^t\left(\int_0^v G(u,v)\,dW_u\right)dv - \int_0^t\left(\int_0^v G(s,v)\,dW_s\right)dv \\
&= W_t.
\end{aligned}$$

Hence we have

$$W_t(\omega) = \tilde{\xi}_t(\omega) + \int_0^t\left[\int_0^v K(u,v)\,d\tilde{\xi}_u(\omega)\right]dv. \tag{9.4.3}$$

The kernels $G(u,v)$ and $K(u,v)$ are square-integrable Volterra kernels on $[0,T] \times [0,T]$, both vanishing a.e. if $u > v$. They are determined uniquely by the factorization

$$S = (I + G^*)(I + G) \quad \text{and} \quad S^{-1} = (I + K)(I + K^*), \tag{9.4.4}$$

where S is the operator determined by the covariance R on $L^2[0,T]$ and G,G^* (transpose of G), K and K^* are Hilbert-Schmidt, Volterra operators on

$L^2[0,T]$. The kernels, corresponding to G and K are given by $G(u,v)$, $K(u,v)$ in formulas (9.4.2) and (9.4.3). The equations in (9.4.4) can be rewritten explicitly as equations of Wiener-Hopf type in terms of the kernels, but this point will not be pursued at the moment. We are concerned now only with the existence of G and K which, of course, is guaranteed by the factorization of S or S^{-1}. From (9.4.2) and (9.4.3) we get the following important fact. For each t,

$$\mathscr{F}_t^{\xi} = \mathscr{F}_t^W. \tag{9.4.5}$$

Setting $\tilde{\gamma}(t,\tilde{\xi}(\omega)) = -\int_0^t K(u,t)\, d\tilde{\xi}_u(\omega)$ let us rewrite (9.4.3) as

$$\tilde{\xi}_t(\omega) = \int_0^t \tilde{\gamma}(s,\tilde{\xi}(\omega))\, ds + W_t(\omega). \tag{9.4.6}$$

Observe that by the definition of the nonanticipative representation $(\tilde{\xi}_t)$, $P\tilde{\xi}^{-1} = \mu_\xi$. Now

$$P\left[\omega\colon \int_0^T \tilde{\gamma}(t,\tilde{\xi}(\omega))^2\, dt < \infty\right] = \mu_\xi\left[x \in C\colon \int_0^T \tilde{\gamma}^2(t,x)\, dt < \infty\right] = 1, \tag{9.4.7}$$

since $\mu_\xi \equiv \mu$ and

$$\mu\left[x \in C\colon \int_0^T \tilde{\gamma}^2(t,x)\, dt < \infty\right] = 1. \tag{9.4.8}$$

The last step follows because

$$E_\mu \int_0^T [\tilde{\gamma}(t,x)]^2\, dt = \int_0^T \int_0^t K^2(u,t)\, du\, dt < \infty. \tag{9.4.9}$$

It now follows from the remark after Theorem 7.3.3 that the Radon-Nikodym derivative $(d\mu_\xi/d\mu)(x)$ relative to $\mathscr{B}_T$ is given by

$$\exp\left[\int_0^T \gamma(s,x)\, dx_s - \frac{1}{2}\int_0^T \gamma^2(s,x)\, ds\right]$$

and also by

$$\exp\left[\int_0^T \tilde{\gamma}(s,x)\, dx_s - \frac{1}{2}\int_0^T \tilde{\gamma}^2(s,x)\, ds\right].$$

Taking conditional expectations $E(d\mu_\xi/d\mu|\mathscr{B}_{t+})$ from the uniqueness of the Radon-Nikodym derivative we obtain for all t,

$$\exp\left[\int_0^t \tilde{\gamma}(s,x)\, dx_s - \frac{1}{2}\int_0^t \tilde{\gamma}^2(s,x)\, ds\right] = \exp\left[\int_0^t \gamma(s,x)\, dx_s - \frac{1}{2}\int_0^t \gamma^2(s,x)\, ds\right] \quad (\mu\text{-a.s.}). \tag{9.4.10}$$

From (9.4.10) for all t,

$$\int_0^t \tilde{\gamma}(s,x)\, dx_s - \int_0^t \gamma(s,x)\, dx_s = \frac{1}{2}\int_0^t \tilde{\gamma}^2(s,x)\, ds - \frac{1}{2}\int_0^t \gamma^2(s,x)\, ds \qquad (\mu\text{-a.s.}).$$

Since the left side of the above relation is a sample-continuous $(\mathscr{B}_{t+}\mu)$-martingale vanishing at $t = 0$ and the right side is sample absolutely continuous, it follows that for all t,

$$\int_0^t \tilde{\gamma}(s,x)\,dx_s = \int_0^t \gamma(s,x)\,dx_s$$

and

$$\int_0^t \tilde{\gamma}^2(s,x)\,ds = \int_0^t \gamma^2(s,x)\,ds \qquad (\mu\text{-a.s.}). \tag{9.4.11}$$

The second equality in (9.4.11) and (9.4.9) yields a property of γ, viz.,

$$\int_0^T E_\mu(\gamma^2(s,x))\,ds < \infty. \tag{9.4.12}$$

Hence from (9.4.11) we have

$$\int_0^T E_\mu[\tilde{\gamma}(s,x) - \gamma(s,x)]^2\,ds = E_\mu\left(\int_0^T [\tilde{\gamma}(s,x) - \gamma(s,x)]\,dx_s\right)^2 = 0. \tag{9.4.13}$$

Denoting Lebesgue measure by L, we have

$$(L \times \mu)(M) = 0, \tag{9.4.14}$$

where

$$M = [(s,x) \in [0,T] \times C: \tilde{\gamma}(s,x) \neq \gamma(s,x)].$$

Since $\mu_\xi \equiv \mu$, $(L \times \mu_\xi)(M) = 0$ from which we have

$$P[\omega \in \Omega: \gamma(s,\xi(\omega)) = \tilde{\gamma}(s,\xi(\omega))] = 1 \tag{9.4.15}$$

for a.e. s. in $[0,T]$. From (9.4.13) we also get the following conclusion:

$$E_\mu\left[\int_0^t \tilde{\gamma}(s,x)\,ds - \int_0^t \gamma(s,x)\,ds\right]^2 \leq T\int_0^T E_\mu[\tilde{\gamma}(s,x) - \gamma(s,x)]^2\,ds = 0.$$

Using once again the equivalence of μ_ξ and μ, we find that for all t in $[0,T]$,

$$\int_0^t \tilde{\gamma}(s,x)\,ds = \int_0^t \gamma(s,x)\,ds \qquad (\mu_\xi\text{-a.s.}), \tag{9.4.16}$$

that is,

$$P\left[\omega: \int_0^t \gamma(s,\xi(\omega))\,ds = \int_0^t \tilde{\gamma}(s,\xi(\omega))\,ds \text{ for all } t\right] = 1. \tag{9.4.17}$$

Let us now deduce all the implications of (9.4.14), (9.4.15), and (9.4.17). First of all, condition 2 may now be written as

$$\xi_t(\omega) = \int_0^t \tilde{\gamma}(s,\xi(\omega))\,ds + W_t(\omega) \qquad (P\text{-a.s.}), \tag{9.4.18}$$

and recalling the form of $\tilde{\gamma}$ [given just prior to (9.4.6)], we have

$$W_t(\omega) = \xi_t(\omega) + \int_0^t\left[\int_0^s K(u,s)\,d\xi_u(\omega)\right]ds \qquad (P\text{-a.s.}). \tag{9.4.19}$$

Equation (9.4.19) can be inverted as we did in passing from (9.4.2) to (9.4.3), but going backward this time, and we obtain

$$\xi_t(\omega) = W_t(\omega) + \int_0^t \left[\int_0^s G(u,s)\, dW_u(\omega) \right] ds, \tag{9.4.20}$$

for all t in $[0,T]$, (P-a.s.).

Comparing (9.4.2) with (9.4.20) we have

$$\xi_t(\omega) = \tilde{\xi}_t(\omega) \quad \text{for all } t, (P\text{-a.s.}). \tag{9.4.21}$$

As a consequence of (9.4.5) and (9.4.21) we obtain the important equality of σ-fields

$$\mathscr{F}_t^{\xi} = \mathscr{F}_t^{W} \quad \text{for every } t. \tag{9.4.22}$$

In the preceding discussion we have produced the explicit solution [given by (9.4.20)] to the stochastic equation under the stated assumptions. In addition, a number of other interesting facts have been established, all of which we now collect in the following theorems.

Theorem 9.4.1. *Let γ be a nonanticipative functional and let (ξ_t), (W_t) be processes defined on $(\Omega,\mathscr{F},P)$ satisfying conditions 1 and 3 and connected by the stochastic equation*

$$\xi_t(\omega) = \int_0^t \gamma(s,\xi(\omega))\, ds + W_t(\omega).$$

Suppose further that the measure $\mu_\xi = P\xi^{-1}$ is Gaussian.

(a) *Then (ξ_t) is expressed in terms of (W_t) by the formula*

$$\xi_t(\omega) = W_t(\omega) + \int_0^t \left[\int_0^s G(u,s)\, dW_u(\omega) \right] ds$$

for all t, (P-a.s.), G being a square integrable Volterra kernel given by the factorization formula (9.4.4).

(b) *(ξ_t) is pathwise identical with the nonanticipative representation $(\tilde{\xi}_t)$ of μ_ξ with respect to the Wiener measure μ on $(\Omega,\mathscr{F},P)$. In other words* $P[\omega\colon \xi_t(\omega) = \tilde{\xi}_t(\omega) \text{ for all } t] = 1$.

(c) *For every t, $\mathscr{F}_t^{\xi} = \mathscr{F}_t^{W}$.*

Theorem 9.4.2. *Let γ be a nonanticipative functional such that*

$$\mu\left[x \in C\colon \int_0^T \gamma^2(t,x)\, dx < \infty \right] = 1. \tag{9.4.23}$$

(a) *Then the stochastic equation*

$$d\xi_t = \gamma(t,\xi)\, dt + dW_t \tag{9.4.24}$$

has a Gaussian weak solution if and only if γ is of the form

$$\gamma(t,x) = \int_0^t G(u,t)\, dx_u \tag{9.4.25}$$

for a.e. (t,x) *with respect to the product measure* $L \times \mu$, *where* $G(u,t)$ *belongs to* $L^2[0,T]^2$ *and* $G(u,t) = 0$ a.e. *for* $u > t$.

(b) *If a Gaussian weak solution exists, then so does a Gaussian strong solution, and the latter is unique.*

PROOF. Most of the work has already been done above and it only remains to make some supplementary remarks.

(a) First of all, the "only if" part follows from the proof of Theorem 9.4.1. To prove the "if" part one needs only to reproduce a familiar argument. Suppose γ is given by (9.4.25). Consider the coordinate Wiener process (x_t) on $(C,\mathscr{B}_T,\mu)$. Let $\tilde{G}$ denote the Volterra operator determined by the kernel $G(u,t)$ on the linear Hilbert space of (x_t) which we have denoted by $L(x;T)$. $\int_0^t \gamma(s,x)\,ds = (\tilde{G}x)_t$. Define the process (y_t) on $(C,\mathscr{B}_T,\mu)$ by $y_t(x) = (I + G)x_t$. Then clearly under μ, (y_t) is Gaussian with zero mean function and covariance R, where the operator S determined by R on $L^2[0,T]$ is given by $S = (I + G^*)(I + G)$. Since G is a Volterra operator, S is a positive, self-adjoint operator satisfying all the requirements that ensure the equivalence of the measures μ_y and μ. Thus μ_y is the desired unique weak solution.

(b) The following argument shows how the existence of a Gaussian strong solution of (9.4.24) is implied by a Gaussian weak solution. Let $(\Omega,\mathscr{F}_T,P)$ be a probability space on which is defined a Wiener martingale $(W_t,\mathscr{F}_t)$. If μ^* is the Gaussian weak solution, $\mu^* \equiv \mu$. μ^* is determined by its covariance function, say, R and its mean function which we assume to be zero without restricting the generality. Now by the work of the last section R defines a nonanticipative representation, which we denote by $(\tilde{\xi}_t)$, of μ^* with respect to μ on the given space $(\Omega,\mathscr{F}_T,P)$. It is given by (9.4.6), where $\tilde{\gamma}(t,\tilde{\xi}(\omega)) = -\int_0^t K(u,t)\,d\tilde{\xi}_u(\omega)$. By (9.4.5) and (9.4.6) $(\tilde{\xi}_t)$ can be taken to be (t,ω)-measurable and adapted to $(\mathscr{F}_t)$. Hence $(\tilde{\xi}_t)$ is a strong solution of (9.4.24). Next, let (ξ_t) be any strong solution of (9.4.24) such that $\mu_\xi = \mu^*$. Because of (9.4.23), condition 3 is satisfied and the work preceding Theorem 9.4.1 shows that (9.4.21) is valid. This proves uniqueness since the nonanticipative representation $(\tilde{\xi}_t)$ is itself unique. The latter follows from the uniqueness of the Gohberg-Krein factorization of S given by $S = (I + G^*)(I + G)$. □

It is possible further to characterize the Gaussian solutions of the stochastic equation (9.4.24) in terms of the canonical representations of Levy, Cramér, and Hida.

Theorem 9.4.3. *Let the assumptions of Theorem 9.4.1 hold. Also assume (9.4.23). If the Gaussian solution of (9.4.24) exists it has a canonical representation of multiplicity 1:*

$$\xi_t = \int_0^t F(t,u\; dW_u, \tag{9.4.26}$$

where $F(t,u)$ *is a square-integrable kernel of the form*

$$F(t,u) = I_{[0,t]}(u) + H(t,u), \qquad H(t,u) = 0 \quad \text{a.e. } for\ u > t$$

and

$$\int_0^T \int_0^t H^2(t,u)\,dt\,du < \infty. \tag{9.4.27}$$

Moreover, the representation is proper canonical in the sense defined by Hida [18].

PROOF. If a Gaussian solution exists, it follows from Theorem 9.4.2(a) that γ is of the form (9.4.25). Let G be the Volterra operator on $L^2[0,T]$ defined by the square-integrable Volterra kernel $G(u,t)$. Let K be the operator given by $(I + K) = (I + G)^{-1}$. Then K is a Hilbert-Schmidt Volterra operator on $L^2[0,T]$ determined by a square-integrable kernel $K(t,u)$ which $= 0$ a.e. if $u > t$. It is then easy to see that we have a solution (on some probability space),

$$\xi_t(\omega) = W_t(\omega) + \int_0^t \left[\int_0^s G(s,u)\,dW_u(\omega) \right] ds \quad \text{for all } t \text{ } (P\text{-a.s.}). \tag{9.4.28}$$

The stochastic integral on the right side can be written in the form

$$\int_0^t \left[\int_u^t G(s,u)\,ds \right] dW_u(\omega) \text{ and so } \xi_t(\omega) = W_t(\omega) + \int_0^t \left[\int_u^t G(s,u)\,ds \right] dW_u(\omega),$$

which is (9.4.26). Next, we observe that exactly as we did earlier, we can invert (9.4.28) to yield $W_t(\omega) = \xi_t(\omega) + \int_0^t [\int_0^s K(u,s)\,d\xi_u(\omega)]\,ds$ for all t (P-a.s.). It follows that for every t,

$$L(\xi;t) = L(W;t), \tag{9.4.29}$$

where $L(\xi;t)$ and $L(W;t)$ are the linear Hilbert spaces introduced in Section 9.1, viz., subspaces of the Hilbert space $L^2(\Omega,\mathscr{F},P)$ spanned by finite linear combinations $\sum_{t_i \le t} c_i \xi_{t_i}$ ($\sum_{t_i \le t} c_i W_{t_i}$, respectively). Equation (9.4.29) defines the proper canonical property of the representation (9.4.26), and the result is established. □

The remainder of this section is a digression from the main subject of interest and is not necessary for its understanding. The work we have been discussing in this chapter is closely related to certain ideas concerning "stochastic infinitesimal equations" advanced by Paul Lévy over twenty years ago. Equation (9.4.24) governing a Gaussian process (ξ_t) is the "normal form" of (ξ_t) in Lévy's terminology. As a prelude to a general theory of stochastic infinitesimal equations Lévy discusses by means of examples Gaussian processes given by an equation of the form (in his terminology)

$$\delta x(t) \sim dt \int_0^t F(t,u)\,dx(u) + \zeta\sqrt{d\omega(t)},$$

ζ being a standard Gaussian random variable and $\omega(t)$ is a positive increasing function of t. For the case $\omega(t) = t$ the results of this chapter provide a complete answer to the following questions raised by him [36, p. 349]:

1. How to pass from (9.4.24) to the covariance function of (ξ_t) and conversely.
2. Given the covariance how to express the process (ξ_t) by means of an infinitesimal equation.

In the light of Theorems 9.4.2 and 9.4.3 and the theory of equivalence and singularity of Gaussian measures, it is instructive to work out completely the following illuminating example of Lévy's.

EXAMPLE 9.4.1. Consider Equation (9.4.24) with $T = 1$ and

$$\gamma(t,x) = \int_0^t G^{(\alpha)}(t,u)\,dx_u \qquad (\mu\text{-a.s.}), \tag{9.4.30}$$

where $\alpha > \frac{1}{2}$ and

$$G^{(\alpha)}(t,u) = \frac{(t-u)^{\alpha-1}}{\Gamma(\alpha)} - \frac{(t-u)^{2\alpha-1}}{\Gamma(2\alpha)} + \frac{(t-u)^{3\alpha-1}}{\Gamma(3\alpha)} - \cdots \tag{9.4.31}$$

for $0 \le u < t$, and $= 0$ otherwise.

For every (t,u) with $u < t$, the series on the right side of (9.4.31) converges. Hence it is easily seen that $G^{(\alpha)}(t,u)$ defined by (9.4.31) is (t,u)-measurable. Also $\int_0^1\int_0^t [G^{(\alpha)}(t,u)]^2\,du\,dt$ is finite as can be directly verified from (9.4.31). Hence from Theorem 9.4.2, Equation (9.4.24) with γ given by (9.4.30) has a Gaussian solution which we denote by (ξ_t^α). Now from Theorem 9.4.3 we can express (ξ_t^α) by

$$\xi_t^\alpha(\omega) = W_t(\omega) + \int_0^t \left[\int_u^t G_1^{(\alpha)}(s,u)\,ds\right] dW_u(\omega), \tag{9.4.32}$$

where $G_1^{(\alpha)}(s,u)$ is the resolvent kernel of $G^{(\alpha)}$. To determine the resolvent operator $G_1^{(\alpha)}$ (defined on $L^2[0,1]$) we proceed as follows. Consider the Riemann-Liouville fractional order integral operator $I_\beta (\beta > 0)$:

$$(I_\beta f)(t) = \frac{1}{\Gamma(\beta)} \int_0^t (t-u)^{\beta-1} f(u)\,du, \qquad f \in L^2[0,1].$$

Then if $\beta' > 0$,

$$\begin{aligned} I_{\beta'}[I_\beta f](t) &= \frac{1}{\Gamma(\beta)\Gamma(\beta')} \int_0^t (t-u)^{\beta'-1} \int_0^u (u-v)^{\beta-1} f(v)\,dv\,du \\ &= \frac{1}{\Gamma(\beta+\beta')} \int_0^t (t-v)^{\beta+\beta'-1} f(v)\,dv = (I_{\beta+\beta'} f)(t), \end{aligned}$$

that is, $I_{\beta'}I_\beta = I_{\beta'+\beta}$. Hence from (9.4.31) we have (the various steps can be justified without difficulty),

$$\begin{aligned}(G^{(\alpha)}f)(t) &= \int_0^t G^{(\alpha)}(t,u)f(u)\,du \\ &= \int_0^t \frac{(t-u)^{\alpha-1}}{\Gamma(\alpha)} f(u)\,du - \int_0^t \frac{(t-u)^{2\alpha-1}}{\Gamma(2\alpha)} f(u)\,du \\ &\quad + \int_0^t \frac{(t-u)^{3\alpha-1}}{\Gamma(3\alpha)} f(u)\,du - \cdots \\ &= (I_\alpha f)(t) - (I_\alpha^2 f)(t) + (I_\alpha^3 f)(t) - \cdots. \end{aligned} \tag{9.4.33}$$

Equation (9.4.33) holds in the sense that $\|G^{(\alpha)}f - T_n f\|_{L^2} \to 0$ as $n \to \infty$, where we write $T_n = I_\alpha - I_\alpha^2 + \cdots + (-1)^{n-1}I_\alpha^n$. Now if we denote by $I_\alpha(t,u)$ the kernel determining the Volterra operator I_α, it follows after some calculation that the series of iterated kernels $I_\alpha(t,u) - I_\alpha^2(t,u) + \cdots$ converges in $L^2([0,1] \times [0,1])$. Hence T_n converges strongly to the Volterra operator $I_\alpha - I_\alpha^2 + I_\alpha^3 - \cdots$, and we obtain $G^{(\alpha)} = I_\alpha - I_\alpha^2 + \cdots + (-1)^{n-1}I_\alpha^n - \cdots$. Hence $I - G^{(\alpha)} = (I + I_\alpha)^{-1}$ and it therefore follows that the resolvent $G_1^{(\alpha)} = I_\alpha$. Substituting in (9.4.32) we get

$$\xi_t^{(\alpha)} = W_t + \int_0^t \left[\int_u^t \frac{(s-u)^{\alpha-1}}{\Gamma(\alpha)}\,ds\right] dW_u, \tag{9.4.34}$$

that is,

$$\xi_t^{(\alpha)} = W_t + \frac{1}{\Gamma(\alpha+1)} \int_0^t (t-u)^\alpha\,dW_u. \tag{9.4.35}$$

For the special case $\alpha = 1$, $G^{(\alpha)}(t,u) = e^{-(t-u)}$ if $u < t$, $= 0$ if $u > t$. We find that the solution to the stochastic equation $d\xi_t = [\int_0^t e^{-(t-u)}\,d\xi_u]\,dt + dW_t$ is given by $\xi_t = W_t + \int_0^t W_u\,du$. Let us now go back to the process (ξ_t^α) given by (9.4.35) and set

$$W_t^{(\alpha)} = \frac{1}{\Gamma(\alpha+1)} \int_0^t (t-u)^\alpha\,dW_u. \tag{9.4.36}$$

It is easy to verify that $(W_t^{(\alpha)})$ for $\alpha > 0$ is sample-continuous. [This can be seen either directly or from (9.4.35) when $\alpha > \frac{1}{2}$.] For $\alpha = 0$, $W_t^{(0)} = W_t$ and for $\alpha > 0$ one can show that $(W_t^{(\alpha)})$ is singular with respect to the Wiener process. Turning to (9.4.35), we obtain an interesting fact: On a suitable probability space $(\Omega,\mathscr{F},P)$ we have (a.s.)

$$W_t^{(\alpha)}(\omega) = \xi_t^{(\alpha)}(\omega) - W_t(\omega) \quad \text{for all } t \in [0,1], \tag{9.4.37}$$

and

$$\mu_{\xi^{(\alpha)}} \equiv \mu, \qquad \mu_{W^{(\alpha)}} \perp \mu. \tag{9.4.38}$$

Thus we have an example here of a sample-continuous Gaussian process singular with respect to Wiener measure which is the sum (sample-pathwise) of two Gaussian processes each of which is equivalent to Wiener measure.

EXAMPLE 9.4.2. (A stochastic differential equation for the pinned Wiener process). A pinned Wiener process (or *Brownian bridge* as it is sometimes called in the statistical literature) is a Gaussian process with mean zero and with covariance $(t \wedge s) - ts/T$, $(0 \leq t, s \leq T)$. If W_t is a standard Wiener process, then $Y_t = W_t - (t/T)W_T$ $(0 \leq t \leq T)$ is a pinned Wiener process. For each t, Y_t depends not only on W_t but also on the "future" value W_T. To obtain a nonanticipative representation of $Y = (Y_t)$, we seek a transformation of $W = (W_t)$ of the form $\hat{W}_t = \int_0^t F(t,u)\,dW_u$. We have to find $F(t,u)$ satisfying

$$\int_0^{t \wedge s} F(t,u)F(s,u)\,du = t \wedge s - \frac{ts}{T}.$$

Let us set $F(t,u) = f(t)g(u)$ and assume $c^{-1} = f(0)g^2(0) \neq 0$. For t fixed, differentiating

$$f(t)f(s) \cdot \int_0^{t \wedge s} g^2(u)\,du = t \wedge s - \frac{ts}{T}$$

with respect to s $(s < t)$, we obtain

$$f(t)f'(s) \cdot \int_0^s g^2(u)\,du + f(t)f(s)g^2(s) = \frac{T-t}{T}$$

Putting $s = 0$ in the above, $f(t) = c(T-t)/T$. Hence substituting for $f(t)$ in the equation for $F(t,u)$, we get

$$c^2 \frac{(T-t)(T-s)}{T^2} \int_0^{t \wedge s} g^2(u)\,du = t \wedge s - \frac{ts}{T}.$$

For $s < t < T$ this becomes

$$c^2 \frac{(T-s)}{T} \int_0^s g^2(u)\,du = s.$$

Hence $g(s) = T/c(T-s)$ and so

$$F(t,s) = c\left(\frac{T-t}{T}\right)\frac{T}{c(T-s)} = \frac{T-t}{T-s}.$$

Thus we have

$$\hat{W}_t = \int_0^t \frac{T-t}{T-s}\,dW_s \quad \text{if } t < T.$$

Define $\hat{W}_T = 0$. Clearly $(\hat{W}_T)$ is the required representation of the pinned Wiener process. It is easy to see that $\hat{W}_t$ satisfies the stochastic differential equation

$$d\hat{W}_t = dW_t - \frac{\hat{W}_t}{T-t}\,dt \qquad (0 \leq t < T).$$

9.5 Vector-Valued Processes

The representation (9.3.5) can be extended to a d-dimensional Gaussian process equivalent to a d-dimensional Wiener process. We give here only the main steps and leave the details to the reader.

Let $X_t = (X_t^1, \ldots, X_t^d)$, $t \in [0,T]$, be a standard Wiener process on $(\Omega,\mathscr{A},Q)$, that is, with mean zero and covariance Γ_Q given by $\Gamma_Q(t,s) = \min(t,s)I$ where I is the d-dimensional identity matrix. Then for $a,b \in \mathbf{R}^d$ we have

$$E[(a,X_t)(b,X_s)] = (\Gamma_Q(t,s)a,b) = \min(t,s) \cdot (a,b).$$

It is convenient to consider the real-valued process (a,X_t) depending on the parameter (t,a) with $t \in [0,T]$ and $a \in \mathbf{R}^d$. Under the probability measure P, (a,X_t) will be assumed to be Gaussian with mean zero and covariance $\Gamma_P(t,a;\, s,b) = (\Gamma_P(t,s)a,\, b)$.

The definition of a nonanticipative representation of the process (X_t,P) with respect to (X_t,Q) is the same as in Section 9.2 except that Y_t is understood to be a d-dimensional process and $L(X;t)$ in (9.2.1) is defined as the closed linear subspace of $L^2(\Omega,\mathscr{A},Q)$ spanned by the family $\{(a,X_s),\, 0 \le s \le t,\, a \in \mathbf{R}^d\}$ or, equivalently, by $\{X_s^i,\, 0 \le s \le t,\, i = 1, \ldots, d\}$. The RKHS $H(\Gamma_Q)$ of the Wiener process (X_t,Q) is the space of all real functions f on $[0,T] \times \mathbf{R}^d$ of the form $f(t,a) = (\int_0^t \phi(u),a)$ where $\phi = (\phi^1, \ldots, \phi^d) \in L^{2,d}[0,T]$. Recall that $L^{2,d}[0,T]$ is the Hilbert space of square-integrable, d-dimensional functions with the inner product

$$(\phi,\psi)_{L^{2,d}} = \sum_{i=1}^{d} \int_0^T \phi^i(u)\psi^i(u)\, du, \quad (\psi = (\psi^1, \ldots, \psi^d) \in L^{2,d}[0,T]).$$

Every element η in $L(X;T)$ is of the form

$$\eta = \int_0^T (\phi(u),dX_u) = \sum_{i=1}^{d} \int_0^T \phi^i(u)\, dX_u^i$$

where $\phi \in L^{2,d}[0,T]$. The map J which sends η into ϕ is an isometry from $L(X;T)$ onto $L^{2,d}[0,T]$ and is the composition of the two isometries J_1 and J_2, where $J_1\colon H(\Gamma_Q) \to L^{2,d}[0,T]$ sends f into ϕ and $J_2\colon L(X;\, T) \to H(\Gamma_Q)$ sends η into f. Here, $f(t,a)$ is as defined above.

Let K be a bounded linear operator on $L^{2,d}[0,T]$ belonging to the class $\mathscr{S}_2$. Then K is determined by a kernel $K = (K^{ij}) \in L^{2,d}[0,T]^2$. For $\phi \in L^{2,d}[0,T]$ we have

$$(K\phi)(u = \int_0^T K(u,v)\phi(v)\, dv,$$

that is

$$(K\phi)^i(u) = \sum_{j=1}^{d} \int_0^T K^{ij}(u,v)\phi^j(v)\, dv,$$

where

$$\sum_{i,j} \int_0^T \int_0^T [K^{ij}(u,v)]^2\, du\, dv < \infty.$$

Suppose that K in $\mathcal{S}_2$ has the eigenchain $\pi = \{P(t)\}$, where the range of the orthoprojector $P(t)$ is the subspace $\{f \in L^{2,d}[0,T]: f(s) = 0 \text{ a.e. for } t < s \leq T\}$. Then the analogue of Lemma 9.3.1 shows that for each i,j

$$K^{ij}(u,v) = 0 \quad \text{a.e. for } u > v.$$

Let $\tilde{W}_+$ be the operator obtained through the factorization of the operator $\tilde{S}$ as explained in Theorem 9.2.1. Then $\tilde{W}_+ \in \mathcal{S}_2$ on $L(X;T)$ and is a Volterra operator having $\tilde{\pi}$ as an eigenchain ($\tilde{\pi}$ consists of the orthoprojectors $\tilde{P}(t)$ whose range is $L(X;t), t \in [0,T]$). Writing $W_+ = J\tilde{W}_+J^{-1}$, it follows that W_+ is a Volterra operator belonging to the class $\mathcal{S}_2$ on $L^{2,d}[0,T]$ and having π as an eigenchain. Denoting by $(W_+^{ij}(u,v))$ the kernel of W_+ it is easy to verify that

$$\tilde{W}_+\eta = \sum_i \int_0^T \left[\int_0^T \left\{ \sum_j W_+^{ij}(u,v)\phi^j(v) \right\} dv \right] dX_u^i. \tag{9.5.1}$$

Define

$$Y(a,t) = (I + \tilde{W}_+)(a,X_t).$$

Taking $\eta = (a,X_t)$ and the corresponding $\phi(\cdot) = \chi_t(\cdot)\, a$ in expression (9.5.1) and noting that $W_+^{ij}(u,v) = 0$ a.e. if $u > v$, we have

$$\begin{aligned}\tilde{W}_+(a,X_t) &= \sum_i \int_0^T \left\{ \int_0^t \sum_j W_+^{ij}(u,v)a^j\, dv \right\} dX_u^i \\ &= \sum_i \int_0^t \left\{ \int_0^v \sum_j a^j W_+^{ij}(u,v)\, dX_u^i \right\} dv.\end{aligned}$$

Hence

$$Y(a,t) = \sum_j a^j \left[X_t^j + \int_0^t \left\{ \int_0^v \sum_i W_+^{ij}(u,v)\, dX_u^i \right\} dv \right]$$

and so

$$Y(a,t) = (a,Y_t), \quad \text{where } Y_t = (Y_t^1, \ldots, Y_t^d),$$

and

$$Y_t^j = X_t^j + \int_0^t \left[\sum_i \int_0^v W_+^{ij}(u,v)\, dX_u^i \right] dv \tag{9.5.2}$$

($j = 1, \ldots, d$). The d-dimensional Gaussian process (Y_t, Q) is the desired non-anticipative representation. Following the notation for the vector-valued stochastic integral introduced in Chapter 3, we may write (9.5.2) in the form

$$Y_t = X_t + \int_0^t \left[\int_0^v W_+(u,v)\, dX_u \right] dv. \tag{9.5.3}$$

As in the one-dimensional case, a (v,ω)-measurable version of the stochastic integral $\int_0^v W_+(u,v)\, dX_u(\omega)$ is chosen, so that $Y = (Y_t)$ defined by (9.5.3) is a continuous process.

Let us now consider the vector-valued stochastic equation treated in Section 7.5. Assume that $\xi = (\xi_t)$ is given by

$$\xi_t(\omega) = \int_0^t \gamma(s,\xi(\omega))\, ds + W_t(\omega) \tag{9.5.4}$$

(a.s.) for all $t \in [0,T]$. Here $(W_t,\mathscr{F}_t,P)$ is a d-dimensional Wiener martingale and γ is the nonanticipative functional defined in Section 7.5 which satisfies the condition

$$P\left[\omega \in \Omega: \int_0^T |\gamma(t,\xi(\omega))|^2 \, dt < \infty\right] = 1. \tag{9.5.5}$$

We shall make the further assumption that the induced measure μ_ξ is Gaussian. From Theorem 7.5.1 it follows that $\mu_\xi \ll \mu$ and we have $\mu_\xi \equiv \mu$ since μ_ξ and μ are Gaussian measures. Hence Theorem 7.5.2 implies that

$$\mu\left[x \in C_d: \int_0^T |\gamma(t,x)|^2 \, dt < \infty\right] = 1. \tag{9.5.6}$$

With the help of the representation (9.5.3) derived above, Theorems 7.5.1 and 7.5.2 and the remark following them, it is easy to obtain the multi-dimensional analogue of Theorem 9.4.1.

Theorem 9.5.1. *Let $\xi = (\xi_t)$ be a Gaussian process given by Equation (9.5.4) and let condition (9.5.5) be satisfied. Then (ξ_t) has the representation*

$$\xi_t(\omega) = W_t(\omega) + \int_0^t \left[\int_0^s G(u,s) \, dW_u(\omega)\right] ds, \tag{9.5.7}$$

that is,

$$\xi_t^i(\omega) = W_t^i(\omega) + \int_0^t \left[\sum_j \int_0^s G^{ij}(u,s) \, dW_u^j(\omega)\right] ds \qquad (i = 1, \ldots, d).$$

(a.s.) *for all t. $G(u,s)$ is a matrix-valued, Volterra kernel corresponding to the operator G given by the factorization $S = (I + G^*)(I + G)$, where S is the operator determined on $L^{2,d}[0,T]$ by the covariance of ξ as in Theorem 9.4.1.*

It also follows that for every t,

$$\mathscr{F}_t^\xi = \mathscr{F}_t^W. \tag{9.5.8}$$

Linear Filtering Theory 10

10.1 Introduction

The most valuable achievements to date of the filtering theory of Chapter 8 belong to the linear theory which forms the subject of the present chapter and which is associated with the names of Kalman and Bucy. The Kalman filter (as this theory has come to be known) has a central place in our discussion of linear filtering not merely because of the fact that it is the precursor of the general nonlinear theory treated in Chapter 8 (and is still its most important special case) but because of its extensive applications in the post-Sputnik era to problems of tracking of satellites, signal detection, stochastic control, and aerospace engineering.

The problems of prediction, smoothing, and filtering were first considered for stationary processes by Kolmogorov and by Wiener. Developed over the last three decades, this theory has acquired a vast literature and has inspired many attempts to extend it, including the work of Kalman and his school. For the understanding of the material of this chapter it will therefore be helpful to consider briefly the point of view of the Kolmogorov-Wiener theory. However, a description of even the most important results of this theory is beyond the scope of this book. The reader will find an account of the general linear theory of filtering and estimation in the survey article [26] by Kailath. Our aim here is to focus attention on the innovation process and on the manner in which the optimal predictor (or filter) is calculated in the prediction theory of stationary processes.

The innovation process ν has been defined in Chapter 8 and shown to be a Wiener martingale. In the filtering problem of discrete parameter stationary processes, the notion of an innovation subspace and the corresponding notion of an innovation process can be introduced in a very natural manner. Since the latter turns out to be a sequence of uncorrelated random variables it is perhaps more appropriate to call it the difference innovation sequence to distinguish it from the innovation process already introduced.

Let $Z_n = (Z_n^1, \ldots, Z_n^d)$ be a zero mean, stationary process taking values in $\mathbf{R}^d$ and such that it has the form

$$Z_n = S_n + N_n,$$

where the "signal" process (S_n) and the "noise" process (N_n) are zero mean and weakly stationary. By a *noise* process we simply mean here that $E(N_m^i N_n^j) = \delta_{ij}\delta_{nm}$. Let us also make the simplifying assumption that $E(S_m^i N_n^j) = 0$ for $i, j = 1, \ldots, d$ and all m and n (that is, $S \perp N$). Let $H(Z;n)$ be the Hilbert space spanned by $\{Z_k^i, -\infty < k \leq n, i = 1, \ldots, d\}$. To simplify the writing we also suppose that (Z_n) is purely nondeterministic, that is, $\bigcap_n H(Z;n) = 0$. Given the observations $\{Z_k, k \leq n\}$, the problem is to find the best linear estimate in the least squares sense of S_n. Such an estimate $\hat{S}_n = (\hat{S}_n^i)$, where $\hat{S}_n^i$ is given by the orthogonal projection of S_n^i onto $H(Z;n)$ and has the property

$$\operatorname{Trace} E[(S_n - \hat{S}_n)(S_n - \hat{S}_n)^*] = \min_{L = (L^i) \in H(Z;n)} \operatorname{Trace} E[(S_n - L)(S_n - L)^*].$$

Let us now define $\xi_n = (\xi_n^1, \ldots, \xi_n^d)$ by $\xi_n^i = Z_n^i - \hat{S}_n^i$ and also the *innovation subspaces* $M_n = H(Z;n) \ominus H(Z;n-1)$. If U is the underlying unitary operator, it is easily seen that $\xi_n = U^n \xi_0$, $M_n = U^n M_0$ and $M_n \perp M_m$, $(n \neq m)$. We shall first show that $\xi_n^i \in M_n$ for each i. Clearly $\xi_n^i \in H(Z;n)$. For $m \leq n-1$ and $j = 1, \ldots, d$,

$$\begin{aligned} E(\xi_n^i Z_m^j) &= E[(S_n^i - \hat{S}_n^i + N_n^i) Z_m^j] \\ &= E(N_n^i Z_m^j) = E[N_n^i(S_m^j + N_m^j)] = 0, \end{aligned}$$

so that $\xi_n^i \perp H(Z;n-1)$, proving $\xi_n^i \in M_n$. We call (ξ_n) the *innovation process* for the filtering problem and ξ_n, the nth innovation vector. Note that $\xi_n^i = U^n \xi_0^i$. It is well known that the dimension of M_n is independent of n and does not exceed d. It is easy to verify that ξ_n^i $(i = 1, \ldots, d)$ are linearly independent so that dimension $(M_n) = d$. Let (ζ_n^i), $(i = 1, \ldots, d)$ be the set obtained by orthonormalizing (ξ_n^i). Then, from the relation $H(Z;n) = \sum_{j=0}^{\infty} \oplus M_{n-j}$, it follows that every element in $H(Z;n)$ has an "expansion" in terms of the orthonormal basis (ζ_{n-j}^i), $(i = 1, \ldots, d, j \geq 0)$. In particular, we obtain

$$\hat{S}_n^i = \sum_{j=0}^{\infty} \sum_{k=1}^{d} f_{n-j}^{ik} \zeta_{n-j}^k,$$

where $\sum_{j=0}^{\infty} \sum_{k=1}^{d} (f_{n-j}^{ik})^2 < \infty$. Letting f_{n-j} be the $d \times d$ matrix with the entries f_{n-j}^{ik}, we have the following representation for the best filter in terms of the innovations:

$$\hat{S}_n = \sum_{j=0}^{\infty} f_{n-j} \zeta_{n-j}.$$

Another consequence is the property $H(Z;n) = H(\xi;n)$ for every n.

Now, in view of the orthogonal sum decomposition of $H(Z;n)$ in terms of the spaces M_{n-j}, one can obtain a representation for $\hat{S}_n$ of the above form starting with an *arbitrary* orthonormal basis in M_n and constructing a basis in $H(Z;n)$ by multiplying by powers of U. The point of the above discussion is that the ξ_s^i's are already defined independently and then shown to be a basis in M_n.

The algorithm for calculating the coefficients f_{n-j}, even when $d = 1$, is based on spectral analysis and involves rather deep results from the theory of Hardy class functions.

It should be noted that for the prediction problem, the definition of the innovation vector is to be suitably modified. We have $S_n = Z_n$ since $N_n = 0$. The appropriate definition then is $\xi_n^i = Z_n^i - \hat{Z}_{n,n-1}^i$, where $\hat{Z}_{n,n-1}^i$ is the projection of Z_n^i on $H(Z;n-1)$.

The situation for continuous parameter processes is more complicated but the basic ideas are the same. For simplicity, let us consider the prediction problem for a purely nondeterministic process (Z_t). The best linear predictor of Z_{t_1} based on $H(Z;t)$, $(t \leq t_1)$ is then given by

$$\hat{Z}_{t_1,t} = \int_{-\infty}^{t} f(t_1 - s)\, d\zeta_s.$$

The process (ζ_s) has orthogonal increments, that is,

$$E(\zeta_b - \zeta_a) = 0, \qquad E(\zeta_b - \zeta_a)(\zeta_{b'} - \zeta_{a'}) = 0$$

if the intervals (a,b) and (a',b') are disjoint and $E[\zeta_b - \zeta_a]^2 = b - a$. A rigorous analogue of the ζ_n's in the discrete case does not now exist. What corresponds is the "process" $d\zeta_t$ or $\dot{\zeta}_t\, dt$, which may be called a *differential innovation process*. The process (ζ_t) itself is akin to the innovation process (ν_t) introduced in Chapter 8 (and of which more will be seen in this chapter).

The difficulties of defining "white noise" directly as a proper process stand in the way of formulating the precise extension to the continuous case of the discrete time filtering problem treated above. Nevertheless, the following heuristic approach shows how one is led to the Wiener-Hopf equation in determining the optimal linear filter. So, let $Z_t = S_t + N_t$ where all the processes are stationary, have zero means and (N_t) is "white noise," $EN_tN_s = \delta(t - s)$, δ being the Dirac function. We also assume $(S_t) \perp (N_t)$. If we set

$$\hat{S}_t = \int_{-\infty}^{t} f(t - u) Z_u\, du,$$

we have the orthogonality condition

$$E(S_t Z_s) = E(\hat{S}_t Z_s) \text{ for } s \leq t.$$

The left-hand side in the above equation equals $\rho_S(t - s)$, where ρ_S is the covariance of (S_t). The right-hand side, which equals

$$\int_{-\infty}^{t} f(t - u) E(Z_u Z_s)\, du,$$

reduces to $\int_{-\infty}^{t-s} f(t-s-u)\rho_S(u)\,du + f(t-s)$, and replacing $t-s$ by t, we obtain the Wiener-Hopf equation

$$\rho_S(t) = f(t) + \int_{-\infty}^{t} f(t-u)\rho_S(u)\,du \qquad (0 < t < \infty).$$

The solution f and hence the complete determination of the optimal filter is then achieved by spectral factorization techniques.

The Kalman theory is concerned with the circle of ideas outlined above but provides a new direction of development. Its novel feature—and as it turns out, a fundamental change in point of view from the Kolmogorov-Wiener theory—is to regard the signal process as the output of a linear, stochastic dynamical system driven by white noise. Then the optimal linear filter also has a stochastic differential structure, thus making it possible to compute the filter recursively. Stationarity is no longer a basic assumption, and the spectral factorization which plays so essential a role in the classical theory is now replaced by the solution of an ordinary differential equation with known initial conditions, though the differential equation itself is non-linear and of Riccati type.

Before proceeding to rigorous definitions, it is instructive to look at some of the examples which motivated Kalman and Bucy in the choice of their model. We retain their notation and the white noise formulation.

The state process is an n-dimensional process x_t satisfying

$$\dot{x}_t = F(t)x_t + G(t)u_t, \tag{10.1.1}$$

where $F(t)$ is an $n \times n$ continuous matrix (nonrandom) function of t, $G(t)$ is an $n \times m$ continuous matrix (nonrandom) function of t and $u(t)$ is m-dimensional white noise. The output is denoted by

$$s_t = H(t)x_t, \tag{10.1.2}$$

where $H(t)$ is a $d \times n$ matrix of continuous functions. The observations are on the process

$$o_t = s_t + v_t = H(t)x_t + v_t, \tag{10.1.3}$$

where v_t is "observation white noise." It is further assumed that the white noises u and v are independent.

Example 10.1.1. Consider the motion of a particle whose acceleration is white noise. It is assumed that its position and velocity can be observed in the presence of additive white noise.

Let $x_t^{(1)}$ = position and $x_t^{(2)}$ = velocity of the particle. Then

$$x_t^{(2)} = \dot{x}_t^{(1)} \quad \text{and} \quad \dot{x}_t^{(2)} = u_t.$$

We obtain the dynamical system [Equation (10.1.1)] if we set

$$x_t = \begin{bmatrix} x_t^1 \\ x_t^2 \end{bmatrix}, \qquad F(t) = \begin{bmatrix} 0 & 1 \\ 0 & 0 \end{bmatrix}, \qquad G = \begin{bmatrix} 0 \\ 1 \end{bmatrix},$$

and u_t is scalar white noise. For the observation process

$$o_t^{(1)} = h_1(t)x_t^{(1)} + v_t^1$$
$$o_t^{(2)} = h_2(t)x_t^{(2)} + v_t^2,$$

where

$$v_t = \begin{bmatrix} v_t^{(1)} \\ v_t^{(2)} \end{bmatrix}$$

is white noise. We thus get Equation (10.1.3) upon setting

$$H(t) = \begin{bmatrix} h_1(t) & 0 \\ 0 & h_2(t) \end{bmatrix}.$$

EXAMPLE 10.1.2. A particle leaves at time $t = 0$ with a fixed, unknown velocity v. Only the position at time t is continually observed in the presence of white noise. The problem is to find the best estimate of position and velocity.

Let $x_t^{(1)}$ = position and $x_t^{(2)}$ = velocity at time t. Then

$$\dot{x}_t^{(1)} = x_t^{(2)}, \qquad x_0^{(2)} = v, \qquad \dot{x}_t^{(2)} = u_t,$$

and observed position

$$o_t = x_t^{(1)} + v_t^{(1)},$$

$v_t^{(1)}$ being white noise. This information can be cast in the form of Equations (10.1.1) and (10.1.3) if we write

$$x_t = \begin{bmatrix} x^{(1)} \\ x_t^{(2)} \end{bmatrix}, \qquad F(t) = \begin{bmatrix} 0 & 1 \\ 0 & 0 \end{bmatrix}, \qquad G(t) = 0,$$

$$H(t) = [1,0] \quad \text{and} \quad v_t = \begin{bmatrix} v_t^{(1)} \\ 0 \end{bmatrix}.$$

EXAMPLE 10.1.3. Again we take $x_t^{(1)}$ = position, $x_t^{(2)} = \dot{x}_t^{(1)}$,

$$x_t = \begin{bmatrix} x_t^1 \\ x_t^2 \end{bmatrix}, \qquad F(t) = \begin{bmatrix} 0 & 1 \\ 0 & 0 \end{bmatrix}, \qquad G(t) = \begin{bmatrix} 0 \\ 1 \end{bmatrix}.$$

If both position and velocity are observed in the presence of white noise, take

$$H(t) = \begin{bmatrix} h_1 & 0 \\ 0 & h_2 \end{bmatrix} \quad \text{and} \quad v_t = \begin{bmatrix} v_t^1 \\ v_t^2 \end{bmatrix}.$$

The basic assumption of this chapter (except where we indicate generalizations) is that the signal or system process $X = (X_t)$ and the observation process $Y = (Y_t)$ together form a Gaussian Markov process governed by a stochastic differential equation. The conditional distribution of the state of the system, that is, X_t given $\mathscr{F}_t^Y$, is easily seen to be Gaussian. The problem then is to describe the behavior of the conditional mean and covariance "dynamically," that is, by deriving differential equations for $\hat{X}_t = E(X_t|\mathscr{F}_t^Y)$ and $P(t) = E[(X_t - \hat{X}_t)(X_t - \hat{X}_t)^*|\mathscr{F}_t^Y]$.

Following Kalman and Bucy's original derivation in [35], other proofs have since been given which are based on a passage from discrete time to the continuous or on a Bayes formula [32, 54]. We shall follow the martingale approach adopted in Chapter 8. However, two proofs will be given: one which deduces the result as a corollary from the equation of the optimal nonlinear filter obtained in Chapter 8, while the second is an independent derivation which exploits the Gaussian features of the model. (See [2]).

10.2 The Stochastic Model for the Kalman Theory

The signal process X_t and the observation process Y_t are given by the stochastic differential equations

$$dX_t = [A_0(t) + A_1(t)X_t + A_2(t)Y_t]\,dt + B(t)\,dW_t$$

and (10.2.1)

$$dY_t = [C_0(t) + C_1(t)X_t + C_2(t)Y_t]\,dt + D(t)\,dW_t$$

$(0 \le t \le T)$, with initial random variable X_0 given to be independent of $\mathscr{F}_T^W$ and $Y_0 = 0$ (a.s.). (X_t) and (Y_t) are m- and n-dimensional processes and (W_t) is a q-dimensional Wiener process $(q = m + n)$. The coefficients A_i, C_i $(i = 0,1,2)$, B and D are assumed to be nonrandom, matrix-valued functions of t of appropriate dimensions.

Let us first assure ourselves that, under suitable conditions on the coefficients, Equation (10.2.1) has a unique solution. For this it is convenient to write $\xi_t = (\xi_t^1, \ldots, \xi_t^q)$, where $\xi_t^i = X_t^i$ for $i = 1, \ldots, m$ and $\xi_t^{m+j} = Y_t^j$ for $j = 1, \ldots, n$. Then it is easy to see that (10.2.1) can be recast in the form

$$d\xi_t = [\bar{A}_0(t) + \bar{A}_1(t)\xi_t]\,dt + \bar{B}(t)\,dW_t \qquad \xi_0 = \zeta, \tag{10.2.2}$$

where $\zeta^i = X_0^i$ $(i = 1, \ldots, m)$ and $\zeta^{m+i} = 0$ (a.s.) $(i = 1, \ldots, n)$. The coefficients $\bar{A}_0$, $\bar{A}_1$, and $\bar{B}$ can easily be expressed in terms of the coefficients of (10.2.1). Linear stochastic equations of this type for one-dimensional processes were studied in Chapter 5. The existence and uniqueness of the solution of (10.2.2) can be deduced easily from Theorem 5.1.3. However, we shall give an independent proof under slightly different conditions on the coefficients and at the same time, obtain the solution explicitly.

Assume that the entries in the coefficients of (10.2.2) are measurable. Further suppose that the entries in A_i and C_i $(i = 0,1,2)$ are integrable over $[0,T]$ and those in B and D are square-integrable. Then, for Equation

(10.2.2), the integrals

$$\int_0^T |\bar{A}_0^i(t)|\,dt, \qquad \int_0^T |\bar{A}_1^{ij}(t)|\,dt, \quad \text{and} \quad \int_0^T [\bar{B}^{ij}(t)]^2\,dt$$

are finite for every i and j.

Now it is easy to verify that (10.2.2) has a unique solution (ξ_t) if and only if $\xi_t = \tilde{\xi}_t + \bar{\xi}_t$, where $(\tilde{\xi}_t)$ and $(\bar{\xi}_t)$ are, respectively, the unique solutions of the following equations:

$$d\tilde{\xi}_t = \bar{A}_1(t)\tilde{\xi}_t\,dt + \bar{B}(t)\,dW_t, \qquad \tilde{\xi}_0 = \zeta \tag{10.2.2a}$$

and

$$\frac{d\bar{\xi}_t}{dt} = \bar{A}_0(t) + \bar{A}_1(t)\bar{\xi}_t, \qquad \bar{\xi}_0 = 0. \tag{10.2.2b}$$

Equation (10.2.2a) has a unique solution given by

$$\tilde{\xi}_t = F(t)\left[\zeta + \int_0^t F(s)^{-1}\bar{B}(s)\,dW_s\right], \tag{10.2.3}$$

where $F(t)$ is the fundamental matrix given by the equation

$$F(t) = I + \int_0^t \bar{A}_1(s)F(s)\,ds \tag{10.2.4}$$

(I is the identity matrix).

Proof. First, we show that (10.2.4) has a solution and that $F(t)^{-1}$ exists. Recursively define

$$F_0(t) \equiv I,\ F_{j+1}(t) = I + \int_0^t \bar{A}_1(s)F_j(s)\,ds \qquad (j \geq 1).$$

From

$$F_{j+1}(t) - F_j(t) = \int_0^t \bar{A}_1(s)[F_j(s) - F_{j-1}(s)]\,ds,$$

we have (in terms of entries of the matrices),

$$\sum_{l,m=1}^{k} |(F_{j+1}(t) - F_j(t))_{lm}| \leq \int_0^t \sum_{l,m} |\bar{A}_1^{lm}(s)| \sum_{l,m} |(F_j(s) - F_{j-1}(s))_{l,m}|\,ds.$$

For all t,

$$\sum_{l,m} |(F_1(t) - F_0(t))_{l,m}| \leq \int_0^t \sum_{l,m} |\bar{A}_1^{lm}(s)|\,ds.$$

Using this with (10.2.4) we have

$$\sum_{l,m=1} |(F_{j+1}(t) - F_j(t))_{l,m}| \leq \frac{1}{j!}\left[\int_0^T \sum_{l,m} |\bar{A}_1^{lm}(s)|\,ds\right]^j$$

It follows that the series $F_0(t) + \sum_{j=0}^{\infty} [F_{j+1}(t) - F_j(t)]$ converges absolutely and uniformly to $F(t)$, say, whose elements are continuous functions of t. Furthermore, for a.e. t, $F(t)$ is differentiable. Denoting the determinant of

$F(t)$ by $[F(t)]$, it can be verified that (for a.e. t),

$$\frac{d}{dt}[F(t)] = \text{Trace}\,\bar{A}_1(t) \cdot [F(t)], \qquad [F(0)] = 1.$$

The existence of $F(t)^{-1}$ now follows from

$$[F(t)] = \exp \int_0^t \text{Trace}\,\bar{A}_1(s)\,ds.$$

From the fact that, for a.e. t,

$$\frac{dF(t)^{-1}}{dt} = -F(t)^{-1}\frac{dF(t)}{dt}F(t)^{-1} = -F(t)^{-1}\bar{A}_1(t),$$

uniqueness of the solution of (10.2.4) follows if we observe that $(d/dt)[F_1(t)^{-1}F_2(t)] = 0$, where F_1 and F_2 are two solutions of (10.2.4) with $F_1(0) = F_2(0) = I$.

The existence of a solution of Equation (10.2.2a) is established by applying the version of Ito's lemma given in Section 4.5 to the process $(\tilde{\xi}_t)$ defined by (10.2.3). The uniqueness is proved as follows. Let $\tilde{\xi}_t^{(1)}(\omega)$ and $\tilde{\xi}_t^{(2)}(\omega)$ be two solutions of (10.2.2a) (which are necessarily sample continuous). Set $\Delta_t(\omega) = \tilde{\xi}_t^{(1)}(\omega) - \tilde{\xi}_t^{(2)}(\omega)$ and denote its ith component by $\Delta_t^i(\omega)$. Then for all t in $[0,T]$ we have

$$\Delta_t(\omega) = \Delta_0(\omega) + \int_0^t \bar{A}_1(s)\,\Delta_s(\omega)\,ds,$$

$$\sum_i |\Delta_t^i(\omega)| \le \sum_i |\Delta_0^i(\omega)| + \int_0^t \sum_{i,j} |\bar{A}_1^{ij}(s)| \sum_i |\Delta_s^i(\omega)|\,ds.$$

Introducing the abbreviated notation $f(t) = \sum_i |\Delta_t^i|, h(s) = \sum_{i,j} |\bar{A}_1^{ij}(s)|$, we have $f(t) \le f(0) + \int_0^t h(s)f(s)\,ds$. Using this bound for $f(s)$ inside the integral, we obtain

$$f(t) \le f(0)\left[1 + \int_0^t h(s)\,ds\right] + \rho_2(t),$$

where $\rho_2(t) = \int_0^t h(s)(\int_0^s h(u)f(u)\,du)\,ds$. Letting $M = \sup_{0 \le t \le T} f(t)$ (note that M depends on ω), we find that $\rho_2(t) \le \frac{1}{2}M(\int_0^t h(s)ds)^2$. Proceeding similarly for any positive integer n, we have [writing $H(t) = \int_0^t h(s)\,ds$]

$$f(t) \le f(0)\left[1 + H(t) + \frac{(H(t))^2}{2!} + \cdots + \frac{(H(t))^n}{n!}\right] + \rho_{n+1}(t),$$

where

$$\rho_{n+1}(t) \le M\frac{(H(t))^{n+1}}{(n+1)!} \le \frac{M(H(T))^{n+1}}{(n+1)!} \to 0 \quad \text{as } n \to \infty.$$

Hence for all t in $[0,T]$, $f(t) \le f(0)\exp H(T)$. Now $f(0) = \sum_i |\Delta_0^i(\omega)| = 0$ if $\tilde{\xi}_t^1(\omega)$ and $\tilde{\xi}_t^2(\omega)$ satisfy the same initial condition in (10.2.2a). Thus we have $f(t) \equiv 0$, that is, $\tilde{\xi}_t^1(\omega) \equiv \tilde{\xi}_t^2(\omega)$ for all t (a.s.), and the uniqueness is proved.

Next, consider Equation (10.2.2b). The unique solution exists and is given by

$$\bar{\xi}_t = F(t)\int_0^t F(s)^{-1}\bar{A}_0(s)\,ds, \tag{10.2.5}$$

where $F(t)$ is the solution of (10.2.4). This can be proved by direct verification. □

Combining (10.2.3) and (10.2.5), we have the following result.

Theorem 10.2.1. *The unique solution of Equation (10.2.2) is given by*

$$\xi_t = F(t)\left[\zeta + \int_0^t F(s)^{-1}\bar{A}_0(s)\,ds + \int_0^t F(s)^{-1}\bar{B}(s)\,dW_s\right], \tag{10.2.6}$$

where $F(t)$ satisfies (10.2.4).

Remark 10.2.1. If X_0 (hence ζ) is a Gaussian random variable independent of (W_t) $(0 \le t \le T)$, it follows immediately from (10.2.6) that (ξ_t), that is, the process (X_t, Y_t), is jointly Gaussian. The assumption that X_0 is Gaussian will be made throughout this chapter.

Remark 10.2.2. From (10.2.6) we can also conclude that the conditional distribution $\mathscr{L}(X_t|\mathscr{F}_t^Y)$, of X_t given $\mathscr{F}_t^Y$ is Gaussian. This can be seen as follows. Fix $t > 0$ and let $D_N = \{0 = s_1^N < \cdots < s_{a_N}^N = t\}$ $(N \ge 1)$ be such that $D_N \subseteq D_{N+1}$ and $\bigcup_N D_N$ is dense in $[0,t]$. Remark 10.2.1 implies that for each $\theta \in \mathbf{R}^m$ and $A \in \mathscr{B}(\mathbf{R})$, the conditional probability $P[(\theta,X_t) \in A|Y_s, s \in D_N]$ is Gaussian. Now since (Y_s) is a continuous process, the σ-fields $\sigma[Y_s,\ s \in D_N] \uparrow \mathscr{F}_t^Y$ as $N \to \infty$. Hence $P[(\theta,X_t) \in A|\mathscr{F}_t^Y]$ is a Gaussian random variable.

Remark 10.2.3. Let $\hat{X}_t = E(X_t|\mathscr{F}_t^Y)$ be the conditional mean and $P(t) = E[(X_t - \hat{X}_t)(X_t - \hat{X}_t)^*|\mathscr{F}_t^Y]$ the conditional covariance matrix of X_t given $\mathscr{F}_t^Y$. We then have the following important and well-known consequence of the Gaussian hypotheses:

$$P(t) = E[(X_t - \hat{X}_t)(X_t - \hat{X}_t)^*] = E(X_t X_t^*) - E(\hat{X}_t \hat{X}_t^*). \tag{10.2.7}$$

Let us write $m(t) = E(X_t)$, $\gamma(t) = E(Y_t)$ and $Y_t' = Y_t - \gamma(t)$. Recalling that $L(Y';t)$ is the Hilbert space spanned by $\{Y_s'^j, 0 \le s \le t, j = 1, \ldots, n\}$ and using Remarks 10.2.1 and 10.2.2, it follows that

$$\begin{aligned}\hat{X}_t^i - m_t^i &= E(X_t^i - m_t^i|\mathscr{F}_t^Y)\\ &= P_{L(Y';t)}(X_t^i - m_t^i),\end{aligned}$$

$[P_{L(Y';t)} =$ orthoprojector with range $L(Y';t)]$. Hence $X_t^i - \hat{X}_t^i \perp L(Y';t)$ and again because we are dealing with Gaussian families of random variables,

$X_t^i - \hat{X}_t^i \perp\!\!\!\perp Y_s'$ for all $s \leq t$. Thus

$$E[(X_t^i - \hat{X}_t^i)(X_t^j - \hat{X}_t^j)|\mathscr{F}_t^Y] = E[(X_t^i - \hat{X}_t^i)(X_t^j - \hat{X}_t^j)],$$

and Equation (10.2.7) is proved.

We shall take the system (10.2.1) as the model for the signal and observation processes though in most treatments of the Kalman theory it is customary to take $A_0 = C_0 = 0$ and $A_2 = C_2 = 0$.

The conditional mean $\hat{X}_t$ is clearly the optimal linear (as well as nonlinear) estimate of X_t given $\mathscr{F}_t^Y$ in the sense that

$$\text{Trace } P(t) \leq \text{Trace } E(X_t - m(t) - \phi_t)(X_t - m(t) - \phi_t)^*$$

for all $\mathscr{F}_t^Y$-measurable, m-dimensional random vectors ϕ_t such that $E(\phi_t) = 0$ and $E|\phi_t|^2 < \infty$.

The problem, as already mentioned in the previous section, is to obtain differential equations (stochastic or otherwise) for $\hat{X}_t$ and $P(t)$.

10.3 Derivation of the Kalman Filter from the Nonlinear Theory

If A is a matrix, let A^{ij} denote the (i,j)th element of $A, A^{\cdot j}$ the vector given by the jth column of A and $|A|^2 = \sum_{i,j}(A^{ij})^2$. The symbol, (,) denotes the inner product in any Euclidean space.

Let us impose the following additional restrictions on the coefficients of (10.2.1). For all $t \in [0,T]$, there is a constant d such that

$$D(t)D^*(t) \geq d > 0; \tag{10.3.1}$$

$$\int_0^T |C_i(t)|^2\, dt < \infty \qquad (i = 0,1,2). \tag{10.3.2}$$

The optimal (nonlinear) filter satisfies the stochastic equations (8.4.22) of Theorem (8.4.4). The model (10.2.1) is a special case of the stochastic differential system given in Section 8.5, and we note that if $f \in C^2(\mathbf{R}^m)$,

$$\begin{aligned}\tilde{A}_t f &= \sum_{i=1}^m [A_0^i(t) + (A_1(t)X_t)^i + (A_2(t)Y_t)^i] \frac{\partial f}{\partial x^i} \\ &\quad + \frac{1}{2} \sum_{i,j=1}^m (B(t)B^*(t))^{ij} \frac{\partial^2 f}{\partial x^i\, \partial x^j}\end{aligned}$$

and

$$\tilde{D}_t^i f = \sum_{j=1}^m B^{ji}(t) \frac{\partial f}{\partial x^j}.$$

The assumptions of Theorem 8.4.4 are easily seen to be satisfied. We have

$$a_s = C_0(s) + C_1(s)X_s + C_2(s)Y_s,$$

$c_s^{-\frac{1}{2}} = (D(s)D^*(s))^{-\frac{1}{2}}$ which, for convenience, we denote by α_s, and

$$h'_s = c_s^{-\frac{1}{2}}a_s = \alpha_s[C_0(s) + C_1(s)X_s + C_2(s)Y_s].$$

Let us first take $f(x) = (\theta,x)$, where θ is an arbitrary vector in $\mathbf{R}^m$. Then

$$E^s[f(X_s)a_s] = E[(\theta,X_s)\{C_0(s) + C_1(s)X_s + C_2(s)Y_s\}|\mathscr{F}_s^Y].$$

The ith component

$$\begin{aligned}(E^s[f(X_s)a_s])^i &= E\left[\sum_k \theta^k X_s^k \left\{C_0^i(s) + \sum_j C_1^{ij}(s)X_s^j + \sum_l C_2^{il}(s)Y_s^l\right\}\Bigg|\mathscr{F}_s^Y\right] \\ &= (\theta,\hat{X}_s)C_0^i(s) + \sum_{j,k}\theta^k C_1^{ij}(s)E(X_s^jX_s^k|\mathscr{F}_s^Y) \\ &\quad + \sum_{l,k}\theta^k C_2^{il}(s)\hat{X}_s^kY_s^l,\end{aligned} \tag{10.3.3}$$

$$\begin{aligned}(E^s[f(X_s)]E^s[a_s])^i &= (\theta,\hat{X}_s)\left[C_0^i(s) + \sum_j C_1^{ij}(s)\hat{X}_s^j + \sum_l C_2^{il}(s)Y_s^l\right] \\ &= (\theta,\hat{X}_s)C_0^i(s) + \sum_{j,k}\theta^k C_1^{ij}(s)\hat{X}_s^j\hat{X}_s^k + (\theta,\hat{X}_s)\sum_l C_2^{il}(s)Y_s^l.\end{aligned} \tag{10.3.4}$$

Hence

$$\begin{aligned}(E^s[f(X_s)a_s] - E^s[f(X_s)]E^s[a_s])^i &= \sum_k \theta^k \sum_j C_1^{ij}(s)\{E(X_s^jX_s^k|\mathscr{F}_s^Y) - \hat{X}_s^j\hat{X}_s^k\} \\ &= \sum_k \theta^k \sum_j C_1^{ij}(s)P^{jk}(s) \\ &= \sum_k \theta^k \sum_j P^{jk}(s)C_1^{*ji}(s) \\ &= \sum_k \theta^k[P(s)C_1^*(s)]^{ki} = (\theta,(P(s)C_1^*(s))^{\cdot i}),\end{aligned} \tag{10.3.5}$$

where $(P(s)C_1^*(s))^{\cdot i}$ is the vector with components $(P(s)C_1^*(s))^{ki}$.

Next from the definition of $\tilde{D}_s$,

$$\tilde{D}_s^i f(X_s) = \sum_{j=1}^m \theta^j B^{ji}(s), \qquad i = 1, \ldots, q,$$

so that

$$E^s[\tilde{D}_s^i f(X_s)] = \sum_{j=1}^m \theta^j B^{ji}(s), \qquad i = 1, \ldots, q.$$

The ith component

$$\begin{aligned}
[b_s E^s(\tilde{D}_s f(X_s))]^i &= \sum_{j=1}^{q} b_s^{ij} E^s(\tilde{D}_s^j f(X_s)) \\
&= \sum_{j=1}^{q} b_s^{ij} \sum_{k=1}^{m} \theta^k B^{kj}(s) \\
&= \sum_{j=1}^{q} D_s^{ij} \sum_{k=1}^{m} \theta^k B^{kj}(s) \\
&= \sum_{k=1}^{m} \theta^k \left(\sum_{j=1}^{q} D_s^{ij} B^{kj}(s) \right) \\
&= \sum_{k=1}^{m} \theta^k \left(\sum_{j=1}^{q} B^{kj}(s) D^{*ji}(s) \right) \\
&= \sum_{k=1}^{m} \theta^k (B(s)D^*(s))^{ki} \\
&= (\theta, (B(s)D^*(s))^{\cdot i}).
\end{aligned} \tag{10.3.6}$$

Hence, the ith component of

$$E^s[f(X_s)a_s] - E^s(f(X_s))E^s(a_s) + b_s E^s(\tilde{D}_s f(X_s)) = (\theta, R(s)^{\cdot i}), \tag{10.3.7}$$

where $R(s) = P(s)C^*(s) + B(s)D^*(s)$. Let us define

$$L(s) = [P(s)C^*(s) + B(s)D^*(s)][D(s)D^*(s)]^{-\frac{1}{2}} = R(s)\alpha(s)$$

and write

$$Q(s) = E^s[f(X_s)a_s] - E^s[f(X_s)]E^s(a_s) + b_s E^s(\tilde{D}_s f(X_s)).$$

The stochastic integral

$$\int_0^t (\alpha(s)Q(s), dv_s) = \sum_k \int_0^t [\alpha(s)Q(s)]^k \, dv_s^k.$$

Now,

$$\begin{aligned}
[\alpha(s)Q(s)]^k &= \sum_i \alpha(s)^{ki}(\theta, R(s)^{\cdot i}) = \left(\theta, \sum_i \alpha(s)^{ki} R(s)^{\cdot i}\right) \\
&= \sum_l \theta^l \left(\sum_i \alpha(s)^{ki} R(s)^{li} \right) = \sum_l \theta^l \sum_i R(s)^{li} \alpha(s)^{ik} \\
&= \sum_l \theta^l [R(s)\alpha(s)]^{lk} = \sum_l \theta^l L(s)^{lk}.
\end{aligned}$$

We have

$$\begin{aligned}
\int_0^t (\alpha(s)Q(s), dv_s) &= \sum_k \int_0^t \sum_l \theta^l L(s)^{lk} \, dv_s^k \\
&= \sum_l \theta^l \int_0^t \sum_k L(s)^{lk} \, dv_s^k.
\end{aligned} \tag{10.3.8}$$

For $f(x) = (\theta,x)$, we have

$$\tilde{A}_s f = (\theta, A_0(s) + A_1(s)X_s + A_2(s)Y_s)$$

and

$$E^s[\tilde{A}_s f] = (\theta, A_0(s) + A_1(s)\hat{X}_s + A_2(s)Y_s). \tag{10.3.9}$$

Equation (8.4.22) then becomes

$$(\theta,\hat{X}_t) = (\theta,\hat{X}_0) + \int_0^t (\theta, A_0(s) + A_1(s)\hat{X}_s + A_2(s)Y_s)ds + \sum_l \theta^l \int_0^t \sum_k L(s)^{lk}\, dv_s^k \tag{10.3.10}$$

where v stands for v' in (8.4.22). It follows that

$$\hat{X}_t = \hat{X}_0 + \int_0^t [A_0(s) + A_1(s)\hat{X}_s + A_2(s)Y_s]\, ds + \int_0^t L(s)\, dv_s, \tag{10.3.11}$$

or

$$d\hat{X}_t = [A_0(t) + A_1(t)\hat{X}_t + A_2(t)Y_t]\, dt + [P(t)C_1^*(t) + B(t)D^*(t)][D(t)D^*(t)]^{-\frac{1}{2}}\, dv_t \tag{10.3.11a}$$

with $\hat{X}_0 = E(X_0)$.

It now remains to derive the differential equation satisfied by $P(t)$. We first obtain the differentials for $E(X_t X_t^*)$ and $E(\hat{X}_t \hat{X}_t^*)$ using Ito's formula and then use the fact

$$P(t) = E(X_t X_t^*) - E(\hat{X}_t \hat{X}_t^*).$$

In Equation (4.4.5), take $\psi(t,x) = x^i x^j$. We then have

$$X_t^i X_t^j = X_0^i X_0^j + \int_0^t \Bigg[\Big(A_0^i(s) + \sum_l A_1^{il}(s)X_s^l + \sum_l A_2^{il}(s)Y_s^l\Big) X_s^j + \Big(A_0^j(s) + \sum_l A_1^{jl}(s)X_s^l + \sum_l A_2^{jl}(s)Y_s^l\Big) X_s^i + (B(s)B^*(s))^{ij}\Bigg] ds + \sum_l \int_0^t (X_s^i B(s)^{jl} + X_s^j B_s^{il})\, dW_s^l.$$

It is easy to see that we may take expectations on both sides. Noting that $E(X_s^k Y_s^l) = E(\hat{X}_s^k Y_s^l)$ $(k = i,j)$, we obtain

$$E(X_t^i X_t^j) = E(X_0^i X_0^j) + \int_0^t \Bigg[A_0^i(s)E(X_s^j) + \sum_l A_1^{il}(s)E(X_s^l X_s^j) + \sum_l A_2^{il}(s)E(\hat{X}_s^j Y_s^l) + A_0^j(s)E(X_s^i) + \sum_l A_1^{jl}(s)E(X_s^l X_s^i) + \sum_l A_2^{jl}(s)E(\hat{X}_s^i Y_s^l) + (B(s)B^*(s))^{ij} \Bigg] ds. \tag{10.3.12}$$

Again applying the Ito formula of Section 4.4 to the stochastic differential given by (10.3.11), we have

$$E(\hat{X}_t^i\hat{X}_t^j) = E(\hat{X}_0^i\hat{X}_0^j) + \int_0^t \Big[A_0^i(s)E(X_s^j) + \sum_l A_1^{il}(s)E(\hat{X}_s^l\hat{X}_s^j)$$

$$+ \sum_l A_2^{il}(s)E(\hat{X}_s^j Y_s^l) + A_0^j(s)E(X_s^i)$$

$$+ \sum_l A_1^{jl}(s)E(\hat{X}_s^l\hat{X}_s^i) + \sum_l A_2^{jl}(s)E(\hat{X}_s^i Y_s^l)$$

$$+ (L(s)L^*(s))^{ij} \Big] ds. \tag{10.3.13}$$

The (i,j)th element of the matrix $P(t)$, $P^{ij}(t) = E(X_t^i X_t^j) - E(\hat{X}_t^i\hat{X}_t^j)$ satisfies the following equation obtained by subtracting Equation (10.3.13) from (10.3.12):

$$P^{ij}(t) = P^{ij}(0) + \int_0^t \Big[\sum_l A_1^{il}(s)P^{lj}(s) + \sum_l A_1^{jl}(s)P^{li}(s)$$

$$+ (B(s)B^*(s))^{ij} - (L(s)L^*(s))^{ij} \Big] ds$$

$$= P^{ij}(0) + \int_0^t [(A_1(s)P(s))^{ij} + (P(s)A_1^*(s))^{ij}$$

$$+ (B(s)B^*(s))^{ij} - (L(s)L^*(s))^{ij}]\, ds. \tag{10.3.14}$$

Hence $P(t)$ satisfies the matrix differential equation

$$\frac{dP(t)}{dt} = A_1(t)P(t) + P(t)A_1^*(t) + B(t)B^*(t) - [P(t)C_1^*(t)$$

$$+ B(t)D^*(t)][D(t)D^*(t)]^{-1}[C_1(t)P(t) + D(t)B^*(t)], \tag{10.3.15}$$

with

$$P(0) = E[(X_0 - EX_0)(X_0 - EX_0)^*].$$

Equations (10.3.11a) and (10.3.15) are called the *Kalman filter* or the *Kalman-Bucy filter.*

10.4 The Filtering Problem for Gaussian Processes

For an independent derivation of the Kalman filter it will be instructive first to solve the linear filtering problem for a model more general than (10.2.1) but which retains the essential Gaussian properties of the latter.

On $(\Omega,\mathscr{A},P)$ the following conditions will be assumed given:

1. $X = (X_t)$ $(0 \leq t \leq T)$, a measurable, sample-continuous, m-dimensional signal process, and $W = (W_t)$, a standard q-dimensional Wiener process.

2. The joint distribution of the process (X,W) is Gaussian.
3. For each s, $\mathscr{G}_s \perp\!\!\!\perp \sigma[W_v - W_u, s \leq u < v \leq T]$. (The σ-field $\mathscr{G}_s$ has been defined in Chapter 8. We could also denote it by $\mathscr{F}_s^{X,W}$.)

The observation process $Y = (Y_t)$ is an n-dimensional Gaussian process which is the unique solution of the stochastic equation

$$dY_t = [C_0(t) + C_1(t)X_t + C_2(t)Y_t]\,dt + D(t)\,dW_t \qquad (Y_0 = 0), \quad (10.4.1)$$

where $C_i(t)$ $(i = 0,1,2)$ and $D(t)$ are matrix-valued functions of t satisfying the conditions stated in Sections 10.2 and 10.3. Equation (10.4.1) is the second member in (10.2.1). However, we are now assuming that X and W are given processes which satisfy conditions 1, 2, and 3. Let us therefore, first show that (10.4.1) does have a unique solution $Y = (Y_t)$ such that Y_t is $\mathscr{F}_t^{X,W}$-measurable. As in the proof of Theorem 8.2.1 we use the method of successive approximations. Define $Y_t^0 \equiv 0$ and

$$Y_t^n = \int_0^t [C_0(s) + C_1(s)X_s + C_2(s)Y_s^{n-1}]\,ds + \int_0^t D(s)\,dW_s$$

for $n \geq 1$. Then

$$|Y_t^n - Y_t^{n-1}|^2 = \left|\int_0^t C_2(s)(Y_s^{n-1} - Y_s^{n-2})\,ds\right|^2.$$

Defining $\rho_n(t) = \sup_{0 \leq s \leq t} E|Y_s^n - Y_s^{n-1}|^2$ and writing $k = \int_0^T |C_2(s)|^2\,ds$, we have

$$\rho_n(t) \leq k \int_0^t \rho_{n-1}(s)\,ds \leq \cdots \leq \frac{k^{n-1}}{(n-1)!}\,\rho_1(t).$$

k being a constant independent of n. Now

$$\rho_1(t) \leq 2t \int_0^t E|C_0(s) + C_1(s)X_s|^2\,ds + 2E\left|\int_0^t D(s)\,dW_s\right|^2 < \infty.$$

Hence Y_t^n converges to a continuous process Y_t which is $\mathscr{F}_t^{X,W}$-measurable and satisfies (10.4.1). Furthermore, Y_t is Gaussian because Y_t^n is a Gaussian for each n.

To prove uniqueness, let (Y_t) and (Y_t') be two solutions of (10.4.1) and let $\rho(t) = \sup_{0 \leq s \leq t} E|Y_s - Y_s'|^2$. Then we have

$$\rho(t) \leq k \int_0^t \rho(s)\,ds,$$

from which it follows that $\rho(t) \equiv 0$. (See the argument of Section 10.2). Letting $\alpha(t) = [D(t)D^*(t)]^{-\frac{1}{2}}$, define

$$W_t' = \int_0^t \alpha(s)D(s)\,dW_s$$

and

$$\begin{aligned} Z_t &= \int_0^t \alpha(s)[C_0(s) + C_1(s)X_s + C_2(s)Y_s]\,ds + W_t' \\ &= \int_0^t h_s\,ds + W_t', \end{aligned}$$

where $h_s = \alpha(s)[C_0(s) + C_1(s)X_s + C_2(s)Y_s]$. It is easy to verify that $\mathscr{F}_t^Z = \mathscr{F}_t^Y$ for every t. Let

$$\begin{aligned}\hat{h}_s &= E[h_s|\mathscr{F}_s^Z] = E[h_s|\mathscr{F}_s^Y] \\ &= \alpha(s)[C_0(s) + C_1(s)\hat{X}_s + C_2(s)Y_s],\end{aligned}$$

where $\hat{X}_s = E(X_s|\mathscr{F}_s^Y)$. Note that $\int_0^T E|h_t|^2\,dt < \infty$ and we have

$$Z_t = \int_0^t \hat{h}_s\,ds + v_t, \tag{10.4.2}$$

where (v_t) is the innovation Wiener process. Let $\xi_t = Z_t - E(Z_t)$. Since $\hat{h}_s(\omega) - E(h_s)$ is (s,ω)-measurable and $(\mathscr{F}_s^\xi)$-adapted (we take a progressively measurable version of $\hat{X}_s$), we have

$$\hat{h}_s(\omega) - E(h_s) = \gamma(s,\xi(\omega)),$$

where $\gamma\colon [0,T] \times C_n[0,T] \to \mathbf{R}^n$ is a nonanticipative functional (see Section 7.4). Hence Equation (10.4.2) takes the form

$$\xi_t(\omega) = \int_0^t \gamma(s,\xi(\omega))\,ds + v_t(\omega). \tag{10.4.3}$$

Thus, (ξ_t) is a Gaussian process satisfying a stochastic equation of the kind studied in Section 7.4. Furthermore, $E\int_0^T |\gamma(s,\xi(\omega)|^2\,ds < \infty$, since $E\int_0^T |\hat{h}_s|^2\,ds \le E\int_0^T |h_s|^2\,ds < \infty$. Then Theorem 9.5.1 yields the following representation for (ξ_t):

$$Z_t(\omega) = E(Z_t) + v_t(\omega) + \int_0^t \left[\int_0^s G(u,s)\,dv_u(\omega)\right] ds \tag{10.4.4}$$

and

$$\mathscr{F}_t^Z = \mathscr{F}_t^v \quad \text{for each } t. \tag{10.4.5}$$

In fact, from the definition of v and (10.4.4) we even have equality of the linear spaces of Z and v, that is,

$$L(Z;t) = L(v;t) \quad \text{for each } t. \tag{10.4.6}$$

Let us recall briefly from Chapter 9 how the square-integrable Volterra kernel $G(u,s)$ in (10.4.4) [or (9.5.7)] is obtained and also establish its uniqueness. Let $R_Z(t,s) = E(\xi_t \cdot \xi_s^*)$ be the covariance of Z. Since μ_ξ, the measure induced by ξ in $C_n[0,T]$ is equivalent to Wiener measure, by the general theory of equivalent Gaussian measures discussed in Chapter 9 the kernel $R_Z(t,s)$ determines an operator S on $L_n^2[0,T]$ which can be factored as

$$S = (I + G^*)(I + G) \tag{10.4.7}$$

where G is a Hilbert-Schmidt, Volterra operator on $L_n^2[0,T]$ having $\mathscr{M}_t = \{f \in L_n^2[0,T] : f(s) = 0 \text{ a.e. for } t \le s \le T\}$ for its chain of invariant subspaces. Then $G(u,s)$ of (10.4.4) is the kernel determined by G. The operator G^* in (10.4.7) denotes the adjoint of G. Writing (10.4.7) in terms of the Hilbert-Schmidt operator $T = I - S$, we have

$$G + G^* + G^*G + T = 0. \tag{10.4.8}$$

Equation (10.4.8) can be explicitly written as an integral equation. Letting $T(t,s)$ be the kernel determined by T and noting that the kernel of G^*, $G^*(t,s) = G(s,t)$, Equation (10.4.8) becomes

$$
\begin{aligned}
G(s,t) + \int_0^s G(r,t)G(r,s)\,dr + T(t,s) &= 0 \qquad (t > s)\\
G(t,s) + \int_0^t G(r,t)G(r,s)\,dr + T(t,s) &= 0 \qquad (t < s).
\end{aligned}
\tag{10.4.9}
$$

Before proceeding further let us comment on the existence and uniqueness of the solution of (10.4.9). The existence of a solution follows from the existence of the factorization (10.4.8) which is a special case of the general factorization result due to Gohberg and Krein (Theorem 9.1.2). Let $P(t)$ be the orthoprojector in $L_n^2[0,T]$ with range $\mathscr{M}_t$ and denote the chain $\{P(t)\}$ by π. The uniqueness of the solution $G(u,s)$ of (10.4.9) follows from the uniqueness of the corresponding operator G occurring in the factorization (10.4.8). To see this, suppose that $S = (I + M^*)(I + M)$ is another factorization of S, where M is a Hilbert-Schmidt, Volterra operator having π for its eigenchain. We have

$$(I + G)(I + M)^{-1} = (I + G^*)^{-1}(I + M^*),$$

or

$$(I + G)(I + M_1) = (I + G_1^*)(I + M^*),$$

where we set $M_1 = (I + M)^{-1} - I$ and $G_1^* = (I + G^*)^{-1} - I$. Hence $G + M_1 + GM_1 = G_1^* + M^* + G_1^*M^* = V$, say. Clearly V is a Hilbert-Schmidt, Volterra operator which has both π and $\pi^\perp$ as eigenchains. This shows $V = 0$, that is, $(I + G)(I + M_1) = I$, or $G = M$. Equation (10.4.9) is equivalent to (10.4.8) and $G(u,s)$, the solution of (10.4.9) is the kernel determined by G. The uniqueness of $G(u,s)$ is thus established. [We note here that two square-integrable kernels differing only on (u,s) sets of Lebesgue measure zero are regarded as identical.]

In (10.4.9) $T(t,s)$ is assumed known because R_Z (and hence S) is given. For the special case when $C_0 = C_2 = 0$ and $X \perp\!\!\!\perp W$, we have

$$R_Z(t,s) = \int_0^t \int_0^s \alpha(u)C_1(u)R_X(u,v)C_1^*(v)\alpha^*(v)\,du\,dv + \min(t,s)\cdot I,$$

where R_X is the covariance of X and I is the identity matrix.

Let us return to the filtering problem. Under assumptions 1 to 3 and (10.4.1), the conditional distribution of X_t given $\mathscr{F}_t^Y$ is Gaussian and hence the problem is to find $\hat{X}_t$ and $P(t)$.

Stochastic Integral for $\hat{X}_t$

Since $\{X_t, Y_t, 0 \le t \le T\}$ and, therefore, $\{X_t - E(X_t), \xi_t, 0 \le t \le T\}$ is a Gaussian system of random variables (the variables of the latter system having zero means), we have [writing $m(t)$ for $E(X_t)$]

$$\begin{aligned}\hat{X}_t - m(t) &= E[X_t - m(t)|\mathscr{F}_t^{\xi}] \\ &= \text{orthogonal projection of } X_t - m(t) \text{ on } L(\xi;t) \\ &= \text{orthogonal projection of } X_t - m(t) \text{ on } L(v;t). \quad (10.4.10)\end{aligned}$$

The last step in (10.4.10) follows from (10.4.6). Since v is a Wiener process, $X_t - m(t)$ is given by a stochastic integral

$$\hat{X}_t - m(t) = \int_0^t \Gamma(u,t)\, dv_u, \tag{10.4.11}$$

where $\int_0^t |\Gamma(u,t)|^2\, du < \infty$ for each t. Defining $\Gamma(u,t) = 0$ for $u > t$, we can choose Γ to be (u,t)-measurable and such that

$$\int_0^T \int_0^T |\Gamma(u,t)|^2\, du\, dt < \infty. \tag{10.4.12}$$

Formula for $P(t)$

From Remark 10.2.3,

$$\begin{aligned}P(t) &= E[(X_t - \hat{X}_t)(X_t - \hat{X}_t)^*] \\ &= E[(X_t - m(t))(X_t - m(t))^*] - E[(\hat{X}_t - m(t))(\hat{X}_t - m(t))^*].\end{aligned}$$

Using (10.4.11) we obtain the following expression for $P(t)$:

$$P(t) = R_X(t,t) - \int_0^t (\Gamma\Gamma^*)(u,t)\, du, \tag{10.4.13}$$

R_X being the covariance of X and Γ^* the matrix adjoint of Γ.

The bridge between formulas (10.4.11) and (10.4.13) and the Kalman filter equations is provided by the Markovian assumption on the signal process (X_t). Let us examine, in some detail, the consequences of this assumption. Only the case of real-valued signal and observation processes will be considered. We shall also assume $C_0 = C_2 = 0$ and that $X \perp\!\!\!\perp W$. Writing $K(s) = \alpha(s)C_1(s)$ and comparing (10.4.4) with (10.4.11), Γ can be defined in terms of G by

$$\begin{aligned}K(t)\Gamma(s,t) &= G(s,t) \quad \text{for } 0 \le s \le t, \\ \Gamma(s,t) &= 0 \qquad\quad \text{for all } s \text{ if } K(t) = 0.\end{aligned} \tag{10.4.14}$$

The argument presented below has to be buttressed by ad hoc assumptions in several places and must be regarded as heuristic. Since our aim here is to clarify the ideas we shall make such assumptions freely whenever necessary.

Suppose that the Gaussian process (X_t) is Markov with respect to the family $(\mathscr{G}_t)$. Then the Gaussian property of (X_t) yields the relation

$$E(X_t|\mathscr{F}^{X_s}) = \Phi(t,s)X_s \qquad \text{(a.s.) for } s \le t. \tag{10.4.15}$$

Taking expectations, we have

$$m(t) = \Phi(t,s)m(s). \tag{10.4.16}$$

Now

$$\begin{aligned} E(\hat{X}_t|\mathscr{F}_s^Z) &= E(X_t|\mathscr{F}_s^Z) \\ &= E[E(X_t|\mathscr{G}_s)|\mathscr{F}_s^Z] \\ &= E[\Phi(t,s)X_s|\mathscr{F}_s^Z] = \Phi(t,s)\hat{X}_s. \end{aligned}$$

Using the integral representation (10.4.11) for $\hat{X}_t$, we get

$$m(t) + \int_0^s \Gamma(u,t)\,dv_u = \Phi(t,s)m(s) + \int_0^s \Phi(t,s)\Gamma(u,s)\,dv_u \qquad \text{(a.s.)}.$$

Comparing with (10.4.16), we have

$$\Gamma(u,t) = \Phi(t,s)\Gamma(u,s) \quad \text{(a.e.) for } 0 \le u \le s.$$

Let us assume that K and Γ are continuous and that $\Gamma(s,s) \neq 0$ for every s. Then

$$\Gamma(s,t) = \Phi(t,s)\Gamma(s,s) \tag{10.4.17}$$

and

$$\Phi(s,s) = 1. \tag{10.4.18}$$

From the Markov property of (X_t) it easily follows that for every u,s,t $(0 \le u \le s \le t)$,

$$\Phi(t,s)\Phi(s,u) = \Phi(t,u).$$

Suppose further that the function $\Phi(\cdot,0)$ is continuous and never 0. Setting $f(t) = \Phi(t,0)$, we then get

$$\Phi(t,s) = f(t)(f(s))^{-1} \qquad (0 \le s \le t).$$

Substituting in (10.4.17), we have

$$\Gamma(s,t) = f(t)(f(s))^{-1}\Gamma(s,s). \tag{10.4.19}$$

From (10.4.11) and (10.4.19) we obtain the representation

$$\hat{X}_t - m(t) = f(t)\int_0^t (f(s))^{-1}\Gamma(s,s)\,dv_s. \tag{10.4.20}$$

Thus $\hat{X}_t - m(t)$ is the product of $f(t)$ and a martingale. Now take differentials on both sides of (10.4.20) and assume the necessary differentiability properties for $f(t)$ and $m(t)$. Then

$$\begin{aligned} d\hat{X}_t - dm(t) &= df(t)\left[\int_0^t (f(s))^{-1}\Gamma(s,s)\,dv_s\right] \\ &\quad + f(t)f(t)^{-1}\Gamma(t,t)\,dv_t \\ &= [\hat{X}_t - m(t)](f(t))^{-1}\,df(t) + \Gamma(t,t)\,dv_t. \end{aligned}$$

Now from (10.4.16), $m(t) = f(t)f(s)^{-1}m(s)$; that is,

$$\frac{m(t)}{f(t)} = \frac{m(s)}{f(s)} \quad \text{for all } 0 \le s \le t \le T.$$

Hence $m(t)/f(t)$ is a constant independent of t in $[0,T]$ and which we assume to be nonzero. It is easy to see then that $dm(t) = m(t)(f(t))^{-1}\,df(t)$, and we have

$$d\hat{X}_t = (f(t))^{-1}f'(t)\hat{X}_t\,dt + \Gamma(t,t)\,dv_t. \tag{10.4.21}$$

Simplifying both sides of the relation

$$E(X_tZ_s) = E(\hat{X}_tZ_s) \qquad (0 \le s \le t), \tag{10.4.22}$$

we have

$$\int_0^s K(u)[E(X_tX_u) - E(\hat{X}_t\hat{X}_u)]\,du = \int_0^s \Gamma(u,t)\,du. \tag{10.4.23}$$

Formally differentiating and putting $s = t$, we obtain

$$\Gamma(t,t) = K(t)P(t). \tag{10.4.24}$$

To find the equation for the conditional variance, we start with the relation

$$\begin{aligned} P(t) &= R_X(t,t) - \int_0^t \Gamma^2(s,t)\,ds \\ &= R_X(t,t) - f^2(t)\int_0^t (f(s))^{-2}\Gamma^2(s,s)\,ds. \end{aligned}$$

Hence differentiating with respect to t, we have

$$\begin{aligned} P'(t) &= R_X'(t,t) - \Gamma^2(t,t) - 2f(t)f'(t)\int_0^t (f(s))^{-2}\Gamma^2(s,s)\,ds \\ &= R_X'(t,t) - \Gamma^2(t,t) - 2f(t)f'(t)[f(t)]^{-2}[R_X(t,t) - P(t)] \\ &= R_X'(t,t) - \Gamma^2(t,t) - 2f'(t)(f(t))^{-1}R_X(t,t) + 2f'(t)(f(t))^{-1}P(t). \end{aligned}$$

Substituting for $\Gamma(t,t)$ from (10.4.24), we finally obtain the differential equation satisfied by $P(t)$,

$$\begin{aligned} P'(t) = {} & [R_X'(t,t) - 2f'(t)(f(t))^{-1}R_X(t,t)] \\ & + 2f'(t)(f(t))^{-1}P(t) - K^2(t)P^2(t). \end{aligned} \tag{10.4.25}$$

Thus the Markov assumption (with a few additional restrictions) leads from (10.4.11) to the differential equations of the Kalman filter. The connection between the Gohberg-Krein factorization and the Riccati equation is given by (10.4.14), (10.4.24), and (10.4.25).

10.5 The Kalman Filter (Independent Derivation)

We start with the model (10.2.1) for the signal and observation processes,

$$\begin{aligned} X_t &= X_0 + \int_0^t [A_0(s) + A_1(s)X_s + A_2(s)Y_s]\,ds + \int_0^t B(s)\,dW_s \\ Y_t &= \int_0^t [C_0(s) + C_1(s)X_s + C_2(s)Y_s]\,ds + \int_0^t D(s)\,dW_s \qquad (0 \le t \le T). \end{aligned} \tag{10.5.1}$$

Let

$$\bar{M}_t = \hat{X}_t - \hat{X}_0 - \int_0^t [A_0(s) + A_1(s)\hat{X}_s + A_2(s)Y_s]\, ds. \tag{10.5.2}$$

Then $(\bar{M}_t, \mathscr{F}_t^Y)$ is an m-dimensional, continuous martingale. If we assume (10.4.6), it follows that

$$\bar{M}_t = \int_0^t L(u)\, dv_u, \tag{10.5.3}$$

where L is an $m \times n$-matrix-valued function on $[0,T]$ such that

$$\int_0^T |L(u)|^2\, du < \infty.$$

Or, we may conclude (10.5.3) from (10.5.2) if we show that $\bar{M}_t^i \in L(v;t)$ for $i = 1, \ldots, m$.

Now for each i, the ith component of $\hat{X}_t - E(X_t)$ is an element of $L(v;t)$. Therefore, we have to verify that the ith component of

$$\int_0^t [A_0(s) + A_1(s)\hat{X}_s + A_2(s)Y_s]\, ds - \int_0^t [A_0(s) + A_1(s)E(X_s) + A_2(s)E(Y_s)]\, ds$$
$$= \int_0^t (A_1(s)[\hat{X}_s - E(X_s)] + A_2(s)[Y_s - E(Y_s)])\, ds \tag{10.5.4}$$

belongs to $L(v;t)$. If A_1 and A_2 are continuous, the problem reduces to the following: Let $\Phi_s(\omega)$ be jointly measurable, $E(\Phi_s) = 0$ for every s, $\Phi_s \in L^2(\Omega,P)$ and $E(\Phi_s^2)$ is continuous. Suppose that $\Phi_s \in L(v;s)$. Then for every t, $\int_0^t \Phi_s\, ds \in L(v;t)$. This is so because the integral $\int_0^t \Phi_s\, ds$ in the L^2-mean (or quadratic mean) sense exists and almost surely equals the Lebesgue integral $\int_0^t \Phi_s(\omega)\, ds$ which exists for almost all ω. In (10.5.4), then, take Φ_s successively, to be the ith component of $A_1(s)[\hat{X}_s - EX_s] + A_2(s)[Y_s - EY_s]$. Next, if $\int_0^T |A_i(s)|\, ds < \infty$ $(i = 1,2)$, we have sequences (A_i^n) of continuous functions such that $\int_0^T |A_i(s) - A_i^n(s)|\, ds \to 0$. It follows that

$$\int_0^t (A_1^n(s)[\hat{X}_s - E(X_s)] + A_2^n(s)[Y_s - EY_s])\, ds \to$$
$$\int_0^t (A_1(s)[\hat{X}_s - E(X_s)] + A_2(s)[Y_s - EY_s])\, ds \quad \text{in } L^1.$$

Denote the left-hand side sequence by η_n and the random vector on the right side by η. Let us consider componentwise convergence. For each $i = 1, \ldots, m$, (η_n^i) is a sequence of zero-mean Gaussian random variables converging in L^1 to η^i. Hence η^i is Gaussian with $E(\eta^i) = 0$. Also $\eta_n^i - \eta^i$ is a Gaussian sequence converging to zero in L^1. Hence $E(\eta^i - \eta_n^i)^2 \to 0$ and so $\eta^i \in L(v;t)$. This establishes the claim that the ith component of the quantity in (10.5.4) and, consequently of $\bar{M}_t$ given by (10.5.2) is an element of $L(v;t)$. It follows also that $(\bar{M}_t)$ is an L^2-martingale and (10.5.3) is proved.

In order to obtain the desired stochastic differential equation for $\hat{X}_t$, it now remains to find $L(u)$ in terms of the quantities given in the problem.

Since $\bar{M}_t$ and v_t are continuous, L^2-martingales with respect to $(\mathscr{F}_t^Y)$, the ($m \times n$-matrix-valued) process $\langle \bar{M}, v \rangle_t$ is given by the L^1-limit as $|\Pi_N| \to 0$

of $\sum_{\Pi_N} E[(\bar{M}_{t_{j+1}} - \bar{M}_{t_j})(v_{t_{j+1}} - v_{t_j})^*|\mathscr{F}^Z_{t_j}]$, where $\Pi_N = \{t_j^N\}$ is a finite partition of $[0,t]$ and $|\Pi_N| = \max_j(t_{j+1}^N - t_j^N)$. (The superscript N is suppressed for convenience of writing.) The convergence is understood to be elementwise in the matrix sequence. A typical term in the above sum equals

$$E\left\{\left[\int_{t_j}^{t_{j+1}} B(s)\,dW_s - (\tilde{X}_{t_{j+1}} - \tilde{X}_{t_j}) + \int_{t_j}^{t_{j+1}} A_1(s)\tilde{X}_s\,ds\right](v_{t_{j+1}} - v_{t_j})^*|\mathscr{F}^Z_{t_j}\right\}, \tag{10.5.5}$$

where $\tilde{X}_u = X_u - \hat{X}_u$. Writing $K(s)$ for $\alpha(s)C_1(s)$, we have

$$\begin{aligned}\sum E\left[\left(\int_{t_j}^{t_{j+1}} B(s)\,dW_s\right)(v_{t_{j+1}} - v_{t_j})^*|\mathscr{F}^Z_{t_j}\right] &= \sum E\left[\left(\int_{t_j}^{t_{j+1}} B(s)\,dW_s\right)\left(\int_{t_j}^{t_{j+1}} K(s)\tilde{X}_s\,ds\right)^*\middle|\mathscr{F}^Z_{tj}\right]\\ &\quad + \sum E\left[\left(\int_{t_j}^{t_{j+1}} B(s)\,dW_s\right)\left(\int_{t_j}^{t_{j+1}} dW'_s\right)^*\middle|\mathscr{F}^Z_{t_j}\right].\end{aligned}$$

The first term on the right-hand side tends to 0 in L^1 as $|\Pi_N| \to 0$. Since $\mathscr{F}^Z_{t_j} \subset \mathscr{G}_{t_j} \perp\!\!\!\perp [W_v - W_u, t_j \le u < v < T]$, the second term is easily seen to converge in L^1 to $\int_0^t B(s)D^*(s)\alpha(s)\,ds$. The sum

$$\sum E\left[\left(\int_{t_j}^{t_{j+1}} A_1(s)\tilde{X}_s\,ds\right)(v_{t_{j+1}} - v_{t_j})^*|\mathscr{F}^Z_{t_j}\right] \to 0$$

in L^1 by an easy argument. As for the sums involving the remaining terms in (10.5.5), we first note that $\sum E[\tilde{X}_{t_{j+1}}(v_{t_{j+1}} - v_{t_j})^*|\mathscr{F}^Z_{t_j}] = 0$, since $\tilde{X}_{t_{j+1}} \perp\!\!\!\perp \mathscr{F}^Z_{t_j}$ and $\tilde{X}_{t_{j+1}} \perp\!\!\!\perp v_s$ for all $s \le t_{j+1}$. The sum remaining to be considered is $\sum E[\tilde{X}_{t_j}(v_{t_{j+1}} - v_{t_j})^*|\mathscr{F}^Z_{t_j}]$. The term in the sum equals

$$E\left[\tilde{X}_{t_j}\left(\int_{t_j}^{t_{j+1}} K(s)\tilde{X}_s\,ds\right)^*\middle|\mathscr{F}^Z_{t_j}\right] + E[\tilde{X}_{t_j}(W'_{t_{j+1}} - W'_{t_j})^*|\mathscr{F}^Z_{t_j}].$$

In the second term, $\tilde{X}_{t_j}$ and $W'_{t_{j+1}} - W'_{t_j} \perp\!\!\!\perp \mathscr{F}^Z_{t_j}$. Also, since $\tilde{X}_{t_j}$ is $\mathscr{G}_{t_j}$-measurable, $\tilde{X}_{t_j} \perp\!\!\!\perp W'_{t_{j+1}} - W'_{t_j}$. Hence the second term vanishes, while the first term equals

$$E\left[\int_{t_j}^{t_{j+1}} \tilde{X}_{t_j}\tilde{X}_s^* K^*(s)\,ds|\mathscr{F}^Z_{t_j}\right] = \int_{t_j}^{t_{j+1}} E[\tilde{X}_{t_j}\tilde{X}_s^*|\mathscr{F}^Z_{t_j}]K^*(s)\,ds.$$

Hence, using the continuity of $E(\tilde{X}_u\tilde{X}_u^*)$, we find that

$$\sum E\tilde{X}_{t_j}\left[\left(\int_{t_j}^{t_{j+1}} K(s)\tilde{X}_s\,ds\right)^*\middle|\mathscr{F}^Z_{t_j}\right] \to \int_0^t P(s)K^*(s)\,ds$$

elementwise in L^1. Thus all the terms in (10.5.5) have been accounted for and we have shown that

$$\langle \bar{M}_1, v\rangle_t = \int_0^t [P(s)C_1^*(s) + B(s)D^*(s)]\alpha(s)\,ds. \tag{10.5.6}$$

On the other hand, from (10.5.3) it follows that

$$\langle \bar{M}_1, v \rangle_t = \int_0^t L(s)\,ds. \tag{10.5.7}$$

Comparing (10.5.6) and (10.5.7), we obtain

$$L(t) = [P(t)C_1^*(t) + B(t)D^*(t)]\alpha(t) \quad \text{for a.e. } t. \tag{10.5.8}$$

From (10.5.2), (10.5.3), and (10.5.8) we obtain the stochastic differential equation satisfied by $\hat{X}_t$:

$$\begin{aligned} d\hat{X}_t &= [A_0(t) + A_1(t)\hat{X}_t + A_2(t)Y_t]\,dt \\ &\quad + [P(t)C_1^*(t) + B(t)D^*(t)][D(t)D^*(t)]^{-\frac{1}{2}}\,dv_t \end{aligned} \tag{10.5.9}$$

with $\hat{X}_0 = E(X_0)$. The differential equation satisfied by $P(t)$, viz.,

$$\begin{aligned} \frac{dP(t)}{dt} &= A_1(t)P(t) + P(t)A_1^*(t) + B(t)B^*(t) \\ &\quad - [P(t)C_1^*(t) + B(t)D^*(t)][D(t)D^*(t)]^{-1}[C_1(t)P(t) + D(t)B^*(t)] \end{aligned} \tag{10.5.10}$$

with initial condition

$$P(0) = E[(X_0 - EX_0)(X_0 - EX_0)^*] \tag{10.5.11}$$

is obtained exactly as in Section 10.3 by applying the Ito formula to $X_tX_t^*$ and $\hat{X}_t\hat{X}_t^*$ and using (10.5.9) and the first equation in (10.2.1).

Uniqueness

Let us now establish the uniqueness of the solution of (10.5.10). More precisely, we show the following: If $P_1(t)$ and $P_2(t)$ are covariance matrices satisfying Equation (10.5.10) for $t \in [0,T]$ and if $P_1(0) = P_2(0) = E(X_0 - EX_0)(X_0 - EX_0)^*$, then $P_1(t) = P_2(t)$ for all t.

Proof. First of all, we need a bound for $\operatorname{Tr} P(t)$ (Tr standing for trace) for any covariance matrix $P(t)$ satisfying (10.5.10). Writing $F(t) = \exp[\int_0^t A_1(s)\,ds]$, a direct verification shows that

$$\begin{aligned} P(t) = F(t)\Big\{P(0) &+ \int_0^t F(s)^{-1}[B(s)B^*(s) - (P(s)C_1^*(s) + B(s)D^*(s)) \\ &\times (D(s)D^*(s))^{-1}(C_1(s)P(s) + D(s)B^*(s))](F(s)^{-1})^*\,ds\Big\}F(t)^* \end{aligned} \tag{10.5.12}$$

For symmetric matrices A and B of the same order, use the notation $A \le B$ (or $B \ge A$) to mean that $B - A$ is nonnegative definite. Note also that $A \le B$ implies $\operatorname{Tr} A \le \operatorname{Tr} B$. The matrix in brackets on the right-hand side of (10.5.12) is $\leq B(s)B^*(s)$. Hence, for $0 \leq t \leq T$,

$$0 \leq \operatorname{Tr} P(t) \leq \operatorname{Tr} F(t)\left\{P(0) + \int_0^t F(s)^{-1}B(s)B^*(s)(F(s)^{-1})^*\,ds\right\}F(t)^*$$

$$\leq H < \infty, \tag{10.5.13}$$

where H is a constant which depends only on the initial condition (10.5.11) and not on $P(t)$ for $t > 0$. Since $P(t)$ is a covariance matrix, $|P^{ij}(t)| \leq (P^{ii}(t)P^{jj}(t))^{\frac{1}{2}} \leq H$, so that $|P(t)|^2 = \sum_{i,j}[P^{ij}(t)]^2 \leq m^2H^2$. Suppose that $P_i(t)$, $(i = 1,2)$ are covariance matrices satisfying (10.5.10) and (10.5.11). Then

$$\begin{aligned} P_1(t) - P_2(t) = \int_0^t [&A_1(s)(P_1(s) - P_2(s)) + (P_1(s) - P_2(s))A_1^*(s) \\ &- \{(P_1(s) - P_2(s))C_1^*(s)(D(s)D^*(s))^{-1}(C_1(s)P_1(s) \\ &+ D(s)B^*(s)) + (P_2(s)C_1^*(s) + B(s)D^*(s)) \\ &\times (D(s)D^*(s))^{-1}C_1(s)(P_1(s) - P_2(s)\}]\,ds \\ = \int_0^t &R(s)\,ds, \quad \text{say.} \end{aligned}$$

Now $|\int_0^t R(s)\,ds|^2 \leq \sum_{i,j}[\int_0^t |R^{ij}(s)|\,ds]^2$. Setting $\Delta(s) = |P_1(s) - P_2(s)|$ and using $|P_k(s)| \leq mH$ for $k = 1,2$, we have

$$|R^{ij}(s)| \leq |R(s)| \leq \Delta(s)h(s),$$

where

$$h(s) = 2[|A_1(s)| + |C_1(s)| \cdot |D(s)D^*(s)|^{-1}(mH|C_1(s)| + |D(s)| \cdot |B(s)|)].$$

Hence it follows that

$$\Delta(t) \leq m\int_0^t \Delta(s)h(s)\,ds.$$

(Note that the last integral is finite since $\Delta(s)$ is continuous and $h \in L^1[0,T]$.) The above inequality immediately yields the conclusion $\Delta(t) = 0$ for each t, that is, $P_1 = P_2$ (see Proposition 5.1.1). □

The stochastic equation (10.5.9) has the same form as (10.4.1). The uniqueness of the solution of the latter equation was established in Section 10.4. Hence, with $P(t)$ already uniquely determined through (10.5.10) and (10.5.11) it follows that (10.5.9) has a unique solution. We have thus proved the following theorem.

Theorem 10.5.1. *For the signal-observation model (10.2.1) assume the conditions stated in Sections 10.2 and 10.3.*

Then the Kalman filter $(\hat{X}_t, P(t))$ is the unique solution of the following system of equations:

$$\begin{aligned} d\eta_t = {} &[A_0(t) + A_1(t)\eta_t + A_2(t)Y_t]\,dt \\ &+ [P(t)C_1^*(t) + B(t)D^*(t)][D(t)D^*(t)]^{-\frac{1}{2}}\,d\nu_t, \end{aligned} \tag{10.5.14}$$

with $\eta_0 = E(X_0)$, *and*

$$P'(t) = A_1(t)P(t) + P(t)A_1^*(t) + B(t)B^*(t) - [P(t)C_1^*(t) + B(t)D^*(t)][D(t)D^*(t)]^{-1}[C_1(t)P(t) + D(t)B^*(t)] \tag{10.5.15}$$

with $P(0) = E[(X_0 - EX_0)(X_0 - EX_0)^*]$.

Remark 10.5.1. The solution of (10.5.14) is shown by direct verification to be

$$\eta_t = F(t)\left[E(X_0) + \int_0^t F(s)^{-1}\{A_0(s) + A_2(s)Y_s\}\,ds + \int_0^t F(s)^{-1}L(s)\,dv_s\right].$$

Remark 10.5.2. It might be of interest to point out two variations in the proof of the derivation of the Kalman equations. For convenience, we take (X_t) and (Y_t) to be real-valued and set $A_0 = A_2 = C_0 = C_2 = 0$. It will be assumed that the appropriate conditions imposed on the model (10.2.1) are still in force.

(a) Proceeding as in Section 10.3, if f is any C^2 function, we have

$$(\tilde{A}_t f)(\omega) = A_1(t)X_t(\omega)f'(X_t(\omega)) + \tfrac{1}{2}\lambda(t)f''(X_t(\omega))$$

and

$$(\tilde{D}_t^i f)(\omega) = B_{1i}(t)f'(X_t(\omega)), \qquad i = 1,2,$$

where $\lambda(t) = B_{11}^2(t) + B_{12}^2(t)$. We need consider only two functions: $f(x) = x$ and $f(x) = x^2$. The first choice readily yields the differential

$$d\hat{X}_t = A_1(t)\hat{X}_t\,dt + \alpha(t)[C_1(t)P(t) + \delta(t)]\,dv_t,$$

where $\delta(t) = D_{11}(t)B_{11}(t) + D_{12}(t)B_{12}(t)$. The variation appears in the next step. If $f(x) = x^2$, the term $E^s[f(X_s)a_s] - E^s[f(X_s)]E^s[a_s]$ in Equation (8.4.26) becomes $C_1(s)[E^s(X_s^3) - E^s(X_s^2)\hat{X}_s]$. Now, since the conditional distribution of X_t given $\mathscr{F}_t^Y$ is Gaussian with mean $\hat{X}_t$ and variance $P(t)$, by a well-known property of the normal distribution we have $E^t(X_t^3) = (\hat{X}_t)^3 + 3\hat{X}_tP(t)$. Hence

$$\begin{aligned} C_1(s)[E^s(X_s^3) - E^s(X_s^2)\hat{X}_s] &= C_1(s)[(\hat{X}_s)^3 + 3\hat{X}_sP(s) - E^s(X_s^2)\hat{X}_s] \\ &= 2C_1(s)\hat{X}_sP(s). \end{aligned}$$

The equation corresponding to $f(x) = x^2$ then reduces to

$$dE^t(X_t^2) = [2A_1(t)E^t(X_t^2) + \lambda(t)]\,dt + \alpha(t)[2C_1(t)\hat{X}_tP(t) + 2\delta(t)\hat{X}_t]\,dv_t.$$

Applying the Ito formula to $(\hat{X}_t)^2$ and using the stochastic differential for $\hat{X}_t$, we observe that the random term (that is, the coefficient of dv_t) in the stochastic differential for $P(t)$ vanishes, and we have

$$dP(t) = \{2A_1(t)P(t) + \lambda(t) - \alpha^2(t)[C_1(t)P(t) + \delta(t)]^2\}\,dt,$$

which is the desired Riccati equation.

(b) Let $M_1(t) = X_t - \int_0^t A_1(s)X_s\,ds$ and $M_2(t) = X_t^2 - \int_0^t [2A_1(s)X_s^2 + \lambda(s)]\,ds$. Then $M_1(t)$ and $M_2(t)$ are square-integrable $(\mathscr{G}_t, P)$-martingales. Using the square-integrable $(\mathscr{F}_t^Y, P)$-martingale $\bar{M}_1(t) = \hat{X}_t - \int_0^t A_1(s)\hat{X}_s\,ds$, we obtain (as we did earlier in this section)

$$\hat{X}_t = \int_0^t A_1(s)\hat{X}_s\,ds + \int_0^t L(s)\,dv_s,$$

where L is to be determined. Next,

$$\bar{M}_2(t) = E(X_t^2|\mathscr{F}_t^Y) - \int_0^t [2A_1(s)E(X_s^2|\mathscr{F}_s^Y) + \lambda(s)]\,ds$$

is also an L^2-martingale relative to $(\mathscr{F}_t^Y)$. Now use the fact that $\mathscr{F}_t^Y = \mathscr{F}_t^v$. From Ito's version of the Cameron-Martin theorem proved in Chapter 6 it follows that

$$\bar{M}_2(t) = E(X_0^2) + \int_0^t \Phi(s)\,dv_s,$$

where on the right-hand side we have the Ito integral (with Φ possibly random). The stochastic differential for $\hat{X}_t$ also yields

$$d(\hat{X}_t^2) = [2A_1(t)\hat{X}_t^2 + L^2(t)]\,dt + 2\hat{X}_t L(t)\,dv_t.$$

Hence

$$dP(t) = [2A_1(t)P(t) + \lambda(t) - L^2(t)]\,dt + [\Phi(t) - 2\hat{X}_t L(t)]\,dv_t.$$

But the term involving dv_t has to vanish because $P(t)$ is nonrandom. The calculation of $L(t)$ is done exactly as before by finding $\langle \bar{M}_1, v\rangle_t$.

11 The Stochastic Equation of the Optimal Filter (Part II)

11.1 Introduction

In Theorems 8.4.3 and 8.4.4 of Section 8.4 the conditional expectation $E^t f(X_t)$ was shown to satisfy a stochastic differential equation for all f in $\mathscr{D}(\tilde{A})$ for which the condition $\int_0^T E|f(X_t)h_t|^2 < \infty$ is fulfilled. However, the filtering problem can be regarded as completely solved if we can derive from (8.4.22) a stochastic differential equation for the conditional probability distribution—or the condition probability density—of X_t given $\mathscr{F}_t^Z$ and if, furthermore, it can be established that the equation has a unique solution. This was achieved in Chapter 10 for the linear theory, and we saw that the general equations of Chapter 8 yielded the Kalman filter. The complete solution of the optimal nonlinear filtering problem presents a much more difficult task.

To see this, let us consider a special case of the model (8.5.1) and (8.5.2), where X_t and Y_t are real-valued processes governed by the differential equations

$$dX_t = A(t,X_t)\,dt + B_1(t,X_t)\,dW_t^1,$$
$$dY_t = X_t\,dt + dW_t^2,$$
$$X_0 = \text{a constant}, \qquad Y_0 = 0, \quad \text{and} \quad (W_t^1) \perp\!\!\!\perp (W_t^2).$$

The coefficients $A(t,x)$ and $B_1(t,x)$ are assumed to satisfy the Lipschitz and growth conditions mentioned in Chapter 8 which guarantee the existence of a unique solution. We see that $(X_t) \perp\!\!\!\perp (W_t^2)$ so that the operator $\tilde{D}_t$ vanishes, that is, $\tilde{D}_t f = 0$ for a C^2 function f. From (8.5.4) it is seen that if $f \in C^2$.

$$\tilde{A}_t f(x) = A(t,x)f'(x) + \tfrac{1}{2}B_1^2(t,x)f''(x).$$

The conditions imposed on $A(t,x)$, $B_1(t,x)$, and X_0 imply that $E|X_t|^n < \infty$ for every t and $\int_0^T E|X_t|^n \, dt < \infty$, $(n \geq 1)$. For simplicity let us now take $A(t,x) = 0$ and $B_1(t,x) = x$. For $f(x) = x^n$, where n is any positive integer, the condition $\int_0^T E|f(X_t)h_t|^2 < \infty$ of Theorem 8.4.3 is obviously fulfilled since $h_t = X_t$. Writing $\alpha_t^{(n)}$ for the conditional moment $E^t(X_t^n)$, $(n \geq 1)$, the stochastic differential equation (8.4.13) takes the form

$$\begin{aligned} d\alpha_t^{(n)} &= \tfrac{1}{2}n(n-1)\alpha_t^{(n)}\, dt + \left[\alpha_t^{(n+1)} - \alpha_t^{(n)}\alpha_t^{(1)}\right] dv_t, \\ \alpha_0^{(n)} &= X_0^n \qquad (n \geq 1). \end{aligned} \tag{11.1.1}$$

The equations in (11.1.1) do not form a closed system; that is, the equation for $\alpha_t^{(n)}$ involves the next higher moment. This example shows that to solve the equation of the optimal nonlinear filter we really have to solve an infinite system of stochastic differential equations. Hence (8.4.13) which governs the conditional expectation $E^t[f(X_t)]$ is to be looked upon as a stochastic differential equation satisfied by a measure-valued stochastic process (in this case, the conditional distribution of X_t given $\mathscr{F}_t^Z$). When the conditional density exists and satisfies certain conditions, a stochastic partial differential equation for it can be deduced from (8.4.13). We consider the latter question in the next section. In general, the problem of verifying the conditions imposed in deriving the existence of a solution, or of showing the uniqueness of the solution of these equations is a formidable one.

A very special case of a stochastic equation for the nonlinear filter is considered in Section 11.5 for which the existence of a unique solution as a measure-valued process is proved. The underlying stochastic model makes the assumption that the signal process is completely independent of the observation noise process. A Bayes formula which is useful in this context is derived in Section 11.3.

11.2 A Stochastic Differential Equation for the Conditional Density

Suppose (X_t,Y_t) is a Markov process satisfying the stochastic differential equation

$$\begin{aligned} dX_t &= A(t,X_t,Y_t)\, dt + B(t,X_t,Y_t)\, dW_t, \\ dY_t &= a(t,X_t,Y_t)\, dt + b(t,Y_t)\, dW_t \end{aligned} \tag{11.2.1}$$

with $Y_0 = 0$ and $X_0 \perp\!\!\!\perp \mathscr{F}_T^W$. The reader is referred to Section 8.5 where the Fleming-Nisio equations (8.5.1) and (8.5.2) are discussed. Here we assume for simplicity that X_t and Y_t are real-valued, that is, $N = 1$ and $M = 2$ in Section 8.5, $B = (B^1,B^2)$ and $W_t = (W_t^1,W_t^2)$ written as a column vector. Lipschitz and growth conditions such as (8.2.7) and (8.2.8) are imposed on the coefficients to ensure the existence of a unique solution to (11.2.1). If f is a C^2-class function of a real variable, it is easy to verify from (8.5.4) and

(8.5.5) that $\tilde{A}_t f$ is a function of (x, y) given by

$$(\tilde{A}_t f)(x,y) = A(t,x,y)f'(x) + \tfrac{1}{2}[(B^1(t,x,y))^2 + (B^2(t,x,y))^2]f''(x) \tag{11.2.2}$$

and

$$([b^1(t,y)\tilde{D}_t^1 + b^2(t,y)\tilde{D}_t^2]f)(x, y) = [b^1(t,y)B^1(t,x,y) + b^2(t,y)B^2(t,x,y)]f'(x).$$

Let us define the operator L_t by

$$(L_t f)(x, y) = c^{-\frac{1}{2}}(t,y)[b^1(t,y)B^1(t,x,y) + b^2(t,y)B^2(t,x,y)]f'(x). \tag{11.2.3}$$

The differential operators $\tilde{A}_t$ and L_t have adjoints given by

$$(\tilde{A}_t^* f)(x,y) = -\frac{\partial}{\partial x}[A(t,x,y)f(x)] + \frac{1}{2}\frac{\partial^2}{\partial x^2}[\{(B^1(t,x,y))^2 + (B^2(t,x,y))^2\}f(x)] \tag{11.2.4}$$

and

$$(L_t^* f)(x) = -\frac{\partial}{\partial x}[c^{-\frac{1}{2}}(t,y)\{b^1(t,y)B^1(t,x,y) + b^2(t,y)B^2(t,x y)\}f(x)] \tag{11.2.5}$$

provided the coefficients A, B^1, B^2, b^1, and b^2 have the appropriate differentiability properties.

Theorem 11.2.1. *In addition to the Lipschitz and growth conditions imposed on (11.2.1) we make the following assumptions:*

A. *The conditional distribution $P[X_t \leq x | \mathscr{F}_t^Y]$ has a probability density (with respect to Lebesgue measure) $\phi_t(x,\omega)$ which is (t,x,ω)-measurable and $\mathscr{F}_t^Y$-adapted for each t and x.*
B. *The derivatives*

$$\frac{\partial}{\partial x}[A(t,x,Y_t(\omega))\phi_t(x,\omega)],$$

$$\frac{\partial}{\partial x}[c^{-\frac{1}{2}}(t,Y_t(\omega))\{b^1(t,Y_t(\omega))B^1(t,x,Y_t(\omega)) + b^2(t,Y_t(\omega))B^2(t,x,Y_t(\omega))\}\phi_t(x,\omega)],$$

and

$$\frac{\partial^2}{\partial x^2}[\{(B^1(t,x,Y_t(\omega)))^2 + (B^2(t,x,Y_t(\omega)))^2\}\phi_t(x,\omega)]$$

exist (a.s.) *for every t in $[0,T]$.*

C. *For each t,*

$$\int_{-\infty}^{\infty} |A_t^*\phi_t(x,\omega)|\,dx < \infty \qquad \text{(a.s.)} \tag{11.2.6a}$$

and

$$\int_0^T \int_{-\infty}^{\infty} |A_t^*\phi_t(x,\omega)|\,dx\,dt < \infty \qquad \text{(a.s.)}. \tag{11.2.6b}$$

For each t,

$$\int_{-\infty}^{\infty} |L_t^*\phi_t(x,\omega) + \phi_t(x,\omega)\{h'(t,x,Y_t(\omega)) - E^t(h_t')(\omega)\}|\,dx < \infty \qquad \text{(a.s.)} \tag{11.2.7a}$$

and

$$E\int_0^T \left[\int_{-\infty}^{\infty} |L_t^*\phi_t(x,\omega) + \phi_t(x,\omega)\{h'(t,x,Y_t(\omega)) - E^t(h_t'(\omega))\}|\,dx\right]^2 dt < \infty. \tag{11.2.7b}$$

Then $\phi_t(x,\omega)$ satisfies the following stochastic integral equation: for every t,

$$\begin{aligned}\phi_t(x,\omega) = \phi_0(x) &+ \int_0^t \tilde{A}_s^*\phi_s(x,\omega)\,ds \\ &+ \int_0^t \{L_s^*\phi_s(x,\omega) + \phi_s(x,\omega)[h'(s,x,Y_s(\omega)) \\ &- \int_{-\infty}^{\infty} h'(s,x',Y_s(\omega))\phi_s(x',\omega)\,dx']\}\,dv_s(\omega)\end{aligned} \tag{11.2.8}$$

for a.e. x, *with probability* 1. *Here* (v_t) *is the Wiener process denoted by* (v_t') *in (8.4.22).*

PROOF. Let $C_0^2(\mathbf{R})$ be the class of twice continuously differentiable functions f on $\mathbf{R}$ such that f (f' and f'') has compact support. We deduce (11.2.8) from the basic equation (8.4.22). Observe that the condition $\int_0^T E|f(X_t)h_t'^2|\,dt < \infty$ of that theorem holds for every $f \in C_0^2(\mathbf{R})$ since h_t' satisfies (8.2.3). The main point of the proof is to justify the interchange of the order of integration in the various terms of (8.4.22). The left-hand side of (8.4.22) is simply $\int_{-\infty}^{\infty} f(x)\phi_t(x,\omega)\,dx$. The first term on the right-hand side is $Ef(X_0) = \int_{-\infty}^{\infty} f(x)p_0(x)\,dx$, where $p_0(x)$ is the probability density (assumed to exist) of $P[X_0 \le x]$. By assumption, $\phi_0(x,\omega)$ is the density of $P[X_0 \le x|\mathscr{F}_0^Y] = P[X_0 \le x]$ (a.s.). So we have the initial condition $\phi_0(x,\omega) = p_0(x)$ for a.e. x with probability 1. Consider the remaining two terms on the right-hand side of (8.4.22). The second term equals

$$\begin{aligned}\int_0^t \left(\int_{-\infty}^{\infty} \tilde{A}_s f(x)\phi_s(x,\omega)\,dx\right) ds &= \int_0^t \left(\int_{-\infty}^{\infty} f(x)\tilde{A}_s^*\phi_s(x,\omega)\,dx\right) ds \\ &= \int_{-\infty}^{\infty} f(x)\left[\int_0^t \tilde{A}_s^*\phi_s(x,\omega)\,ds\right] dx \qquad \text{(a.s.)},\end{aligned}$$

the last equality being justified by condition (11.2.6b) and the Fubini theorem. The third term, which may be written as

$$\int_0^t \left(\int_{-\infty}^{\infty} f(x) \left[\phi_s(x,\omega) \left\{ h'(s,x,Y_s(\omega)) - \int_{-\infty}^{\infty} h'(s,x',Y_s(\omega))\phi_s(x',\omega')\,dx' \right\} + L_s^* \phi_s(x,\omega) \right] dx \right) dv_s(\omega),$$

equals

$$\int_{-\infty}^{\infty} f(x) \left(\int_0^t \left[L_s^* \phi_s(x,\omega) + \phi_s(x,\omega) \left\{ h'(s,x,Y_s(\omega)) - \int_{-\infty}^{\infty} h'(s,x',Y_s(\omega))\phi_s(x',\omega)\,dx' \right\} \right] dv_s(\omega) \right) dx$$

(a.s.) for every t. The justification of the last step is as follows. Set

$$H_s(x,Y_s(\omega)) = L_s^* \phi_s(x,\omega) + \phi_s(x,\omega)\{h'(s,x,Y_s(\omega)) - E^s(h_s')\}.$$

We then have to show that

$$\int_0^t \left[\int_{-\infty}^{\infty} f(x) H_s(x,Y_s(\omega))\,dx \right] dv_s(\omega) = \int_{-\infty}^{\infty} f(x) \left[\int_0^t H_s(x,Y_s(\omega))\,dv_s(\omega) \right] dx \quad \text{(a.s.)}. \tag{11.2.9}$$

Denoting by $I_t^1(\omega)$ and $I_t^2(\omega)$ the two sides of (11.2.9), it suffices to show that

$$E(I_t^1 M_t) = E(I_t^2 M_t) \tag{11.2.10}$$

for every $M_t(\omega)$ of the form $E(M_t) + \int_0^t \Psi_s(\omega)\,dv_s(\omega)$, where $\Psi_s(\omega)$ is (s,ω)-measurable, $(\mathscr{F}_s^Y)$-adapted and $E\int_0^t \Psi_s^2(\omega)\,ds < \infty$. Let us first note that $I_t^k(\omega)$ $(k = 1,2)$ is (t,ω)-measurable, $(\mathscr{F}_t^Y)$-adapted, and we have $EI_t^k = 0$, $E(I_t^k)^2 < \infty$. Now

$$E(I_t^1 M_t) = \int_0^t E\left[\int_{-\infty}^{\infty} f(x) H_s(x,Y_s(\omega))\,dx\,\Psi_s(\omega) \right] ds \tag{11.2.11}$$

where, by (11.2.7b),

$$\begin{aligned}
&\int_0^t E \int_{-\infty}^{\infty} |f(x)\Psi_s(\omega)H_s(x,Y_s(\omega))|\,dx\,ds \\
&\le \int_0^t (E(\Psi_s^2))^{\frac{1}{2}} \left\{ E\left(\int_{-\infty}^{\infty} |f(x)H_s(x,Y_s(\omega))|\,dx \right)^2 \right\}^{\frac{1}{2}} ds \\
&\le \left[\int_0^t E(\Psi_s^2)\,ds \int_0^t E\left(\int_{-\infty}^{\infty} |f(x)H_s(x,Y_s(\omega))|\,dx \right)^2 ds \right]^{\frac{1}{2}} \\
&< \infty.
\end{aligned}$$

Hence by the Fubini theorem we may interchange the order of integration in the right-hand side of (11.2.11) and obtain

$$\int_{-\infty}^{\infty} f(x)E\left[\int_0^t H_s(x,Y_s(\omega))\Psi_s(\omega)\,ds\right]dx = E(I_t^2 M_t).$$

Thus (11.2.9) is proved. Equation (8.4.22) has now been reduced to the following form. For $t \in [0,T]$,

$$\int_{-\infty}^{\infty} f(x)\left[\phi_t(x,\omega) - p_0(x) - \int_0^t \tilde{A}_s^*\phi_s(x,\omega)\,ds - \int_0^t H_s(x,Y_s(\omega))\,dv_s(\omega)\right]dx = 0 \tag{11.2.12}$$

(a.s.) every $f \in C_0^2(\mathbf{R})$. Denoting by $Q_t(x,\omega)$ the expression in the brackets in (11.2.12), we see from conditions (11.2.6a) and (11.2.7a) that for each t, $\int_{-\infty}^{\infty} |Q_t(x,\omega)|\,dx < \infty$ (a.s.). Now $L^1(\mathbf{R})$ is a separable Banach space and there exists a countable set in $C_0^2(\mathbf{R})$ which is dense in $L^1(\mathbf{R})$. Hence it can be shown from (11.2.12) that for every t,

$$P[\omega\colon Q_t(x,\omega) = 0 \text{ for a.e. } x] = 1, \tag{11.2.13}$$

and (11.2.8) is proved. □

Remark 11.2.1. The stronger conclusion that (11.2.8) holds (a.s.) for every (t,x) can be obtained if we impose further restrictions which ensure, besides the (a.s.) continuity in (t,x) of $\phi_t(x,\omega)$ and $\int_0^t \tilde{A}_s^*\phi_s(x,\omega)\,ds$, the existence of a (t,x)-continuous version of the stochastic integral $\int_0^t H_s(x,Y_s(\omega))\,dv_s(\omega)$. For further details on this question we refer the reader to [51].

11.3 A Bayes Formula for Stochastic Processes

An important special case of the filtering model of Chapter 8 occurs when the signal process (X_t) and the observation "noise" (W_t) are completely independent. A Bayes formula will be derived in this case for the conditional expectation $E(g|\mathscr{F}_t^Z)$ where g is $\mathscr{F}_T^X$-measurable and integrable. As mentioned in Chapter 8 the formula can be used to obtain the stochastic differential equation of the optimal nonlinear filter when $(X_t) \perp\!\!\!\perp (W_t)$. Other applications will be made in the next two sections. We shall first prove some lemmas. Statement (b) of Lemma 11.3.1 is used in Theorem 11.3.1, but Lemma 11.3.1 is given in full because of its interest.

Lemma 11.3.1. *Let (W_t), $(0 \le t \le T)$ be a Wiener process and (f_t) be a measurable process defined on a complete probability space $(\Omega,\mathscr{F},P)$. Suppose that*

$\int_0^T f_t^2\,dt < \infty$ (a.s.) *and* $(f_t) \perp\!\!\!\perp (W_t)$. *Then*

$$E\exp\left[\int_0^T f_s\,dW_s - \frac{1}{2}\int_0^T f_s^2\,ds\right] = 1. \tag{11.3.1}$$

We have the following further conclusions:

(a) *Suppose* $(W_t,\mathscr{F}_t,P)$ *is a Wiener martingale,* $(\mathscr{F}_t)$ *being some increasing, right-continuous family such that* $\mathscr{F}_0$ *contains all P-null sets. Then*

$$\tilde{W}_t = W_t - \int_0^t f_s\,ds \tag{11.3.2}$$

is a Wiener martingale with respect to $(\tilde{\mathscr{F}}_t,\tilde{P})$, *where* $\tilde{\mathscr{F}}_t = \mathscr{F}_T^f \vee \mathscr{F}_t$ *and*

$$d\tilde{P} = \exp\left[\int_0^T f_s\,dW_s - \frac{1}{2}\int_0^T f_s^2\,ds\right]dP.$$

(b) *If* $(W_t,\mathscr{F}_t,P)$ *is a Wiener martingale and* f_t *is a nonrandom, Lebesgue measurable function of t such that* $\int_0^T f_t^2\,dt < \infty$, *then Equation (11.3.1) holds and* $\tilde{W}_t$ *given by (11.3.2) is a Wiener martingale with respect to* $(\mathscr{F}_t,\tilde{P})$.

PROOF. In proving (11.3.1) there is no loss of generality in assuming that (f_t) and (W_t) are given on a product space $(\Omega,\mathscr{F}^0,P) = (\Omega_1 \times \Omega_2, \mathscr{F}_1 \times \mathscr{F}_2, P_1 \times P_2)$, where writing $\omega = (\omega_1,\omega_2)$, we set $f_t(\omega) = f_t(\omega_1)$ and $W_t(\omega) = W_t(\omega_2)$. The σ-field $\mathscr{F}$ is the completion of $\mathscr{F}_1 \times \mathscr{F}_2$ with respect to $P_1 \times P_2$. We then have, using Fubini's theorem,

$$\begin{aligned}
&E\exp\left[\int_0^T f_s\,dW_s - \frac{1}{2}\int_0^T f_s^2\,ds\right]\\
&\quad = \int_{\Omega_1}\int_{\Omega_2} \exp\left[\int_0^T f_s(\omega_1)\,dW_s(\omega_2)\right.\\
&\qquad \left. - \frac{1}{2}\int_0^T f_s^2(\omega_1)\,ds\right] P_1(d\omega_1)P_2(d\omega_2)\\
&\quad = \int_{\Omega_1} \exp\left[-\frac{1}{2}\int_0^T f_s^2(\omega_1)\,ds\right]\\
&\qquad \times \left\{\int_{\Omega_2} \exp\left[\int_0^T f_s(\omega_1)\,dW_s(\omega_2)\right] P_2(d\omega_2)\right\} P_1(d\omega_1) = 1,
\end{aligned}$$

since for fixed ω_1, $\int_0^T f_s(\omega_1)\,dW_s(\omega_2)$ is a Gaussian random variable under P_2 with zero mean and variance $\int_0^T f_s^2(\omega_1)\,ds$.

(a) Note that since $\mathscr{F}_T^f \perp\!\!\!\perp \mathscr{F}_t$ for every t, $(W_t,\tilde{\mathscr{F}}_t,P)$ is a Wiener martingale and f_t is $\tilde{\mathscr{F}}_t$-adapted. The conclusion now follows from (11.3.1) and Girsanov's theorem. Statement (b) is an immediate consequence of (a). □

Remark 11.3.1. Recalling the notation of Chapter 7, it follows from Lemma 11.3.1 that $P_{\tilde{W}} \equiv \mu$, (that is, $P_{\tilde{W}} = P\tilde{W}^{-1} \equiv \tilde{P}\tilde{W}^{-1} = \mu$, Wiener measure). The next two lemmas are simple measure theoretic facts.

Lemma 11.3.2. *On the probability space $(\Omega,\mathscr{F},P)$ let g be an integrable random variable, measurable with respect to $\mathscr{F}^x$, a sub σ-field of $\mathscr{F}$. Let $Q(\cdot,\omega)$ be a version of the conditional probability $Q(A,\omega) = E(I_A|\mathscr{F}^x)(\omega)$ (a.s.) for $A \in \mathscr{F}^z \subseteq \mathscr{F}$. Then ϕ_g defined by*

$$\phi_g(A) = \int g(\omega)Q(A,\omega)P(d\omega) \qquad (A \in \mathscr{F}^z) \tag{11.3.3}$$

is a finite signed measure on $(\Omega,\mathscr{F}^z)$ which is absolutely continuous with respect to P^z, the restriction of P to $\mathscr{F}^z$. Also,

$$E(g|\mathscr{F}^z) = \frac{d\phi_g}{dP^z} \qquad (P^z\text{-a.s.}). \tag{11.3.4}$$

PROOF. It is obvious from the integrability of g that ϕ_g is a finite signed measure dominated by P. To complete the proof observe that

$$E[I_A E(g|\mathscr{F}^z)] = \phi_g(A) \quad \text{for } A \in \mathscr{F}^z.$$

Now

$$\begin{aligned} E(gI_A) &= E[E(gI_A|\mathscr{F}^x)] \\ &= E[gE(I_A|\mathscr{F}^x)] \\ &= \int g(\omega)Q(A,\omega)P(d\omega) = \phi_g(A). \end{aligned}$$

Hence

$$\phi_g(A) = E[E(gI_A|\mathscr{F}^z)] = E[I_A E(g|\mathscr{F}^z)] = \int_A E(g|\mathscr{F}^z)\,dP. \qquad \square$$

The notation $\mathscr{F}^x$ and $\mathscr{F}^z$ in the above lemma and in Lemma 11.3.3 is only meant to suggest future applications and does not refer to σ-fields generated by processes X and Z.

Lemma 11.3.3. *Let the following conditions be satisfied:*

(i) *The conditional probability $Q(\cdot,\omega)$ is regular; that is, $Q(\cdot,\omega)$ is a probability measure on $\mathscr{F}^z$ for each ω.*
(ii) *The σ-field $\mathscr{F}^z$ is generated by a countable family (that is, $\mathscr{F}^z$ is separable).*
(iii) *There exists a probability measure λ on $(\Omega,\mathscr{F}^z)$ such that $Q(\cdot,\omega) \ll \lambda$ (P^z-a.s.).*

Then the following conclusions hold:

$$P^z \ll \lambda\,[\mathscr{F}^z]. \tag{11.3.5}$$

There exists $q(\omega,\omega')$ which is $\mathscr{F}^z \times \mathscr{F}^x$-measurable such that for P-a.a. ω'

$$q(\omega,\omega') = \frac{dQ(\cdot,\omega')}{d\lambda}(\omega) \qquad (\lambda\text{-a.s.}) \tag{11.3.6}$$

and

$$0 < \int q(\omega,\omega')P(d\omega') < \infty \qquad (P^z\text{-a.s.}). \tag{11.3.7}$$

If g is $\mathscr{F}^x$-measurable and integrable, then

$$E(g|\mathscr{F}^z)(\cdot) = \frac{\int g(\omega')q(\cdot,\omega')P(d\omega')}{\int q(\cdot,\omega')P(d\omega')}, \qquad (P^z\text{-a.s.}). \tag{11.3.8}$$

PROOF. Let $q(\omega,\omega')$ be the Radon-Nikodym derivative $(dQ(\cdot,\omega')/d\lambda)(\omega)$, which exists in view of condition (iii). For each $\omega', q(\cdot,\omega')$ is $\mathscr{F}^z$-measurable. We further assume that $q(\omega,\omega')$ is jointly measurable since the existence of such a version of q is ensured by condition (ii). We then have, for $A \in \mathscr{F}^z$,

$$P^z(A) = E(I_A) = E\{E(I_A|\mathscr{F}^x)\} = \int Q(A,\omega)P(d\omega),$$

and (11.3.4) follows from (iii). From Lemma 11.3.2, $\phi_g \ll \lambda$, and so

$$\frac{d\phi_g}{d\lambda} = \frac{d\phi_g}{dP^z} \cdot \frac{dP^z}{d\lambda} \qquad (\lambda\text{-a.s.}). \tag{11.3.9}$$

Let $A_0 = \{\omega:(dP^z/d\lambda)(\omega) = 0\}$. Then $P^z(A_0) = 0$, so that $0 < dP^z/d\lambda < \infty$ (P^z-a.s.). Letting B_0 be the set on which (11.3.9) fails to hold, we get

$$\frac{d\phi_g}{d\lambda}(\omega) = \frac{(d\phi_g/d\lambda)(\omega)}{(dP^z/d\lambda)(\omega)} \quad \text{for } \omega \notin A_0 \cup B_0, \tag{11.3.10}$$

and hence (P^z-*a.s.*) because $P^z(A_0 \cup B_0) = 0$. From (11.3.3) and (11.3.6) and using Fubini's theorem, it follows that

$$\phi_g(A) = \int_A \left[\int g(\omega')q(\omega,\omega')P(d\omega') \right] \lambda(d\omega).$$

Noting that the quantity in brackets is $\mathscr{F}^z$-measurable, we obtain

$$\frac{d\phi_g}{d\lambda}(\omega) = \int g(\omega')q(\omega,\omega')P(d\omega').$$

Taking $g = 1$ in the above formula, we get $(dP^z/d\lambda)(\omega) = \int q(\omega,\omega')P(d\omega')$, and (11.3.8) then follows from (11.3.4) and (11.3.10). □

The above lemma can be written in the following form, which is suitable when $\mathscr{F}^z$ is generated by an observation process Z for which the corresponding signal process $X = (X_t)$ and the noise $W = (W_t)$ are stochastically independent.

(i) Let $X = (X_t(u))$ be defined on a probability space $(\Omega^x,\mathscr{F}^x,P^x)$ as a jointly measurable process satisfying the condition.

$$\int_0^T X_s^2(u)\,ds < \infty \qquad (P^x\text{-a.s.}).$$

Now take $\mathscr{F}^x = \mathscr{F}^{X,0}$ $(=\mathscr{F}_T^{X,0})$, where $(\mathscr{F}_t^{X,0})$ is the natural family of X (see Chapter 1, page 5).

(ii) $W = (W_t(v))$ is a standard Wiener process on $(\Omega^w, \mathscr{F}^w, P^w)$, where $\mathscr{F}^w = \mathscr{F}^{W,0} = \mathscr{F}_T^{W,0}$.

(iii) Let $Z_t(\omega) = \int_0^t X_s(\omega)\, ds + W_t(\omega)$ $(0 \le t \le T)$, where $\omega = (u,v)$, $X_t(\omega) = X_t(u)$ and $W_t(\omega) = W_t(v)$. Let $\Omega = \Omega^x \times \Omega^w$, $\mathscr{G} = \mathscr{F}^{X,0} \times \mathscr{F}^{W,0}$, and $P = P^x \times P^w$. We shall also write $\tilde{\mathscr{F}}_t^{X,0} = \mathscr{F}_t^{X,0} \times \{\varnothing, \Omega^w\}$ and $\tilde{\mathscr{F}}_t^{W,0} = \{\varnothing, \Omega^x\} \times \mathscr{F}_t^{W,0}$. Then, $\mathscr{G}_t = \tilde{\mathscr{F}}_t^{X,0} \vee \tilde{\mathscr{F}}_t^{W,0}$ and, in particular, $\mathscr{G} = \tilde{\mathscr{F}}^{X,0} \vee \tilde{\mathscr{F}}^{W,0}$. Since the context will preclude any possible confusion we shall henceforth use the same notation $\mathscr{F}^{X,0}(\mathscr{F}^{W,0})$ to denote $\tilde{\mathscr{F}}^{X,0}(\tilde{\mathscr{F}}^{W,0})$. We thus have in (iii) a model for the processes Z, X, and W defined on $(\Omega, \mathscr{G}, P)$ such that $X \perp\!\!\!\perp W$. Let $\mathscr{F}^z = \mathscr{F}^{Z,0}$. Denote by P^z the restriction of P to $\mathscr{F}^z$. Consider the model (iii) and let $u' \in \Omega^x$ be fixed. It is easy to see that $Q(\cdot, \omega')$, the conditional probability (relative to $\mathscr{F}^x$), on the σ-field $\mathscr{F}^z$ is the restriction to $\mathscr{F}^z$ of the product measure $\varepsilon_{u'} \times P^w$, $\varepsilon_{u'}$ being the probability measure with mass concentrated at $\{u'\}$. Furthermore, $Q(\cdot, \omega')$ depends on ω' only through u' so that we write $Q(\cdot, u')$ for $Q(\cdot, \omega')$.

Since $X \perp\!\!\!\perp W$ it is obvious that (W_t) is a Wiener process under the probability law $\varepsilon_{u'} \times P^w$. Also it is easily verified that (a.s.) with respect to $\varepsilon_{u'} \times P^w$,

$$Z_t(\omega) = \int_0^t X_s(u')\, ds + W_t(\omega) \quad \text{for all } t \in [0,T].$$

Let us now define $\hat{\mathscr{F}}_t^Z = \mathscr{F}_t^{Z,0} \vee \{\text{all } Q(\cdot, u')\text{-null sets in } \mathscr{F}_T^{Z,0}\}$. It then follows that (W_t) is $(\hat{\mathscr{F}}_t^Z)$-adapted, a Wiener process under $Q(\cdot, u')$ and is moreover, a $(\hat{\mathscr{F}}_t^Z, Q(\cdot, u'))$-martingale.

We now apply Lemma 11.3.1 to conclude that

$$\exp\left[-\int_0^T X_s(u')\, dW_s(\omega) - \frac{1}{2} \int_0^T X_s^2(u')\, ds \right] Q(d\omega, u') \tag{11.3.11}$$

is Wiener measure over $\hat{\mathscr{F}}^Z$ (and hence over $\mathscr{F}^Z$). Since Wiener measure over $\mathscr{F}^z$ is unique, it follows that (11.3.11) is independent of u' and we denote it by $\lambda(d\omega)$. Hence condition (iii) of Lemma 11.3.3 is satisfied. For the Radon-Nikodym derivative q of $Q(\cdot, u')$ with respect to λ, we take the $\mathscr{F}^z$-measurable version given by

$$q(\omega, \omega') = q(\omega, u') = \exp\left[\int_0^T X_s(u')\, dZ_s(\omega) - \frac{1}{2} \int_0^T X_s^2(u')\, ds \right]. \tag{11.3.12}$$

This choice is justified by (11.3.11) because of the $\mathscr{F}^z \times \mathscr{F}^x$-measurability of (11.3.12). The desired Bayes formula now follows from (11.3.7) and (11.3.8) of Lemma 11.3.3.

Theorem 11.3.1. *Let X, W, and Z be given as in* (i), (ii), *and* (iii). *Let g on Ω be integrable and* $\mathscr{F}^x$-measurable. *Then, writing $g(u') = g(\omega')$ $(\omega' = (u', v'))$, we* *have*

$$E(g|\mathscr{F}^z)(\omega) = \frac{\int_{\Omega^x} g(u')q(\omega,u')P^x(du')}{\int_{\Omega^x} q(\omega,u')P^x(du')}, \tag{11.3.13}$$

where

$$0 < \int_{\Omega^x} q(\omega,u')P^x(du') < \infty \qquad (P^z\text{-a.s.}) \tag{11.3.14}$$

and $q(\omega,u')$ *is given by (11.3.12).*

The next result provides us with another formula for the conditional expectation in (11.3.13).

Theorem 11.3.2. *Assume the conditions of Lemma 11.3.3. Further suppose that*

$$\int_0^T E(X_t^2)\,dt < \infty. \tag{11.3.15}$$

Then we obtain the formula

$$E(g|\mathscr{F}^z)(\omega) = \int_{\Omega_x} g(u)\exp\left[\int_0^T (X_s(u) - \hat{X}_s(\omega))\,d\nu_s(\omega) - \frac{1}{2}\int_0^T (X_s(u) - \hat{X}_s(\omega))^2\,ds\right]P^x(du), \tag{11.3.16}$$

where $\hat{X}_t(\omega) = E(X_t|\mathscr{F}_t^Z)(\omega)$ *and* $\nu = (\nu_t)$ *is the innovation process.*

Proof. Condition (11.3.15) implies the existence of $\hat{X}_t$ for each t and the finiteness (a.s.) of $\int_0^T (\hat{X}_t)^2\,dt$.

Writing (iii) in the form

$$Z_t(\omega) = \int_0^t \hat{X}_s(\omega)\,ds + \nu_t(\omega), \tag{11.3.17}$$

we again obtain from Girsanov's theorem that

$$\exp\left[-\int_0^T \hat{X}_s\,d\nu_s - \frac{1}{2}\int_0^T \hat{X}_s^2\,ds\right]dP^z = d\lambda. \tag{11.3.18}$$

Since, by Lemma 11.3.2, $dP^z/d\lambda$ is the denominator in the right-hand side of (11.3.13), substituting from (11.3.18) in (11.3.13) and using (11.3.17) in the numerator in the right side of (11.3.13), we get (11.3.16). □

11.4 Equality of the Sigma Fields $\mathscr{F}_t^Z$ and $\mathscr{F}_t^\nu$

The existence of the innovation process (ν_t) was established in Chapter 8 under very general conditions, and the question of the equality of the σ-fields $\mathscr{F}_t^\nu$ and $\mathscr{F}_t^Z$ for each t was raised in that chapter. It was shown in Chapter 9 and 10 that the equality does hold in the case of linear filtering. For nonlinear filtering, that is, when the processes $Z = (Z_t)$ and $X = (X_t)$ are connected by the stochastic equation of Chapter 8, it follows from Tsirel'son's result proved in Section 7.7 that, in general, equality does not hold. However, for the important special case when noise and signal processes are independent we have the following positive result. The proof (see [4]) is based on the Bayes formula derived in the previous section.

Theorem 11.4.1. *Let the processes Z, X, and W be as in* (i) *to* (iii) *of Section 11.3. Further suppose that for some constant K,*

$$|X_t(u)| \leq K \quad \text{for all } t \text{ and a.e. } u(P^x). \tag{11.4.1}$$

Then, for each t in $[0,T]$.

$$\mathscr{F}_t^Z = \mathscr{F}_t^\nu \tag{11.4.2}$$

Proof. Let $H = (H_t)$ be a progressively measurable process relative to $(\mathscr{F}_t^Z)$ and uniformly bounded, $|H_t(\omega)| \leq M$, (M depending on H). Then the process $R_t^H(u,\omega)$ defined by

$$R_t^H(u,\omega) = \int_0^t X_s(u)\,d\nu_s(\omega) - \frac{1}{2}\int_0^t X_s^2(u)\,ds + \int_0^t X_s(u)H_s(\omega)\,ds \tag{11.4.3}$$

has the following properties. It is measurable with respect to $\mathscr{F}^X \times \mathscr{F}_t^Z$ and is sample-continuous in t and for $a > 0$.

$$\int \exp\left[a \sup_{s \leq T} |R_s^H(u,\omega)|\right] P(d\omega) \leq C < \infty \tag{11.4.4}$$

for each $u \in \Omega_x$, where C is a constant not depending on u but possibly depending on K, M and a. Note that the process $\hat{X} = (\hat{X}_t)$ belongs to the class of processes H defined above. Furthermore, it is easy to see that, given H, there exist positive constants c_1 and c_2 (not depending on u, ω, or t) such that

$$c_1 \exp R_t^{\hat{X}}(u,\omega) \leq \exp R_t^H(u,\omega) \leq c_2 \exp R_t^{\hat{X}}(u,\omega) \tag{11.4.5}$$

for every u, ω and $t \in [0,T]$. From (11.4.5) and conclusion (11.3.14) of Lemma 11.3.3 it follows that

$$0 < \int_{\Omega_x} \exp[R_t^H(u,\omega)]P^x(du) < \infty \qquad (P^z\text{-a.s.}). \tag{11.4.6}$$

It can be seen that the quantity in (11.4.6) is continuous in t. Let

$$T_t(H,\omega) = \frac{\int X_t(u) \exp R_t^H(u,\omega) P^x(du)}{\int \exp R_t^H(u,\omega) P^x(du)}. \tag{11.4.7}$$

Then $T_t(H,\cdot)$ is $\mathscr{F}_t^Z$-measurable and for all t,

$$|T_t(H,\omega)| \leq K \qquad \text{(a.s.)}. \tag{11.4.8}$$

For $\alpha \in [0,1]$, let

$$R_t^{H,\alpha}(u,\omega) = R_t^{\alpha H + (1-\alpha)\hat{X}}(u,\omega).$$

Then

$$\begin{aligned}\frac{d}{d\alpha} R_t^{H,\alpha}(u,\omega) &= \int_0^t X_s(u)[H_s(\omega) - \hat{X}_s(\omega)]\,ds,\\ \frac{d}{d\alpha}\exp[R_t^{H,\alpha}(u,\omega)] &= \exp[R_t^{H,\alpha}(u,\omega)]\cdot\int_0^t X_s(u)[H_s(\omega) - \hat{X}_s(\omega)]\,ds.\end{aligned} \tag{11.4.9}$$

Write $T_t^\alpha(H,\omega) = T_t(\alpha H + (1-\alpha)\hat{X},\omega)$. $T_t^\alpha(H,\omega)$ is given by the formula (11.4.7) with H replaced by $\alpha H + (1-\alpha)\hat{X}$. It is easily verified that

$$\left|\frac{d}{d\alpha}\exp R_t^{H,\alpha}\right| \le \text{constant}\left\{\exp[R_t^H] + \exp[R_t^{\hat{X}}]\right\},$$

so from (11.4.4) we may interchange integration and differentiation with respect to α in the numerator and denominator of the formula (11.4.7) for $T_t^\alpha(H,\omega)$. We have

$$\begin{aligned}\frac{d}{d\alpha}T_t^\alpha(H,\omega) &= \frac{\int X_t(u)\exp[R_t^{H,\alpha}(u,\omega)]\cdot\left\{\int_0^t X_s(u)[H_s(\omega)-\hat{X}_s(\omega)]\,ds\right\}P^x(du)}{\int \exp[R_t^{H,\alpha}(u,\omega)]P^x(du)}\\ &\quad - T_t^\alpha(H,\omega)\cdot\frac{\int \exp[R_t^{H,\alpha}(u,\omega)]\left\{\int_0^t X_s(u)[H_s(\omega)-\hat{X}_s(\omega)]\,ds\right\}P^x(du)}{\int \exp[R_t^{H,\alpha}(u,\omega)]P^x(du)}.\end{aligned} \tag{11.4.10}$$

Hence

$$\left|\frac{d}{d\alpha}T_t^\alpha(H,\omega)\right| \le 2K^2\int_0^t |H_s(\omega) - \hat{X}_s(\omega)|\,ds \tag{11.4.11}$$

and

$$|T_t^1(H,\omega) - T_t^0(H,\omega)| \le \int_0^1\left|\frac{d}{d\alpha}T_t^\alpha(H,\omega)\right|d\alpha \le 2K^2\int_0^t |H_s(\omega) - \hat{X}_s(\omega)|\,ds.$$

Since $T_t^1(H,\omega) = T_t(H,\omega)$ and $T_t^0(H,\omega) = T_t(\hat{X},\omega) = \hat{X}_t(\omega)$, we obtain the inequality

$$|T_t(H,\omega) - \hat{X}_t(\omega)| \le 2K^2\int_0^t |H_s(\omega) - \hat{X}_s(\omega)|\,ds. \tag{11.4.12}$$

We now use (11.4.12) for a successive approximation of $\hat{X}_t(\omega)$ as follows. Set $H_s^0(\omega) = 0$ for all $s \in [0,T]$. Define $H_t^1(\omega) = T_t(H^0,\omega)$ and so recursively, $H_t^{n+1}(\omega) = T_t(H^n,\omega)$. It is clear from the definition of $T_t(H,\omega)$ that all the processes $H^n = (H_t^n)$ are measurable and adapted to $(\mathscr{F}_t^\nu)$. Furthermore from

(11.4.12) (writing $A = 2K^2$),

$$\begin{aligned}|H_t^{n+1}(\omega) - \hat{X}_t(\omega)| &\leq A \int_0^t |H_s^n(\omega) - \hat{X}_s(\omega)|\, ds \\ &\leq A^2 \int_0^t \left\{ \int_0^s |H_{s'}^{n-1}(\omega) - \hat{X}_{s'}(\omega)|\, ds' \right\} ds \\ &= A^2 \int_0^t (t-s)|H_s^{n-1}(\omega) - \hat{X}_s(\omega)|\, ds, \\ &\leq \frac{A^3}{2} \int_0^t (t-s)^2 |H_s^{n-2}(\omega) - \hat{X}_s(\omega)|\, ds.\end{aligned}$$

Proceeding similarly, we arrive at the bound

$$|H_t^n(\omega) - \hat{X}_t(\omega)| \leq \frac{K(AT)^n}{n!}, \qquad t \in [0,T]. \tag{11.4.13}$$

Hence the sequence of processes (H_t^n) converges uniformly in t (a.s.) to $\hat{X}_t$, which proves that $\hat{X}_t(\cdot)$ is $\mathscr{F}_t^\nu$-measurable for each t. The conclusion $\mathscr{F}_t^Z = \mathscr{F}_t^\nu$ follows from (11.3.17). □

Remark 11.4.1. Suppose the processes (Z_t), (X_t) and (W_t) are defined for $t \in \mathbf{R}^+$. Then Theorems 11.3.1 and 11.3.2 continue to hold provided that in condition (i) of Theorem 11.3.1 it is assumed that

$$E \int_0^t X_s^2(u)\, ds < \infty.$$

Writing

$$q_t(\omega,u') = \exp\left[\int_0^t X_s(u')\, dZ_s(\omega) - \frac{1}{2} \int_0^t X_s^2(u')\, ds \right],$$

we then have

$$E(g|\mathscr{F}_t^Z)(\omega) = \frac{\int g(u) q_t(\omega,u) P^x(du)}{\int q_t(\omega,u) P^x(du)}, \tag{11.4.14}$$

where

$$0 < \int q_t(\omega,u) P^x(du) < \infty \qquad (P^z\text{-a.s.}) \tag{11.4.15}$$

for each $t \in \mathbf{R}_+$. Here P^z is the restriction of P to $\vee_{t \in \mathbf{R}_+} \mathscr{F}_t^Z$. Similarly we have

$$\begin{aligned}E(g|\mathscr{F}_t^Z)(\omega) = \int_{\Omega^x} g(u) \exp\bigg[&\int_0^t (X_s(u) - \hat{X}_s(\omega))\, d\nu_s(\omega) \\ &- \frac{1}{2} \int_0^t (X_s(u) - \hat{X}_s(\omega))^2\, ds \bigg] P^x(du).\end{aligned} \tag{11.4.16}$$

Also, in Theorem 11.4.1 we have

$$\mathscr{F}_t^Z = \mathscr{F}_t^\nu \quad \text{for each } t \in \mathbf{R}_+. \tag{11.4.2a}$$

Remark 11.4.2. Formulas (11.4.14) and (11.4.16) are, in fact, true for *all t* (a.s.). This has been shown, by Meyer, by defining suitable versions of q_t and of the process in the denominator of the right-hand side of (11.4.14).

Remark 11.4.3. An important consequence of (11.4.2a) is

$$\mathscr{F}_{t+}^{Z} = \mathscr{F}_{t}^{Z} = \mathscr{F}_{t-0}^{Z} \quad \text{for each } t. \tag{11.4.17}$$

11.5 Solution of the Filter Equation

In this section we shall prove that the optimal filter $E^t(f)$ is the unique solution of a stochastic equation related to (8.4.22). We have to impose the severe restriction of boundedness on the signal process.

Let $X = (X_t)$ be a right-continuous temporally homogeneous Markov process whose state space S is a compact, separable, Hausdorff space. Let the semigroup $P_t(t \geq 0)$ associated with the transitional probabilities $P_t(x,E)$ be a Feller semigroup, that is,

$$P_t f(x) = \int_S P_t(x,dy) f(y)$$

maps $C(S)$ into itself for all $t > 0$ and satisfies $\lim_{t \to +0} P_t f(x) = f(x)$ uniformly in S for all $f \in C(S)$, $C(S)$ being the space of all real continuous functions on S. Let A denote the infinitesimal generator of (P_t) and $D(A)$, the domain of A. Assume that $D(A) \subset C(S)$.

We shall now apply the results of the previous sections to study the filtering problem for the particular stochastic model to be introduced below. This model is studied in great detail in [37]. Let us make the following identification of the spaces introduced in Theorem 11.3.1.

Let Ω^x be the space of all functions from $\mathbf{R}_+$ into S which are right-continuous with left limits, and let $(X_t(u))$ be given by $X_t(u) = u(t)$, $u \in \Omega^x$.

Let $\mathscr{F}_t^{X,0}$ be the natural σ-field of (X_t). A family of Markovian measures $P_x^{(1)}$, $x \in S$ is defined on $(\Omega^x, \mathscr{F}^x)$, $(\mathscr{F}^x = \vee \mathscr{F}_t^{X,0})$ by

$$P_x^{(1)}[X_{t_1} \in E_1, \ldots, X_{t_n} \in E_n] = \int \cdots \int_{E_1 \times \cdots \times E_n} P_{t_1}(x,dx_1) \cdots P_{t_n - t_{n-1}}(x_{n-1},dx_n),$$

where $0 < t_1 < \cdots < t_n$ and E_i, $i = 1, \ldots, n$ are Borel sets in S. Let $M(S)$ be the set of all probability measures over the Borel sets of S, endowed with the weak topology. Then $M(S)$ is a compact, separable Hausdorff space. For $m \in M(S)$, define $P_m = \int P_x m(dx)$. The signal process $X = (X_t)$ is then a Markov process on $(\Omega^x, \mathscr{F}^x, P_m)$. In the notation of Section 11.3 and the Bayes formula, $P_m = P^x$. Let Ω^w be the space of all real-valued continuous functions v on $\mathbf{R}_+$, and let $W_t(v) = v(t)$. Suppose that P^w is Wiener measure on $\mathscr{F}^w = \vee_t \mathscr{F}_t^{W,0}$. Now write $\Omega = \Omega^x \times \Omega^w$, $\mathscr{G}_t = \mathscr{F}_t^{X,0} \times \mathscr{F}_t^{W,0}$, $\mathscr{G} = \mathscr{F}^x \times \mathscr{F}^w$ and $P = P_m \times P^w$. As in Section 11.3, we identify $\tilde{\mathscr{F}}_t^{X,0}(\tilde{\mathscr{F}}_t^{W,0})$ with $\mathscr{F}_t^{X,0}(\mathscr{F}_t^{W,0})$,

$\mathscr{G}_t$ with $\tilde{\mathscr{F}}_t^{X,0} \vee \tilde{\mathscr{F}}_t^{W,0}$ and $\mathscr{G}$ with $\tilde{\mathscr{F}}^x \vee \tilde{\mathscr{F}}^w$, where

$$\tilde{\mathscr{F}}^x = \mathscr{F}^x \times \{\varnothing,\Omega^w\} \qquad \text{and} \qquad \tilde{\mathscr{F}}^w = \{\varnothing,\Omega^x\} \times \mathscr{F}^w.$$

Define $X_t(\omega) = X_t(u)$ and $W_t(\omega) = W_t(v)$, where $\omega = (u,v)$. Let $h \in C(S)$ be fixed, $h_s(\omega) = h(X_s(\omega))$ and

$$Z_t(\omega) = \int_0^t h_s(\omega)\, ds + W_t(\omega). \tag{11.5.1}$$

We shall first show that there is a (t,ω)-measurable version of the Bayes formula for the conditional expectation $E^t(f)$. To avoid confusion, we denote by $E_0^t(f)$ the particular version of $E^t(f)$ obtained in the following lemma.

Lemma 11.5.1. *There exists a function $\rho(t,u,\omega)$ which is $\mathscr{B}[0,T] \times \mathscr{F}^{X,0} \times \mathscr{G}$-measurable such that for each $t \in [0,T]$,*

$$\rho(t,u,\omega) = \exp\left[\int_0^t [h_s(u) - \hat{h}_s(\omega)]\, dv_s(\omega) - \frac{1}{2}\int_0^t [h_s(u) - h_s(\omega)]^2\, ds\right]$$

$((P^x \times P)$-a.s.). *Furthermore,*

$$E_0^t(f,\omega) = \int_{\Omega^x} f(X_t(u))\rho(t,u,\omega)P^x(du) \tag{11.5.2}$$

defines a (t,ω)-measurable process on $(\Omega,\tilde{\mathscr{G}},P)$ taking values in $M(S)$ such that for each t in $[0,T]$ and $f \in C(S)$,

$$E_0^t(f,\omega) = E[f(X_t)|\mathscr{F}_t^{Z,0}](\omega) \qquad (P\text{-a.s.}).$$

Proof. Let (h_s^n) be a sequence of simple processes such that

$$E_{P^x} \int_0^T |h_s(u) - h_s^n(u)|^2\, ds \to 0.$$

As in the proof of property 8 of Section 2.3, there is a subsequence $(J_t^{n'}(u,\omega))$ such that

$$\sup_{0 \le t \le T} \left|\int_0^t h_s(u)\, dv_s(\omega) - J_t^{n'}(u,\omega)\right| \to 0 \qquad ((P^x \times P)\text{-a.s.}),$$

where $J_t^n(u,\omega) = \sum_i h_{s_i}^n(u)[v_{s_{i+1}\wedge t}(\omega) - v_{s_i \wedge t}(\omega)]$ is a finite sum which is $\mathscr{B}[0,T] \times \mathscr{F}^{X,0} \times \mathscr{G}$-measurable and continuous in t for $(P^x \times P)$-a.a. (u,ω). Since all the $h^{n'}$ are simple, the set

$$\Lambda = \{(u,\omega)\colon J_t^{n'}(u,\omega) \text{ converges uniformly in } t\}$$

belongs to $\mathscr{F}^{X,0} \times \mathscr{G}$ and we have $(P^x \times P)(\Lambda) = 1$. Define

$$J_t(u,\omega) = \begin{cases} \lim\limits_{n' \to \infty} J_t^{n'}(u,\omega) & \text{if this limit exists.} \\ 0 & \text{otherwise.} \end{cases}$$

Then $J_t(u,\omega)$ is measurable in all three variables, and for each $t \in [0,T]$,

$$J_t(u,\omega) = \int_0^t h_s(u)\,dv_s(\omega) \qquad ((P^x \times P)\text{-a.s.}).$$

We have thus shown the existence of a (t,u,ω) measurable version of the stochastic integral process $\int_0^t h_s(u)\,dv_s(\omega)$ which is continuous in t for a.a. (u,ω). Next, from the joint measurability of $h_s(u)$ and $\hat{h}_s(\omega)$ we obtain versions of $\int_0^t h_s(u)\hat{h}_s(\omega)\,ds$ and $\int_0^t h_s^2(u)\,ds$ which are $\mathscr{B}[0,T] \times \mathscr{F}^{X,0} \times \mathscr{G}$-measurable. In what follows, it is these measurable versions which will be considered. Define

$$R(t,u,\omega) = \exp\left[\int_0^t h_s(u)\,dv_s(\omega) + \int_0^t h_s(u)\hat{h}_s(\omega)\,ds - \frac{1}{2}\int_0^t h_s^2(u)\,ds\right]. \quad (11.5.3)$$

Then $\int_{\Omega^x} R(t,u,\omega)P^x(du) > 0$ for all (t,ω). The set $N = \{(t,\omega)\colon \int_{\Omega^x} R(t,u,\omega)P^X(du) = \infty\}$ is $\mathscr{B}[0,T] \times \mathscr{G}$-measurable and $P(N_t) = 0$ for each t, where N_t is the t section of N. The last fact holds because for each fixed t,

$$0 < \int_{\Omega^x} R(t,u,\omega)P^x(du) < \infty \quad \text{for a.a. } \omega.$$

Hence $(L \times P)(N) = 0$ where L is Lebesgue measure. The function ρ is defined by

$$\rho(t,u,\omega) = \frac{R(t,u,\omega)}{\int_{\Omega^x} R(t,u,\omega)P^X(du)} \quad \text{for } (t,u,\omega) \quad (11.5.4)$$

for which $(t,\omega) \notin N$, and $=1$ otherwise. Then ρ is $\mathscr{B}[0,T] \times \mathscr{F}^{X,0} \times \mathscr{G}$-measurable. From the definition of R given above and the identification of the denominator given in Theorem 11.3.2 it follows that for each t,

$$\rho(t,u,\omega) = \exp\left\{\int_0^t [h_s(u) - \hat{h}_s(\omega)]\,dv_s(\omega) - \frac{1}{2}\int_0^t [h_s(u) - \hat{h}_s(\omega)]^2\,ds\right\}. \quad (11.5.5)$$

Let us now define

$$E_0^t(f,\omega) = \int_{\Omega^x} f(X_t(u))\rho(t,u,\omega)P^x(du) \quad (11.5.6)$$

for each (t,ω) and $f \in C(S)$. Clearly (11.5.6) provides a $\mathscr{B}[0,T] \times \mathscr{G}$-measurable function. From the definition of $\rho(t,u,\omega)$ we find that for each (t,ω) there is precisely one probability measure in $M(S)$ which we denote by $E_0^t(\omega)$ such that $E_0^t(\omega)(f) = E_0^t(f,\omega)$. Hence (E_0^t) is an $M(S)$-valued stochastic process and the map $(t,\omega) \to E_0^t(\omega)$ is $\mathscr{B}(M(S))/\mathscr{B}[0,T] \times \mathscr{G}$-measurable, where $\mathscr{B}(M(S))$ is the Borel σ-field in the weak topology of $M(S)$. □

Remark 11.5.1. From Lemma 11.5.1 and the right-continuity of the paths of the Markov process (X_t), it follows that for $f \in C(S)$, $E_0^t(f,\omega)$ given by (11.5.6) is right-continuous in t, (a.s.). This implies that for a.a. ω, $E_0^t(\omega)$ is right-continuous in the weak topology of $M(S)$.

PROOF. Let $t_n \to t$ $(t_n > t)$. Then we have

$$\begin{aligned}\int_{\Omega^x} |f(X_{t_n}(u))\rho(t_n,u,\omega) - f(X_t(u))\rho(t,u,\omega)| P^x(du) \\ \leq \|f\| \cdot \int_{\Omega^x} |\rho(t_n,u,\omega) - \rho(t,u,\omega)| P^x(du) \\ + \int_{\Omega^x} \rho(t,u,\omega) |f(X_{t_n}(u)) - f(X_t(u))| P^x(du).\end{aligned}$$

Since $X_t(u)$ is a right-continuous function of t for (P^x)-a.a. u and $f \in C(S)$, the second term on the right-hand side of the above inequality tends to 0 by the dominated convergence theorem. The first term goes to 0 because $\rho(t_n,u,\omega) \to \rho(t,u,\omega)$ from formula (11.5.5) and $\rho(t_n,u,\omega)$, $\rho(t,u,\omega)$ are non-negative functions such that $\int_{\Omega^x} \rho(t_n,u,\omega)P^x(du) = \int_{\Omega^x} \rho(t,u,\omega)P^x(du) = 1$ for all ω and n. Hence we have $E_0^{t_n}(f,\omega) \to E_0^t(f,\omega)$ from (11.5.6).

Let us now turn to the stochastic differential equation satisfied by $E^t(f)$. From Theorem 8.4.3 we obtain the following equation:

$$E^t(f) = E^0(f) + \int_0^t E^s(Af)\, ds + \int_0^t [E^s(fh) - E^s(f)E^s(h)]\, dv_s \tag{11.5.7}$$

for $f \in D(A)$. It follows that for each f in $D(A)$, $E^t(f)$ has a sample-continuous version. Since (P_t) is a Feller semigroup $D(A)$ is dense in $C(S)$. (See [10], Vol. 1, p. 72, Theorem 2.8). Using this fact, it is easy to see that there exists a sample-continuous version of $E^t(f)$ for each $f \in C(S)$. Furthermore, let Δ be a countable dense subspace of $C(S)$ and let $\Omega_0 = \{\omega: \xi_t(f_1 + f_2)(\omega) = \xi_t(f_1)(\omega) + \xi_t(f_2)(\omega)$, $\xi_t(af) = a\xi_t(f)$ and $|\xi_t(f)| \leq \|f\|$, for all $t \in \mathbf{R}_+$, all a rational, and all $f, f_1, f_2 \in \Delta\}$. Here $\xi_t(f)$ denotes a continuous version of $E^t(f)$ and $\|f\| = \sup_{x \in S} |f(x)|$. Clearly Ω_0 is measurable and $P(\Omega_0) = 1$. For $\omega \in \Omega_0$ and $f \in C(S)$, defining $\xi_t(f)(\omega)$ as the limit of $\xi_t(f_n)(\omega)$ where $\|f_n - f\| \to 0$, $f_n \in \Delta$, we see that $\xi_t(f)(\omega)$ is continuous in t. For $\omega \notin \Omega_0$, set $\xi_t(f)(\omega) = 0$ for all t. Then it follows that $\xi_t(f)$ thus defined is a continuous stochastic process taking values in the dual of $C(S)$, that is, an $M(S)$-valued, stochastic process. □

It is this version of the process that we have in mind when we refer to $E^t(f)$ from now on. The corresponding $M(S)$-valued process will be denoted by $E^t(\omega)$.

The stochastic equation which will concern us in this section is given in the next resul·.

Theorem 11.5.1. *For all $f \in C(S)$, $E^t(f)$ satisfies the equation*

$$E^t(f) = E^0(P_t f) + \int_0^t [E^s((P_{t-s}f)h) - E^s(P_{t-s}f)E^s(h)]\, dv_s. \tag{11.5.8}$$

PROOF. Let $t > s \geq 0$. Note that

$$E[f(X_t)|\mathscr{G}_s] = P_{t-s}f(X_s) \tag{11.5.9}$$

To prove (11.5.8) it is sufficient to verify that

$$E([E^t(f) - E^0(P_t f)](Y_t - Y_0)) = \int_0^t E([E^s((P_{t-s}f)h) - E^s(P_{t-s}f)E^s(h)]\Phi_s)\,ds \quad (11.5.10)$$

holds for all square-integrable martingales Y_t of the form

$$Y_t - Y_0 = \int_0^t \Phi_s\,dv_s \quad (11.5.11)$$

for it follows from Theorem 11.4.1 that $\mathscr{F}_t^Z = \mathscr{F}_t^v$, so that every square-integrable $(\mathscr{F}_t^Z,P)$-martingale Y_t has the representation (11.5.11), where Φ_s is (s,ω)-measurable, $(\mathscr{F}_s^Z)$-adapted and $E\int_0^t \Phi_s^2\,ds < \infty$ for each t. Since

$$v_t = W_t + \int_0^t [h(X_s) - E^s(h)]\,ds,$$

the left-hand side of (11.5.10) equals

$$E[E^t(f)(Y_t - Y_0)] = E\left[f(X_t)\int_0^t \Phi_s\,dW_s\right] + E\left[f(X_t)\int_0^t \Phi_s(h(X_s) - E^s(h))\,ds\right].$$

The first member on the right-hand side is 0 since (X_t) and (W_t) are independent. This is easily verified by first considering $\Phi_s(\omega)$ of the form $\sum_j \Phi_{s_j}(\omega)I_{(s_j,s_{j+1}]}(s)$. For then

$$\begin{aligned} E\left[f(X_t)\int_0^t \Phi_s\,dW_s\right] &= E\left[f(X_t)\left\{\sum_j \Phi_{s_j}(W_{s_{j+1}} - W_{s_j})\right\}\right] \\ &= \sum_j E[f(X_t)\Phi_{s_j}E\{(W_{s_{j+1}} - W_{s_j})|\mathscr{F}^{X_t} \vee \mathscr{F}_{s_j}^Z\}] \\ &= 0 \end{aligned}$$

since the conditional expectation vanishes because

$$W_{s_{j+1}} - W_{s_j} \perp\!\!\!\perp \mathscr{F}^{X_t} \vee \mathscr{F}_{s_j}^Z.$$

Using (11.5.9) the second member can be written as

$$\begin{aligned} &\int_0^t E\{(E[f(X_t)|\mathscr{G}_s]h(X_s) - E[f(X_t)|\mathscr{G}_s]E^s(h))\Phi_s\}\,ds \\ &\quad = \int_0^t E\{[(P_{t-s}f(X_s))h(X_s)) - P_{t-s}f(X_s)E^s(h)]\Phi_s\}\,ds. \end{aligned}$$

Since Φ_s is $\mathscr{F}_s^Z$-measurable, the last member coincides with the right-hand side of (11.5.10) and the proof is complete. □

Let (β_t), $t \in \mathbf{R}^+$ be a Wiener process defined on a probability space $(\Omega,\mathscr{G},P)$, and let $\pi_0(f) = Ef(X_0)$ [that is, π_0 is a given element of $M(S)$]. An

$M(S)$-valued stochastic process (π_t) is said to be a solution of the equation

$$\pi_t(f) = \pi_0(P_t f) + \int_0^t [\pi_s((P_{t-s}f)h) - \pi_s(P_{t-s}f)\pi_s(h)]\, d\beta_s \quad (11.5.12)$$

if it satisfies the following conditions:

(a) $\pi_t(f)(\omega)$ is (t,ω)-measurable for each $f \in C(S)$.
(b) $\pi_s \perp\!\!\!\perp \sigma[\beta_v - \beta_u, s \le u < v \le s']$ for every s,s' $(0 \le s \le s')$.
(c) $\pi_t(f)(\omega)$ satisfies (11.5.12) (a.s.) for each t.

Theorem 11.5.2. *Let $X = (X_t)$ $(0 \le t \le T)$ be a Feller Markov process with stationary transition probabilities and with state space, a compact separable, Hausdorff space S. Let the model of the observation process be given by (11.5.1) where $h \in C(S)$.*

Then the $M(S)$-valued optimal filter (E^t) $(0 \le t \le T)$ is the unique solution of (11.5.12). Furthermore, the paths $t \to E^t(\omega)$ are continuous for a.e. ω, and $E^t(\cdot)$ is $\mathscr{F}_t^y$-measurable for every t.

Proof. Let us first prove the uniqueness of the solution. Suppose π_t and π_t' are two solutions of (11.5.12) with the same initial condition. Write $\theta_t(f) = E(|\pi_t(f) - \pi_t'(f)|^2)$. Then

$$\theta_t(f) \le 2E(|\pi_t(f)|^2 + |\pi_t'(f)|^2) \le 4\|f\|^2. \quad (11.5.13)$$

On the other hand, it is easy to verify by direct computation that

$$\begin{aligned}\theta_t(f) \le {} & 3\int_0^t E[|\pi_s(hP_{t-s}f) - \pi_s'(hP_{t-s}f)|^2]\, ds \\ & + 3\int_0^t E[(\pi_s(h))^2 |\pi_s(P_{t-s}f) - \pi_s'(P_{t-s}f)|^2]\, ds \\ & + 3\int_0^t E[(\pi_s'(P_{t-s}f)^2 |\pi_s(h) - \pi_s'(h)|^2]\, ds,\end{aligned}$$

so that we obtain

$$\theta_t(f) \le 3\int_0^t \{\theta_s(hP_{t-s}f) + \|h\|^2\theta_s(P_{t-s}f) + \|f\|^2\theta_s(h)\}\, ds. \quad (11.5.14)$$

Noting that $\|P_{t-s}f\| \le \|f\|$ and substituting (11.5.13) in the right-hand side of (11.5.14), we have

$$\theta_t(f) \le 4(3\|h\|)^2\|f\|^2 t.$$

Substitute the above estimate for $\theta_t(f)$ again in the right-hand side of (11.5.14) and repeat this procedure n times. Then

$$\theta_t(f) \le 4(3\|h\|)^{2n}\|f\|^2 \cdot \frac{t^n}{n!}. \quad (11.5.15)$$

Making n tend to infinity we have $\theta_t(f) = 0$ for all $t > 0$ and each $f \in C(S)$. This proves uniqueness.

To prove the existence of the solution of (11.5.12), we argue as follows. When ν_t is the innovation process obtained from an observation process (Z_t) given by (11.5.1) we have shown in Theorem 11.5.1 that the filter $E^t(f)$, based on the σ-fields $(\mathscr{F}_t^Z)$ satisfies (11.5.8). We have also shown above that $E^t(f)$ has a continuous, $M(S)$-valued version which is $(\mathscr{F}_t^\nu)$-adapted, the last fact following from the equality $\mathscr{F}_t^Z = \mathscr{F}_t^\nu$ proved in Section 11.4. Thus, the $M(S)$-valued process $E^t(\omega)$ given by $\int_S f(x)E^t(dx,\omega) = E^t(f,\omega)$ is actually given by a causal functional $\Psi_t(\nu(\omega))$ taking values in $M(S)$, that is, there exists an $M(S)$-valued functional $\Psi_t(\eta)$, where $\eta \in C([0,T],\mathbf{R})$ (we use the symbol η in place of x here because x denotes a point of S), which is $B[0,T] \times \mathscr{B}_T$-measurable as a function of (t,η) and has the following properties: for each t,

(i) $\Psi_t(\cdot)$ is $\mathscr{B}_{t+}$ measurable.
(ii) $P[\omega: E^t(\omega) = \Psi_t(\nu(\omega))] = 1$.

A detailed proof of this statement is left to the reader. The main steps in the argument are as follows. Since the process (E^t) takes values in the compact, separable, Hausdorff space $M(S)$, the reasoning in Theorem 2.7.2 remains valid and we obtain a functional $\Psi_t'(\eta)$ having the property (i) and such that

$$L \times P[(t,\omega): E^t(\omega) \neq \Psi_t'(\nu(\omega))] = 0. \tag{11.5.16}$$

It follows that for a.a. ω, $E^t(\omega) = \Psi_t'(\nu(\omega))$ for a.e. t. Let us introduce the notation $\Psi_t'(f,\nu(\omega))$ to denote the value of the functional $\Psi_t'(\nu(\omega))$ at f, $[f \in C(S)]$. Then for a.a. ω, $E^t(f,\omega) = \Psi_t'(f,\nu(\omega))$ for a.e. t. Hence, for each t, the right-hand side of Equation (11.5.7) equals

$$\pi_0(f) + \int_0^t \Psi_s'(Af,\nu(\omega))\,ds + \int_0^t [\Psi_s'(fh,\nu(\omega)) - \Psi_s'(f,\nu(\omega))\Psi_s'(h,\nu(\omega))]\,d\nu_s(\omega)$$

(a.s.) for $f \in D(A)$. Let us write $\Psi_t(f,\nu(\omega))$ for the above expression, where the meaning of $\Psi_t(f,\eta)$ is clear from the context. It can also be shown that $\Psi_t(f,\eta)$ is defined for all $f \in C(S)$ and that $\Psi_t(f,\eta)$ is the evaluation of $\Psi_t(\eta)$ at f, $\Psi_t(\eta)$ being an $M(S)$-valued functional with the property (i). Finally, it is clear that $E^t(\omega) = \Psi_t(\nu(\omega))$ (a.s.) for all t and (ii) is proved. Hence $\Psi_t(\nu(\omega))$ is a solution of (11.5.8). Now if a Wiener process (β_t) is given on a probability space $(\Omega',\mathscr{G}',P')$, then $\pi_t(\omega') = \Psi_t(\beta(\omega'))$, $(\omega' \in \Omega')$ is a solution of (11.5.12). □

Remark 11.5.2. From Lemma 11.5.1, Remark 11.5.1 and Theorem 11.5.2 it follows that $(E_0^t(\omega))$ given by the Bayes formula (11.5.2) is continuous in t for a.a. ω and is the unique solution of the stochastic Equation (11.5.12).

Notes

Chapter 1

A standard reference for the introductory material on stochastic processes is Doob [8]. Examples 1.2.1 and 1.2.2 are from Doob's earlier papers referred to in [8]. Proposition 1.1.4 is found in Meyer [41], which also contains a discussion of stopping times. A proof of Theorem 1.2.1, as well as of Theorem 1.2.2, is given in Loève [70].

Chapter 2

The use of the Haar family in the construction of the Wiener process is well known. See [55], where further references are given. The reference for Remark 2.1.2 is Ito and Nisio [63]. The strong Markov property was first systematically investigated by Dynkin and Yushkevich (see Dynkin [10]). Property 2(iii) of Section 2.3 is Lévy's Hölder condition and Property 4 is the Blumenthal zero-one law. The treatment of square integrable martingales given here is based on Meyer [43, 44, 45] and Kunita-Watanabe [36]. The proof of Theorem 2.4.4 is due to Murali Rao [50]. Theorem 2.7.2 is from Liptser and Shiryaev [40].

Chapter 3

Section 3.1 is based on the ideas of Dellacherie [7] and Courrège [5]. The proofs of the theorems stated in 3.1 are to be found in [7]. The process (U_t^*)

of Theorem 3.1.4 is called the dual predictable projection of (U_t) by Dellacherie [7]. Section 3.2 is based on Meyer [43, 44] and Courrège [5]. A full discussion of Ito's stochastic integral is given in Ito [20]. Lemma 3.3.1 is from Gihman and Skorohod [15]. Lemma 3.3.3 is given in Friedman [13].

Chapter 4

The proof of Theorem 4.2.1 is due to Meyer [44]. The variant of Ito's formula given in Theorem 4.2.2 is due to Neveu [73].

Chapter 5

The proof of Theorem 5.1.1 is along the lines of Liptser and Shiryaev [40]. Propositions 5.4.1 and 5.4.2 are given in Meyer [46].

Section 5.5. A right-continuous Markov process with the Feller property satisfies the strong Markov property. See Friedman [13]. The treatment of diffusion processes in Section 5.6 follows Ito's lecture notes [20].

The existence of weak solutions is discussed in Skorohod's book [52], though the term "weak solution" is used by later writers. See the papers by Ershov [11] and Yamada and Watanabe [77], and Liptser and Shiryaev's book [40]. For Theorem 5.6.2, see the recent book by N. Krylov [66]. Stroock and Varadhan's work cited in Remark 5.6.2 is contained in their paper [76]. Theorem 5.7.1 appears to be new.

Chapter 6

The idea of using tensor products of Hilbert spaces in deriving the homogeneous chaos for the Wiener process goes back to Segal [75], Ito [62], and Kakutani [65]. The development in the present chapter is taken from Kallianpur [28]. For reproducing kernel Hilbert spaces and their properties, see Aronszajn [56]. The results of Section 6.2 are due to Ito [21]. Theorem 6.6.2 was proved by Le Page in his thesis [69]. Theorem 6.6.4 is Cameron and Martin's [3]. See also [21]. Theorems 6.7.1 and 6.7.2 are based on [21]. The proof of Theorem 6.7.3 seems to be new.

The following is an extension of Theorem 6.7.2.

Theorem 6.7.2 (a). *Let (M_t) $(0 \le t \le T)$ be a separable martingale on $(\Omega, \mathscr{F}_t^W, P)$ with $M_0 = 0$. Then M_t is continuous and there is a unique, measurable, and $(\mathscr{F}_t^W)$-adapted process $g_s(\omega)$ such that*

$$\int_0^T g_s^2 \, ds < \infty, \qquad P\text{-a.s.},$$

and

$$M_t = \int_0^T g_s \, dW_s \qquad \textit{for all } t.$$

PROOF. For $N \geq 1$, set $\phi_N(u) = \begin{cases} N & \text{if } u > N \\ u & \text{if } |u| \leq N. \\ -N & \text{if } u < -N \end{cases}$

Then $M_t^N = E[\phi_N(M_T)|\mathscr{F}_t^W]$ is a bounded $(\mathscr{F}_t^W,P)$-martingale with $M_0^N = 0$. By Theorem 6.7.2, M_t^N has a continuous version which we work with from now on. From Doob's martingale inequality, $P[\sup_{0 \leq t \leq T|}M_t^N - M_{t|} > \lambda] \leq 2\lambda^{-1} \int_{[|M_T|>N]}|M_T| \, dP \to 0$ for $\lambda \downarrow 0$, proving that M_t is continuous (P-a.s.). Introducing the stopping times $\tau_n = \inf\{t: 0 \leq t \leq T, |M_t| \geq N\}, =$ T if the set $\{\cdots\}$ is empty and, applying Theorem 6.7.2 to the bounded $(\mathscr{F}_t^W,P)$-martingale $\tilde{M}_t^n = M_{t \wedge \tau_n}$, we obtain $\tilde{M}_t^n = \int_0^t g_s^n \, dW_s$, g^n being measurable, $\mathscr{F}_t^W$-adapted, and such that $E \int_0^T (g_s^n)^2 \, ds < \infty$. For $m \geq n$, from the relation $\tilde{M}_{t \wedge \tau_n}^m = \tilde{M}_t^n$ it follows that $g_t^m(\omega) = g_t^n(\omega)$ for a.e. t in $[0,\tau_n(\omega)]$, for P – almost all ω. Define $g_t(\omega) = \lim_{m \to \infty} g_t^m(\omega)$ if it exists, $= 0$ otherwise. The process g is measurable and $\mathscr{F}_t^W$-adapted, and for every $n \geq 1$, $g_t^n(\omega) = g_t(\omega)$ for a.e. $t \in [0,\tau_n(\omega)]$, P-a.s. Let $\langle M \rangle_t$ be the quadratic variation process of the continuous martingale (M_t). Then

$$\langle \tilde{M}^n \rangle_t = \int_0^t (g_s^n)^2 \, ds = \int_0^t g_s^2 I_{[0,\tau_n]}(s) \, ds, \ P\text{-a.s.}$$

Noting that $\tau_n \uparrow T$ P-a.s., since M_t is continuous we have

$$\begin{aligned} \int_0^T g_s^2(\omega) \, ds &= \lim_{n \to \infty} \int_0^T g_s^2(\omega) I_{[0,\tau_n(\omega)]}(s) \, ds \\ &= \lim_{n \to \infty} \langle M^n \rangle_T(\omega) = \langle M \rangle_T(\omega) < \infty, \ P\text{-a.s.} \end{aligned}$$

From the fact that $\int_0^T [g_s I_{[0,\tau_n]}(s) - g_s]^2 \, ds \to 0$ P-a.s., it follows that

$$\int_0^t g_s I_{[0,\tau_n]}(s) \, dW_s \to \int_0^t g_s \, dW_s$$

in probability. Hence $M_t = \int_0^t g_s \, dW_s$, P-a.s. The uniqueness of g follows easily.

The most general result in this direction was recently obtained by Dudley: Every measurable, real-valued random variable on $(\Omega,\mathscr{F}_1^W,P)$ can be represented as an Ito stochastic integral $\int_0^1 g_s \, dW_s$, where the process g is measurable and $(\mathscr{F}_t^W)$-adapted with $\int_0^1 g_s^2 \, ds < \infty$, P-a.s. See Dudley [59].

Chapter 7

We follow Neveu in the proof of Theorem 7.1.1 (Neveu [73]). Girsanov's theorem (Theorem 7.1.3) is the central result of this chapter and an indispensable tool in nonlinear filtering. The reference is Girsanov [16]. When f_t

is nonrandom and belongs to $L^2[0,T]$, the result goes back to Cameron and Graves and to Maruyama (see Cameron and Graves [58]).

The following comments pertain to Theorem 7.1.4 and the two paragraphs after Remark 7.3.2. Let x_t be the coordinate process and μ the standard Wiener measure on $(C,\mathscr{B}_T)$. Suppose μ_1 is a second probability measure on $(C,\mathscr{B}_T)$ with $\mu_1 \ll \mu$. Denote by L_t the Radon-Nikodym derivative $d\mu_1/d\mu$ on $\bar{\mathscr{B}}_t$, where $\bar{\mathscr{B}}_t$ is the σ-field defined on p. 175. Noting that $L_0 = 1$, μ-a.s., and applying Theorem 6.7.2(a) (see the notes to Chapter 6), we see that $L_t = 1 + \int_0^t b_s(x)\,dx_s$, where $b_s(x)$ is a measurable, $\bar{\mathscr{B}}_t$-adapted process such that $\int_0^T b_s^2(x)\,ds < \infty$, μ-a.s. Thus the assumptions of Theorem 7.1.4 are satisfied. Using the notation on p. 169 we then have the following

Corollary to Theorem 7.1.4. *$\int_0^T [\hat{L}_s(x)b_s(x)]^2\,ds < \infty$, μ_1-a.s. and the process $y_t(x) = x_t - \int_0^t \hat{L}_s(x)b_s(x)\,ds$ is a Wiener martingale on $(C,\bar{\mathscr{B}}_t,\mu_1)$.*

For Theorem 7.1.4, see also Lipster-Shiryaev [40].

For a proof of Theorem 7.2.3, see Novikov [48]. Sections 7.3 and 7.4 are based on Ershov's results [11]. Explosion times are treated in McKean's book [71] and the illuminating Example 7.6.1 is also due to him. From Theorem 7.4.2 and Example 7.6.1, it follows that the stochastic equation (7.4.4) with $\gamma(t,x) = x_t^3$ has no weak solution on $[0,T]$ for some T. Tsirel'son's example is taken (in a slightly modified form) from his paper [53]. It is also given in [40].

Chapter 8

The innovation (Wiener) process has been known to many authors in the early development of filtering theory, notably Liptser and Shiryaev and Kailath (see the remarks in Kailath's articles [24, 26]). The definition of the innovation process given in Section 8.1 and the ensuing discussion is based on Meyer's seminar article [42]. For a discussion of $[\ ,\ ]_t$ and $\langle\ ,\ \rangle_t$ see Meyer ([45] I, [42]). The remainder of the chapter leading to the derivation of the stochastic equation for the optimal nonlinear filter is taken from Fujisaki-Kallianpur-Kunita [14]. However, Theorems 8.4.1 and 8.4.2 given here are more general. See also Liptser-Shiryaev [40] and the references to their work given there. There has been much further work, in recent years, on filtering and the integral representation of martingales (generalizing Theorem 8.3.1). The reader may consult Grigelionis [60, 61], Jacod and Yor [64], and Brémaud and Yor [57].

The methods of this chapter can be used to obtain stochastic equations for $E[f(X_s)|\mathscr{F}_t^Z]$, where $s < t$ (interpolation problem). Two types of equations can be obtained: a "forward" equation with respect to t, keeping s fixed and a "backward" equation with respect to s, keeping t fixed. Similarly, one can derive equations for $E[f(X_t)|\mathscr{F}_s^Z]$ where $s < t$ (extrapolation problem). For details, see Liptser and Shiryaev [40].

Some additional remarks on the filtering problem may be in order. Consider the model of the observation process (Z_t) given in Theorem 8.1.5. Writing $\mu_1 = PZ^{-1}$, from Theorem 7.3.1 we have $\mu_1 \ll \mu$. The corollary to Theorem 7.1.4 (see Notes on Chapter 7) provides a causal functional $f_s(x) = \hat{L}_s(x)b_s(x)$ for which $\int_0^T f_s^2(x)\,ds < \infty$ μ_1-a.s. and $Y_t(x) = x_t - \int_0^t f_s(x)\,ds$ is a Wiener martingale on $(C,\bar{\mathscr{B}}_t,\mu_1)$. It is easy to see that since (f_s) is $\bar{\mathscr{B}}_s$-adapted and because $\mu_1 \ll \mu$ it follows that $f_s(Z(\omega))$ is $\mathscr{F}_s^Z$-adapted. Also,

$$Y_t(Z(\omega)) = Z_t(\omega) - \int_0^t f_s(Z(\omega))\,ds$$

is an $(\mathscr{F}_t^Z,P)$-Wiener martingale. Now, by Theorem 2.7.1 we can choose $\hat{h}_t(\omega)$, a version of $E(h_t|\mathscr{F}_t^Z)(\omega)$, to be jointly measurable and $\mathscr{F}_t^{Z,0}$-adapted. Denoting the innovation process by (v_t) we see that $Y_t(Z(\omega)) - v_t(\omega) = \int_0^t [\hat{h}_s(\omega) - f_s(Z(\omega))]\,ds$ is a continuous martingale (vanishing at $t = 0$) with paths of bounded variation. It follows that the set $\Lambda = \{\omega: \hat{h}_s(\omega) = f_s(Z(\omega))$ for a.e. s in $[0,T]\}$ has P-measure one. A consequence of this fact is that

$$\int_0^T \hat{h}_s^2(\omega)\,ds < \infty \quad P\text{-a.s.}$$

Next, recall from Theorem 7.3.3 that

$$\frac{d\mu_1}{d\mu}(x) = \begin{cases} \exp[N_T(x) - \frac{1}{2}A_T(x)] & \text{for } x \in \Gamma \\ 0 & \text{for } x \notin \Gamma \end{cases}$$

where $\Gamma = \{x \in C: \int_0^T f_s(x)^2\,ds < \infty\}$. In [40], Section 4.2.9, Liptser and Shiryaev have shown how to define $N_T(x)$ as a stochastic integral of f_s which we write as $\int_0^T f_s(x)\,dx_s$ for $x \in \Gamma$, and $A_T(x)$ may be written $\int_0^T f_s^2(x)\,ds$. Suppose P_0 is a second probability measure on $\mathscr{F}_T^{Z,0}$ such that (Z_t) is a Wiener process under P_0. Then $P \ll P_0[\mathscr{F}_T^{Z,0}]$ and, using the integral notation to express $N_T(Z(\omega))$, we have

$$\frac{dP}{dP_0}(\omega) = \begin{cases} \exp\left[\int_0^T f_s(Z(\omega))\,dZ_s(\omega) - \frac{1}{2}\int_0^T f_s^2(Z(\omega))\,ds\right] & \text{for } \omega \in Z^{-1}\Gamma \\ 0 & \text{for } \omega \notin Z^{-1}\Gamma. \end{cases}$$

Further note that on the set $\Lambda \cap Z^{-1}\Gamma$, the exponent in the above formula becomes

$$\int_0^T \hat{h}_s(\omega)\,dZ_s(\omega) - \frac{1}{2}\int_0^T \hat{h}_s^2(\omega)\,ds.$$

(Compare Theorem 2 of [24].)

Chapter 9

The notation and terminology of Section 1 is taken from Gohberg and Krein's book [17]. The term "orthoprojector" stands for orthogonal projection operator. The material of this chapter is presented in a form that brings out its close connection with Ershov's results discussed in Chapter 7. See the book of Riesz and Nagy [74] for the fact used in Lemma 9.1.3.

The congruence ϕ has been defined in Section 6.3 and denoted there by ψ_1. Theorem 9.2.1 is taken from Kallianpur and Oodaira [31]. It was obtained, independently, by Kailath and Duttweiler [27]. The representation in Section 9.3 is from [31]. It was derived earlier by Hitsuda [19] who used martingale techniques and Girsanov's theorem. In Section 9.4, since the measures μ_ξ and μ are Gaussian, they are either equivalent (i.e., mutually absolutely continuous) or singular. Since $\mu_\xi \ll \mu$, we conclude $\mu_\xi \equiv \mu$. There is an extensive literature on the equivalence-singularity dichotomy. Some of the references to it are given in [31].

Chapter 10

The proof of Theorem 10.2.1 follows Liptser and Shiryaev [40]. To show that (10.2.2a) has a solution, use (10.2.3) and (10.2.4) and apply the Ito formula to each component of the m-vector $[F(t) - I] \cdot [\zeta + \int_0^t F^{-1}(s)\bar{B}(s)\, dW_s]$.

The martingale method for the linear filtering problem was introduced, independently of the developments in Chapter 8, by Balakrishnan [1,2]. The independent derivation of the Kalman-Bucy equations in Section 10.5 makes essential use of the equality of the σ-fields $\mathscr{F}_t^\nu$ and $\mathscr{F}_t^Y$, whereas the proof given in Section 10.3, obtained as a deduction from the nonlinear theory, does not assume this equality but contains it as a consequence. The following is a brief argument to show that the kernel $\Gamma(u,t)$ in (10.4.11) can be chosen to be measurable in (u,t): The linear space M of (t,ω)-measurable real functions $\eta_t(\omega)$ such that $\|\eta\|^2 = \int_0^T E(\eta_t^2)\, dt < \infty$ and $\eta_t \in L(\nu,T)$ for a.e. t is a Hilbert space. It suffices to show that the linear manifold spanned by $\tilde{\eta}_t(\omega) = I_{[a,b)}(t)[\nu_d(\omega) - \nu_c(\omega)]$ (where $a \le b$ and $c \le d$) is dense in M. Suppose $\eta \in M$ such that $\int_a^b E[\eta_t(\nu_d - \nu_c)]\, dt = 0$ for every $[a,b)$ and $c \le d$. Then, since ν_t is continuous, it follows that for a.e. t, $\eta_t(\omega) = 0$, P-a.s., and so $\|\eta\| = 0$.

The proof of (10.4.5) and (10.4.6) given in 10.4 which invokes Theorem 9.5.1 makes a detour via the theory of Gaussian solutions of stochastic equations. A shorter, more direct derivation can be given along the following lines for the simplest Kalman-Bucy model (only real-valued processes are considered),

$$dX_t = A(t)X_t\, dt + B(t)\, dW_t^1,$$
$$dY_t = C(t)X_t\, dt + dW_t^2, \qquad Y_0 = 0, \qquad (W_t^1) \perp\!\!\!\perp (W_t^2),$$

X_0 independent of both Wiener processes with $EX_0 = 0$. Also assume the coefficients to be continuous functions on $[0,T]$. As in Lemma 10.1 of [40], we show that $\bar{X} = \hat{X}_t - E(X_t)$ satisfies

$$\text{(i)} \qquad \bar{X}_t = \int_0^t G(t,s)C(s)\bar{X}_s\, ds + \int_0^t G(t,s)\, d\nu_s,$$

where $G(t,s)$ can be taken to be (t,s)-measurable with $\int_0^T \int_0^t G^2(t,s)\,ds\,dt < \infty$. The latter condition follows from the independence of (X_t) and (W_t^2) and the finiteness of $\int_0^T E(X_t^2)\,dt$. Considering (t,ω)-measurable versions of $\hat{X}_t(\omega)$ and $\eta_t(\omega) = \int_0^t G(t,s)\,dv_s(\omega)$, we have $\hat{X}.(\omega)$ and $\eta.(\omega)$ in $L^2[0,T]$ for P-a.a. ω, and (i) becomes $(I - V)\bar{X}.(\omega) = \eta.(\omega)$. Hence (for P-a.a. ω) $\bar{X}.(\omega) = (I - V)^{-1}\eta.(\omega)$, from which it follows that $\bar{X}_t \in L(\eta;t) \subseteq L(v;t)$. The conclusion holds for every t because of the P-quadratic mean continuity of the $(\hat{X}_t)$ process.

The papers of Wonham [54] and Kallianpur and Striebel [33, 34] are among those that derive the linear filtering equations without recourse to martingale theory. The connection between the linear filtering problem and the Wiener-Hopf equation is discussed in detail in the literature. See Kailath's survey paper [26], where references are given.

For a time-invariant system, i.e., when the matrices $A_1(t)$, $B(t)$, $C_1(t)$, and $D(t)$ in Eq. (10.2.1) are constants independent of t (assume, in addition, that $A_0 = A_2 = C_0 = C_2 = 0$) the asymptotic behavior of $P(t)$ as $t \to \infty$ has been studied by Wonham. (See also Balakrishnan [1].) Set $DD^* = I$. If the matrix A_1 is stable (i.e., all its eigenvalues have strictly negative real parts), and if the matrix $\int_0^\infty e^{A_1^* t} C_1^* C_1 e^{A_1 t}\,dt$ is non-singular, then $P(\infty) = \lim_{t\to\infty} P(t)$ exists and satisfies the equation

$$A_1 P(\infty) + P(\infty)A_1^* + BB^* - P(\infty)C_1^* C_1 P(\infty) = 0.$$

Furthermore, $A_1 - P(\infty)C_1^* C_1$ is stable.

Chapter 11

The proof of Theorem 11.2.1 is along the lines given in Liptser and Shiryaev [40]. Section 11.3 and the Bayes Formula (Theorem 11.3.1) is taken from Kallianpur and Striebel [32]. The proof given in the text, as well as Theorem 11.3.2 is from Meyer [42]. To establish (11.3.18), note that P^z is equivalent to λ on $\mathscr{F}^z$ by (11.3.1). Hence, from the argument used in the proof of Theorem 7.3.2 (page 178), the left-hand side of (11.3.18) is a probability measure under which (Z_t) is a Wiener process by Girsanov's theorem. Theorem 11.4.1 is due to Clark [4]. The central result of Section 11.5 is Kunita's theorem (Theorem 11.5.2), which establishes the uniqueness of the strong solution of Eq. (11.5.12). The proof we give differs somewhat from that given in his paper [37]. It uses the equality of the σ-fields $\mathscr{F}_t^z$ and $\mathscr{F}_t^v$ (available from Clark's theorem), whereas the method of successive approximation is adopted in [37]. In [37] Kunita also studies the asymptotic behavior of the nonlinear filtering error.

The problem of showing the uniqueness of the solution of the most general stochastic equation for the optimal nonlinear filter is still a subject of research. The papers by Rozovskii [51] and Krylov and Rozovskii [68] contain some of the most recent results. In this connection, mention must be made of the early work of Mortensen [47].

Krylov has recently obtained the following result on the equality of the σ-fields (generated by the observation and innovation processes), which generalizes to the case where the signal process and the noise are not independent (see Krylov [67]).

Let W_t, $t \in [0,T]$ be a standard Wiener process in $\mathbf{R}^{d_3}$ defined on a complete probability space $(\Omega,\mathcal{F},P)$. Consider the system of equations

$$\text{(I)} \qquad \begin{aligned} dX_t &= b_1(t,X_t,Y_t)\,dt + \sigma_1(t,X_t,Y_t)\,dW_t, \\ dY_t &= b_2(t,X_t,Y_t)\,dt + \sigma_2(t,Y_t)\,dW_t \qquad (0 \le t \le T), \\ X_0 &= \xi_0,\ Y_0 = \eta_0, \qquad \text{where } \xi_0,\eta_0 \perp\!\!\!\perp (W_t). \end{aligned}$$

The coefficients $b_i(t,x,y)$ are d_i-vector-valued functions; $\sigma_i(t,x,y)$ are $d_i \times d_3$ matrix-valued functions ($i = 1,2$). The following conditions are imposed.

(i) b_i and σ_i are Borel functions of t for each (x,y), satisfying a Lipschitz condition in the pair (x,y) with a constant not depending on t;

(ii) b_i and σ_i are bounded.

(iii) For $t \in [0,T]$, $y \in \mathbf{R}^{d_2}$, $\sigma_1(t,x,y)$ has continuous derivatives with respect to x of the first three orders which are bounded in (t,x,y), of which the first- and second-order derivatives satisfy Lipschitz conditions in y uniformly with respect to x.

(iv) For $t \in [0,T]$, $y \in \mathbf{R}^{d_2}$, $b_1(t,x,y)$ and $b_2(t,x,y)$ have continuous derivatives of the first two orders with respect to x bounded in (t,x,y), of which the first-order derivative of $b_1(t,x,y)$ satisfies a Lipschitz condition in y uniformly with respect to (t,x).

(v) There exists a positive constant K and a function $g \in L^2(\mathbf{R}^{d_1})$ such that

$$|b_2(t,x,y)| \le g(x)$$

and

$$\int |b_2(t,x,y_1) - b_2(t,x,y_2)|^2\,dx \le K^2|y_1 - y_2|^2,$$

for each t in $[0,T]$, y_1, $y_2 \in \mathbf{R}^{d_2}$.

Let $a_2(t,y) = \sigma_2(t,y)\sigma_2^*(t,y)$ (* denotes adjoint).

(vi) There exists $\gamma > 0$ such that, for all λ, $y \in \mathbf{R}^{d_2}$ and $t \in [0,T]$,

$$\sum_{i,j=1}^{d_2} a_2^{ij}(t,y)\lambda^i\lambda^j \ge \gamma|\lambda|^2.$$

Hence the positive symmetric square root $a_2^{1/2}$ of a_2 has an inverse $a_2^{-1/2}$, which is a bounded function of (t, y).

(vii) $|\sigma_1^*(t,x,y)\lambda|^2 - |a_2^{-1/2}(t,y)\sigma_2(t,y)\sigma_1^*(t,x,y)\lambda|^2 \ge \gamma|\lambda|^2$ for $t \in [0,T]$, x, $\lambda \in \mathbf{R}^{d_1}$ and $y \in \mathbf{R}^{d_2}$.

Under the assumptions made above, the system (I) has a unique solution (X_t,Y_t) and the innovation process

$$\bar{W}_t = \int_0^t a_2^{-1/2}(s,Y_s)\,dY_s - \int_0^t a_2^{-1/2}(s,Y_s)E[b_2(s,X_s,Y_s)|\mathcal{F}_s^Y]\,ds$$

is an $(\mathcal{F}_t^Y,P)$-martingale. We need a final set of conditions on the initial random variables ξ_0,η_0.

(viii) The conditional distribution of ξ_0 given η_0 has a continuous density π_0 such that

$$\pi_0 \in W_p^3(\mathbf{R}^{d_1}) \cap W_2^2(\mathbf{R}^{d_1}),$$

and $E\|\pi_0\|_{W_2^2}^2 + E\|\pi\|_{W_p^3}^p < \infty$, where $p > d_1$ and $W_n^m(\mathbf{R}^{d_1})$ is the Sobolev space of all real-valued functions f on $\mathbf{R}^{d_1}$ which, together with the kth order generalized derivatives $D^k f$ ($|k| \leq m$), belong to $L^n(\mathbf{R}^{d_1})$ with the norm $\|f\|_{W_n^m} = \sum_{|k| \leq m} \|D^k f\|_{L^n}$.

Theorem (Krylov). *Suppose the conditions* (i)–(viii) *are satisfied. Then* $\mathcal{F}_t^{\eta_0, \overline{W}} = \mathcal{F}_t^Y$ *for each* $t \in [0,T]$.

References

What follows is not intended to be a complete bibliography. Only those articles and books are listed which have been useful in the organization of the material in this book. The reader will find an extensive bibliography in references [26] and [40].

[1] Balakrishnan, A. V., *Stochastic Differential Systems*. Lecture Notes in Economics and Mathematical Systems 84, Springer-Verlag, Berlin, 1973.

[2] Balakrishnan, A. V., A martingale approach to linear recursive state estimation. *SIAM J. Control Optim.* **10**, 754–766, 1972.

[3] Cameron, R. H. and Martin, W. T., The orthogonal development of non-linear functionals in series of Fourier-Hermite functionals. *Ann. Math.* **48**, 385–392, 1947.

[4] Clark, J. M. C., *Conditions for One-to-one Correspondence between an Observation Process and its Innovation*. Technical Report, Centre for Computing and Automation, Imperial College, London, 1969.

[5] Courrège, P., Intégrales stochastiques et martingales de carré intégrable. *Sémin. Brelot-Choquet-Deny*, 7-e année, 1962/63.

[6] Cramér, H., On some classes of non-stationary processes. *Proc. Fourth Berkeley Symp. Math. Statist. Probability, Vol. II.* University of California Press, Berkeley and Los Angeles, 1961, pp. 57–78.

[7] Dellacherie, C., *Capacités et Processus Stochastiques*. Springer-Verlag, Berlin, 1972.

[8] Doob, J. L., *Stochastic Processes*. John Wiley, New York, 1953.

[9] Duncan, T. E., Evaluation of likelihood functions. *Inf. Control* **13**, 62–74, 1968.

[10] Dynkin, E. B., *Markov Processes, Vols. 1 & 2.* Springer-Verlag, Berlin, 1965.

[11] Ershov, M. P., On the absolute continuity of measures corresponding to diffusion type processes. *Theory Probab. Its Appl.* **17**, 169–174, 1972.

[12] Fleming, W. H. and Nisio, M., On the existence of optimal stochastic controls. *J. Math. Mech.* **15**, 777–794, 1966.

[13] Friedman, A., *Stochastic Differential Equations and Applications, Vol. 1.* Academic, New York, 1975.

[14] Fujisaki, M., Kallianpur, G., and Kunita, H., Stochastic differential equations for the nonlinear filtering problem. *Osaka J. Math.* **9**, 19–40, 1972.

[15] Gihman, I. I. and Skorohod, A. V., *Stochastic Differential Equations.* Springer-Verlag, New York, 1972.

[16] Girsanov, I. V., On transforming a certain class of stochastic processes by absolutely continuous substitution of measures. *Theory Probab. Its Appl.* **5**, 285–301, 1960.

[17] Gohberg, I. C. and Krein, M. G., *Theory and Applications of Volterra Operators in Hilbert Space.* Am. Math. Soc. Transl., Vol. 24, Providence, RI, 1970.

[18] Hida, T., Canonical representations of Gaussian processes and their applications. *Mem. Coll. Sci. Univ. Kyoto Ser. A* **33**, 109–155, 1960.

[19] Hitsuda, M., Representation of Gaussian processes equivalent to Wiener process. *Osaka J. Math.* **5**, 299–312, 1968.

[20] Ito, K., *Lectures on Stochastic Processes.* Tata Inst. Fundamental Research, Bombay, 1961.

[21] Ito, K., Multiple Wiener integrals. *J. Math. Soc. Japan* **3**, 157–169, 1951.

[22] Ito, K. and Nisio, M., On stationary solutions of a stochastic differential equation. *J. Math. Kyoto Univ.* **4**, 1–75, 1964.

[23] Kadota, T. T. and Shepp, L. A., Conditions for the absolute continuity between a certain pair of probability measures. *Z. Wahrscheinlichkeitstheorie Verw. Gebiete* **16**, 250–260, 1970.

[24] Kailath, T., The structure of Radon-Nikodym derivatives with respect to Wiener and related measures. *Ann. Math. Statist.* **42**, 1054–1067, 1971.

[25] Kailath, T., An innovations approach to least squares estimation. *IEEE Trans. Autom. Control* **AC-13**, 646–655, 1968.

[26] Kailath, T., A view of three decades of linear filtering theory. *IEEE Trans. Inf. Theory* **IT-20**, 146–181, 1974.

[27] Kailath, T. and Duttweiler, D., An RKHS approach to detection and estimation problems—Part III: Generalized innovations representations and a likelihood-ratio formula. *IEEE Trans. Inf. Theory* **IT-18**, 730–745, 1972.

[28] Kallianpur, G., The role of reproducing kernel Hilbert spaces in the

study of Gaussian processes. In P. Ney ed.: *Advances in Probability, Vol. 2*, Marcel Dekker, New York, 1970, pp. 49–83.
[29] Kallianpur, G., Zero-one laws for Gaussian processes. *Trans. Am. Math. Soc.* **149**, 199–211, 1970.
[30] Kallianpur, G., A stochastic equation for the optimal non-linear filter. P. R. Krishnaiah, ed.: *Multivariate Analysis IV*. North-Holland, Amsterdam–New York–Oxford, 1977, pp. 267–281.
[31] Kallianpur, G. and Oodaira, H., Non-anticipative representations of equivalent Gaussian processes. *Ann. Probab.* **1**, 104–122, 1973.
[32] Kallianpur, G. and Striebel, C., Estimation of stochastic processes with additive white noise observation errors. *Ann. Math. Statist.* **39**, 785–801, 1968.
[33] Kallianpur, G. and Striebel, C., Stochastic differential equations in statistical estimation problems. In P. R. Krishnaiah, ed.: *Multivariate Analysis II*. Academic, New York and London, 1969, pp. 367–388.
[34] Kallianpur, G. and Striebel, C., Stochastic differential equations in the estimation of continuous parameter stochastic processes. *Theory Probab. Its Appl.* **14**, 567–594, 1969.
[35] Kalman, R. E. and Bucy, R. S., New results in linear filtering and prediction theory. *Trans. ASME Ser. D J. Basic Eng.* **83**, 95–108, 1961.
[36] Kunita, H. and Watanabe, S., On square integrable martingales. *Nagoya Math. J.* **30**, 209–245, 1967.
[37] Kunita H., Asymptotic behavior of the non-linear filtering errors of Markov processes. *J. Multivar. Anal.* **1**, 365–393, 1971.
[38] Lévy, P., A special problem of Brownian motion and a general theory of Gaussian random functions. *Proc. Third Berkeley Symp. Math. Statist. Probab. II*, University of California Press, Berkeley and Los Angeles, 1956, 133–175.
[39] Lévy, P., *Random Functions: General Theory with Special Reference to Laplacian Random Functions*. University of California Press, Berkeley and Los Angeles, 1963.
[40] Liptser, R. Sh. and Shiryayev, A. N., *Statistics of Random Processes I*. Springer-Verlag, New York, 1977.
[41] Meyer, P. A., *Probability and Potentials*. Blaisdell, Waltham, MA, 1966.
[42] Meyer, P. A., Sur un problème de filtration. *Séminaire de Probabilités VII, Université de Strasbourg*. Lecture Notes 321. Springer-Verlag, Berlin–Heidelberg–New York, 1973, pp. 223–247.
[43] Meyer, P. A., Square integrable martingales, a survey. *Martingales, A Report on a Meeting at Oberwolfach, May 17–23, 1970*. Lecture Notes 190. Springer-Verlag, Berlin–Heidelberg–New York, 1971, pp. 32–37.
[44] Meyer, P. A., Un cours sur les intégrales stochastique (Chapitre I). *Séminaire de Probabilités X, Université de Strasbourg*. Lecture Notes 511. Springer-Verlag, Berlin–Heidelberg–New York, 1976, 245–400.

[45] Meyer, P. A., Integrales stochastiques I, II, III, IV. *Séminaire de Probabilités I, Université de Strasbourg*. Lecture Notes 39. Springer-Verlag, Berlin–Heidelberg–New York, 1967 pp. 72–162.
[46] Meyer, P. A., *Processus de Markov*. Lecture Notes 26. Springer-Verlag, Berlin–Heidelberg–New York, 1967.
[47] Mortenson, R. E., *Optimal Control of Continuous-time Stochastic Systems*. Report No. ERL-66-1, Electronics Research Laboratory, College of Engineering, University of California, Berkeley, 1966.
[48] Novikov, A. A., On an identity for stochastic integrals. *Theory Probab. Its Appl.* **17**, 717–720, 1972.
[49] Orey, S., *Radon-Nikodym Derivatives of Probability Measures: Martingale Methods*. Dept. Found. Sci., Tokyo Univ. of Education, 1974.
[50] Rao, M., On decomposition theorems of Meyer. *Math. Scand.* **24**, 66–78, 1969.
[51] Rozovskii, B. L., On stochastic partial differential equations. *Math. USSR-Sb.* **25**, 295–322, 1975.
[52] Skorohod, A. V., *Studies in the Theory of Random Processes*. Addison-Wesley, Reading, MA, 1965.
[53] Tsirelson, B. S., An example of a stochastic differential equation having no strong solution. *Theory Probab. Its Appl.* **20**, 416–418, 1975.
[54] Wonham, W. M., Random differential equations in control theory. In A. T. Bharucha-Reid, ed.: *Probabilistic Methods in Applied Mathematics, Vol. 2*. Academic, New York and London, 1970, pp. 131–212.
[55] Zimmerman, G. J., *Topic in Multiparameter Processes*. Ph.D. Thesis, University of Minnesota, 1970.

Additional References

[56] Aronszajn, N. Theory of reproducing kernels. *Trans. Am. Math. Soc.* **68**, 337–404, 1950.
[57] Bremaud, P. and Yor, M., Changes of filtrations and probability measures. *Z. Wahrscheinlichkeitstheorie Verw. Gebiete* **45**, 269–295, 1978.
[58] Cameron, R. H. and Graves, R. E., Additive functionals on a space of continuous functions. I. *Trans. Am. Math. Soc.* **70**, 160–176, 1951.
[59] Dudley, R. M., Wiener functionals as Ito integrals. *Ann. Probab.* **5**, 140–141, 1977.
[60] Grigelionis, B., On stochastic equations of nonlinear filtering of random processes. *Litovsk. Mat. Sb.* **12**, 37–81, 1972 (in Russian).
[61] Grigelionis, B., On the stochastic integral representation of square integrable martingales. *Litovsk. Mat. Sb.* **14**, 53–68, 1974 (in Russian).
[62] Ito, K., Spectral type of the shift transformation of differential pro-

cesses with stationary increments. *Trans. Am. Math. Soc.* **81**, 253–263, 1956.

[63] Ito, K., and Nisio, M., On the convergence of sums of independent Banach space valued random variables. *Osaka J. Math.* **5**, 35–48, 1968.

[64] Jacod, J. and Yor, M., Étude des solutions extrémales et représentation intégrale des solutions pour certains problèmes de martingales. *Z. Wahrscheinlichkeitstheorie Verw. Gebiete* **38**, 83–125, 1977.

[65] Kakutani, S., Spectral analysis of stationary Gaussian processes. In *Proc. Fourth Berkeley Symp. Mathematical and Statistics and Probability II.* University of California Press, Berkeley and Los Angeles, 1963, pp. 239–247.

[66] Krylov, N. V., *Upravliaemie Protsessi Diffuzionnovo Tipa.* (*Controlled Processes of Diffusion Type.*) Izdatelstvo "Nauka," Moscow, 1977.

[67] Krylov, N. V., On the equality of sigma-algebras in the filtering problem for diffusion processes. *Teor. Veroiatnost. i Primenen.* **24**, 771–780, 1979 (in Russian).

[68] Krylov, N. V. and Rozovskii, B. L., On conditional distributions of diffusion processes. *Izv. Akad. Nauk USSR Ser. Mat.* **42**, 356–378, 1978 (in Russian).

[69] Le Page, R. D., *Estimation of Parameters in Signals of Known Form and an Isometry Related to Unbiased Estimation.* Ph.D. Thesis, University of Minnesota, 1967.

[70] Loève, M., *Probability Theory II, 4th ed.* Springer-Verlag, New York, 1978.

[71] McKean, H. P., *Stochastic Integrals.* Academic, New York and London, 1969.

[72] Neveu, J., *Processus Aléatoires Gaussiens.* Séminaire de Mathématiques Supérieures, Presses de l'Université de Montreal, 1968.

[73] Neveu, J., Martingales, *Notes Partielles d'un Cours de 3-ème Cycle.* Paris, 1970–1971.

[74] Riesz, F. and Nagy, B. Sz., *Functional Analysis.* Ungar, New York, 1955.

[75] Segal, I. E., Tensor algebras over Hilbert spaces. I. *Trans. Am. Math. Soc.* **81**, 106–134, 1956.

[76] Stroock, D. and Varadhan, S. R. S., Diffusion processes with continuous coefficients, I. *Comm. Pure Appl. Math.* **22**, 345–400, 1969.

[77] Yamada, T. and Watanabe, Sh., On the uniqueness of solution of stochastic differential equations. *J. Math. Kyoto Univ.* **11**, 155–167, 1971.

Index of Commonly Used Symbols

Index

Applications of Mathematics

Vol. 1
Deterministic and Stochastic Optimal Control
By **W.H. Fleming** and **R.W. Rishel**
1975. ix, 222p. 4 illus. cloth
ISBN 0-387-**90155**-8

Vol. 2
Methods of Numerical Mathematics
By **G.I. Marchuk**
1975. xii, 316p. 10 illus. cloth
ISBN 0-387-**90156**-6

Vol. 3
Applied Functional Analysis
By **A.V. Balakrishnan**
1976. x, 309p. cloth
ISBN 0-387-**90157**-4

Vol. 4
Stochastic Processes in Queueing Theory
By **A.A. Borovkov**
1976. xi, 280p. 14 illus. cloth
ISBN 0-387-**90161**-2

Vol. 5
Statistics of Random Processes I
General Theory
By **R.S. Lipster** and **A.N. Shiryayev**
1977. x, 394p. cloth
ISBN 0-387-**90226**-0

Vol. 6
Statistics of Random Processes II
Applications
By **R.S. Lipster** and **A.N. Shiryayev**
1978. x, 339p. cloth
ISBN 0-387-**90236**-8

Vol. 7
Game Theory
Lectures for Economists and Systems Scientists
By **N.N. Vorob'ev**
1977. xi, 178p. 60 illus. cloth
ISBN 0-387-**90238**-4

Vol. 8
Optimal Stopping Rules
By **A.N. Shiryayev**
1978. x, 217p. 7 illus. cloth
ISBN 0-387-**90256**-2

Vol. 9
Gaussian Random Processes
By **I.A. Ibragimov** and **Y.A. Rosanov**
1978. x, 275p. cloth
ISBN 0-387-**90302**-X

Vol. 10
Linear Multivariable Control: A Geometric Approach
By **W.M. Wonham**
1979. xi, 326p. 27 illus. cloth
ISBN 0-387-**90354**-2

Vol. 11
Brownian Motion
By **T. Hida**
1980. xvi, 325p. 13 illus. cloth
ISBN 0-387-**90439**-5

Vol. 12
Conjugate Direction Methods in Optimization
By **M. Hestenes**
1980. x, 325p. 22 illus. cloth
ISBN 0-387-**90455**-7

Vol. 14
Controlled Diffusion Processes
By **N.V. Krylov**
1980. approx. 320p. cloth
ISBN 0-387-**90461**-1